Springer Series in Advanced Manufacturing

For further volumes:
http://www.springer.com/series/7113

R. Venkata Rao · Vimal J. Savsani

Mechanical Design Optimization Using Advanced Optimization Techniques

Springer

R. Venkata Rao
Mechanical Engineering Department
S. V. National Institute of Technology
Ichchhanath, Surat
Gujarat 395007
India

Vimal J. Savsani
Department of Mechanical Engineering
B. H. Gardi College of Engineering and Technology
Rajkot
India

ISSN 1860-5168
ISBN 978-1-4471-2747-5 e-ISBN 978-1-4471-2748-2
DOI 10.1007/978-1-4471-2748-2
Springer London Heidelberg New York Dordrecht

British Library Cataloguing in Publication Data
A catalogue record for this book is available from the British Library

Library of Congress Control Number: 2011945824

Printed on acid-free paper

Springer is part of Springer Science+Business Media (www.springer.com)

Dedicated to my parents (Lakshmi Narayana and Jayamma), dearest wife (Sujatha Rao), and beloved daughter (Jaya Lakshmi)

R. Venkata Rao

Preface

This book describes research works that explore different advanced optimization techniques such as GA, PSO, ABC, HEA, DE, AIA, BBO and GEM for mechanical design. This book also includes the modifications of PSO, HEA and ABC to increase the effectiveness of the existing PSO, HEA and ABC techniques. The modified techniques are validated through application to unconstrained and constrained benchmark functions as well as to mechanical design optimization problems. Also new hybrid optimization techniques combining ABC with PSO, BBO, DE and GA are developed and are validated through benchmark functions and mechanical design problems. Moreover, a new efficient and effective optimization technique named as "Teaching–Learning-Based Optimization (TLBO)" is developed for the global optimization problems. The advantage of this new technique is that it does not need any algorithm parameters for it to working and so it eliminates the disadvantages of many existing optimization techniques which need tuning of algorithm parameters.

The algorithms and computer codes for various advanced optimization techniques included in this book will be very useful to the readers. This book is expected to be very useful to the industrial product designers for realizing a product as it presents advanced optimization techniques to make their tasks easier, logical, efficient and effective. This book is intended for designers, practitioners, managers, institutes involved in design-related projects, applied research workers, academics and graduate students in mechanical and industrial design engineering. As such, this book is expected to become a valuable reference for those wishing to do research on the use of advanced optimization techniques for solving single/multi-objective combinatorial design optimization problems.

We are grateful to Anthony Doyle, Claire Protherough and Grace Quinn of Springer-Verlag, London, for their support and help in producing this book. I wish to thank various researchers and the publishers of international journals for giving us the permission to reproduce certain portions of their published research works. Our special thanks are due to the Director, Registrar and the colleagues at S.V. National Institute of Technology, Surat, India.

While every attempt has been made to ensure that no errors (printing or otherwise) enter the book, the possibility of these creeping into the book is always there. We will be grateful to the readers if these errors are pointed out. Suggestions for further improvement of the book will be thankfully acknowledged.

Surat, September 2011

Prof. Dr. Venkata Rao
Dr. V. J. Savsani

Contents

Chapter 1
Introduction

Mechanical design includes an optimization process in which designers always consider certain objectives such as strength, deflection, weight, wear, corrosion, etc. Depending on the requirements. However, design optimization for a complete mechanical assembly leads to a complicated objective function with a large number of design variables. So it is a good practice to apply optimization techniques for individual components or intermediate assemblies than a complete assembly. For example, in an automobile power transmission system, optimization of gearbox is computationally and mathematically simpler than the optimization of complete system.

Analytical or numerical methods for calculating the extreme values of a function have been applied to engineering computations for a long time. Although these methods may perform well in many practical cases, they may fail in more complex design situations. In real design problems, the number of design parameters can be very large and their influence on the value to be optimized (the goal function) can be very complicated, having nonlinear character. The goal function may have many local extrema, whereas the designer is interested in the global extremum. Such problems cannot be handled by classical methods (e.g. gradient methods) at all, or they only compute local extrema. In these complex cases, advanced optimization algorithms offer solutions to the problems because they find a solution near to the global optimum within reasonable time and computational costs.

The optimization techniques can be classified into two distinct types as given below:

(a) *Traditional optimization techniques*: These are deterministic algorithms with specific rules for moving from one solution to the other. These algorithms have been in use for quite some time and have been successfully applied to many engineering design problems. Examples include nonlinear programming, geometric programming, quadratic programming, dynamic programming, generalized reduced gradient method, etc.

R. V. Rao and V. J. Savsani, *Mechanical Design Optimization Using Advanced Optimization Techniques*, Springer Series in Advanced Manufacturing,
DOI: 10.1007/978-1-4471-2748-2_1,

(b) *Advanced optimization techniques*: These techniques are stochastic in nature with probabilistic transition rules. These techniques are comparatively new and gaining popularity due to certain properties which the deterministic algorithm does not have. The examples include Genetic Algorithm (GA), Differential Evolution (DE), Particle Swarm Optimization (PSO), Harmony Elements Algorithm (HEA), Biogeography Based Optimization (BBO), Artificial Bee Colony (ABC), Artificial Immune Algorithm (AIA), etc.

Although, traditional mathematical programming techniques had been employed to solve optimization problems in mechanical design, these techniques have following limitations:

- Traditional techniques do not fare well over a broad spectrum of problem domains.
- Traditional techniques are not suitable for solving multi-modal problems as they tend to obtain a local optimal solution.
- Traditional techniques are not ideal for solving multi-objective optimization problems.
- Traditional techniques are not suitable for solving problems involving large number of constraints.

Considering the drawbacks of the traditional optimization techniques, attempts are being made to optimize the mechanical design optimization problems by using evolutionary optimization techniques. Most commonly used evolutionary optimization technique is GA. However, GA provides a near optimal solution for a complex problem having large number of variables and constraints. This is mainly due to difficulty in determination of optimum controlling parameters such as population size, crossover rate and mutation rate. Therefore, the efforts must be continued to use more recent optimization techniques to modify the existing algorithms and to develop hybrid algorithms which are more powerful, robust and able to provide accurate solution.

The research work reported in this book is therefore carried out keeping in view the following objectives:

- To provide the applications of various recently developed advanced optimization techniques to mechanical design problems such as gear design, bearing design, spring design, clutch design, robot gripper design, etc.
- To modify the existing advanced optimization techniques so as to overcome their limitations.
- To develop new hybrid optimization techniques by hybridization of two existing advanced optimization techniques so as to combine their benefits.
- To develop a new optimization technique that is effective over the existing optimization techniques.

This book is organized as follows: Chap. 2 presents the details of existing optimization algorithms used in this book, the modifications incorporated in the existing algorithms and the hybrid algorithms. Chapter 3 presents the applications of existing advanced optimization algorithms to the mechanical design problems. Chapter 4 presents the applications of modified optimization algorithms to constrained and unconstrained benchmark functions and mechanical design optimization problems. Chapter 5 presents the applications of hybrid algorithms to the constrained and unconstrained benchmark functions and mechanical design optimization problems. Chapter 6 presents the development and application of a new optimization technique, called TLBO, to the constrained and unconstrained benchmark functions and mechanical design optimization problems. Chapter 7 presents the applications of the TLBO and other optimization techniques to the design optimization of some thermal equipment. Chapter 8 presents the general conclusions of the research work reported in the book. Appendix 1 presents some additional demonstrative examples of TLBO and Appendix 2 presents the sample codes for the selected optimization algorithms.

The next chapter presents the details of different advanced optimization algorithms used in this book.

Chapter 2
Advanced Optimization Techniques

Many difficulties such as multi-modality, dimensionality and differentiability are associated with the optimization of large-scale problems. Traditional techniques such as steepest decent, linear programing and dynamic programing generally fail to solve such large-scale problems especially with nonlinear objective functions. Most of the traditional techniques require gradient information and hence it is not possible to solve non-differentiable functions with the help of such traditional techniques. Moreover, such techniques often fail to solve optimization problems that have many local optima. To overcome these problems, there is a need to develop more powerful optimization techniques and research is going on to find effective optimization techniques since last three decades.

Some of the well-known population-based optimization techniques developed during last three decades are: Genetic Algorithms (GA) [16] which works on the principle of the Darwinian theory of the survival-of-the fittest and the theory of evolution of the living beings; Artificial Immune Algorithms (AIA) [14] which works on the principle of immune system of the human being; Ant Colony Optimization (ACO) [10] which works on the principle of foraging behavior of the ant for the food; Particle Swarm Optimization (PSO) [20] which works on the principle of foraging behavior of the swarm of birds; Differential Evolution (DE) [35] which is similar to GA with specialized crossover and selection method; Harmony Search (HS) [15] which works on the principle of music improvisation in a music player; Bacteria Foraging Optimization (BFO) [27] which works on the principle of behavior of bacteria; Shuffled Frog Leaping (SFL) [12] which works on the principle of communication among the frogs, Artificial Bee Colony (ABC) [18] which works on the principle of foraging behavior of a honey bee; Biogeography-Based Optimization (BBO) [34] which works on the principle of immigration and emigration of the species from one place to the other; Gravitational Search Algorithm (GSA) [29] which works on the principle of gravitational force acting between the bodies and Grenade Explosion Method (GEM) [1] which works on the principle of explosion of

R. V. Rao and V. J. Savsani, *Mechanical Design Optimization Using Advanced Optimization Techniques*, Springer Series in Advanced Manufacturing,
DOI: 10.1007/978-1-4471-2748-2_2, © Springer-Verlag London 2012

grenade. These algorithms have been applied to many engineering optimization problems and proved effective to solve some specific kind of problems.

All the above-mentioned algorithms are nature inspired population-based optimization methods, but they have some limitations in one or the other aspect. Due to this fact, more research is required to test algorithms for different problems to check their suitability for a wide variety of problems. Research is continued to enhance the existing algorithms to improve their performance. Enhancement is done either (a) by modifying the existing algorithms or (b) by hybridizing the existing algorithms. Enhancement due to modifications in the existing algorithms is reported in GA [22, 23, 28], PSO [5, 7, 25, 42], ACO [32, 45], ABC [19, 26], etc. Enhancement can also be done by combining the strengths of different optimization algorithms, known as hybridization of algorithms. Hybridization is an effective way to make the algorithm efficient and it combines the properties of different algorithms. Some of such hybridized algorithms can be found in Hui et al. [17], Wen [39], Ying [43], Yannis and Magdalene [41], Shahla et al. [31], Tung and Erwie [36], Dong et al. [8], etc.

Brief discussion of the algorithms, their modifications and hybridizations used in this book is presented in the following subsections.

2.1 Genetic Algorithm

Genetic Algorithm (GA) works on the theory of Darvin's theory of evolution and the survival-of-the fittest [16]. Genetic algorithms guide the search through the solution space by using natural selection and genetic operators, such as crossover, mutation and the selection.

GA encodes the decision variables or input parameters of the problem into solution strings of a finite length. While traditional optimization techniques work directly with the decision variables or input parameters, genetic algorithms usually work with the coding. Genetic algorithms start to search from a population of encoded solutions instead of from a single point in the solution space. The initial population of individuals is created at random. Genetic algorithms use genetic operators to create Global optimum solutions based on the solutions in the current population. The most popular genetic operators are (1) selection, (2) crossover and (3) mutation. The newly generated individuals replace the old population, and the evolution process proceeds until certain termination criteria are satisfied.

2.1.1 Selection

The selection procedure implements the natural selection or the survival-of-the fittest principle and selects good individuals out of the current population for generating the next population according to the assigned fitness. The existing selection operators can be broadly classified into two classes: (1) proportionate schemes, such as roulette-wheel selection and stochastic universal selection and (2) ordinal schemes, such as tournament selection and truncation selection. Ordinal schemes have grown more and more popular over the recent years, and one of the most popular ordinal selection operators is tournament selection. After selection, crossover and mutation recombine and alter parts of the individuals to generate new solutions.

2.1.2 Crossover

Crossover, also called the recombination operator, exchanges parts of solutions from two or more individuals, called parents, and combines these parts to generate new individuals, called children, with a crossover probability. There are a lot of ways to implement a recombination operator. The well-known crossover operators include one-point crossover. When using one-point crossover, only one crossover point is chosen at random, for example let there be two parent string A_1 and A_2 as:

$$\begin{aligned} A_1 &= 1 \quad 1 \quad 1 \quad 1 \mid 1 \quad 1 \\ A_2 &= \mathbf{0} \quad \mathbf{0} \quad \mathbf{0} \quad \mathbf{0} \mid \mathbf{0} \quad \mathbf{0} \end{aligned} \tag{2.1}$$

Then, one-point crossover recombines A_1 and A_2 and yields two offsprings A_{-1} and A_{-2} as:

$$\begin{aligned} A_{-1} &= 1 \quad 1 \quad 1 \quad 1 \mid 1 \quad 1 \\ A_{-2} &= \mathbf{0} \quad \mathbf{0} \quad \mathbf{0} \quad \mathbf{0} \mid \mathbf{1} \quad \mathbf{1} \end{aligned} \tag{2.2}$$

2.1.3 Mutation

Mutation usually alters some pieces of individuals to form perturbed solutions. In contrast to crossover, which operates on two or more individuals, mutation operates on a single individual. One of the most popular mutation operators is the bitwise mutation, in which each bit in a binary string is complemented with a mutation probability. For example,

$$\begin{aligned} A &= 1 \quad 1 \quad 1 \quad 1 \mid 1 \quad 1 \\ A_{-1} &= \mathbf{0} \quad \mathbf{0} \quad \mathbf{0} \quad \mathbf{0} \mid \mathbf{0} \quad \mathbf{1} \end{aligned} \tag{2.3}$$

The step-by-step implementation of GA is explained as follows:

Step 1: Initialize GA parameters which are necessary for the algorithm. These parameters include population size which indicates the number of individuals, number of generations necessary for the termination criterion, crossover probability, mutation probability, number of design variables and respective ranges for the design variables. If binary version of GA is used then string length is also required as the algorithm parameter.

Step 2: Generate random population equal to the population size specified. Each population member contains the value of all the design variables. This value of design variable is randomly generated in between the design variable range specified. In GA, population means the group of individuals which represents the set of solutions.
Step 3: Obtain the values of the objective function for all the population members. The value of the objective function so obtained indicates the fitness of the individuals. If the problem is a constrained optimization problem then a specific approach such as static penalty, dynamic penalty and adaptive penalty is used to convert the constrained optimization problem into the unconstrained optimization problem.
Step 4: This step is for the selection procedure to form a mating pool which consists of the population made up of best individuals. The commonly used selection schemes are roulette-wheel selection, tournament selection, stochastic selection, etc. The simplest and the commonly used selection scheme is the roulette-wheel selection, where an individual is selected for the mating pool with the probability proportional to its fitness value. The individual (solution) having better fitness value will have more number of copies in the mating pool and so the chances of mating increases for the more fit individuals than the less fit ones. This step justifies the procedure for the survival of the fittest.
Step 5: This step is for the crossover where two individuals, known as parents, are selected randomly from the mating pool to generate two new solutions known as off-springs. The individuals from the population can go for the crossover step depending upon the crossover probability. If the crossover probability is more, then more individuals get chance to go for the crossover procedure. The simplest crossover operator is the single point crossover in which a crossover site is determined randomly from where the exchange of bits takes place. The crossover procedure is explained through Eqs. 2.1 and 2.2.
Step 6: After crossover, mutation step is performed on the individuals of population depending on the mutation probability. The mutation probability is generally kept low so that it does not make the algorithm unstable. In mutation, a random site is selected from the string of individuals and it is flapped as explained through Eq. 2.3.
Step 7: Best obtained results are saved using elitism. All elite members are not modified using crossover and mutation operators but can be replaced if better solutions are obtained in any iteration.
Step 8: Repeat the steps (from step 3) until the specified number of generations or termination criterion is reached.

2.2 Artificial Immune Algorithm

The immune system defends the body against harmful diseases and infections. B cells recognize the antigens which enter into the body. B cells circulate through the blood. Each antigen has a particular shape that is recognized by the receptors present on the B cell surface. B cells synthesize and carry antibodies on their surfaces molecules that act like detectors to identify antigens. A B cell with better fitting receptors and binding more tightly the antigen replicate more and survive longer. This process of amplifying, by using proliferation, only those cells that produce a useful B cell type is called clonal selection [11, 21, 30, 38]. Clones are not perfect, but they are subjected to somatic permutations that result in children having slightly different antibodies from the parent. Clonal selection guarantees that only good B cells (i.e., with higher affinity with the antigen) can be cloned to represent the next generation [21]. However, clones with low affinity with antigen do not divide and will be discarded or deleted. Hence, the clonal selection enables the body to have sufficient numbers of antigen-specific B cells to build up an effective immune response. Mapping between the immune system and an optimization problem is done as follows. The immune response represents solutions and antigens represent the problem to solve. More precisely, B cells are considered as artificial agents that roam around and explore an environment. In other words, the optimization problem is described by an environment of antigens. The positive and negative selection mechanism is used to eliminate useless or bad solutions.

The AIA starts with a random population of antibodies [21]. Affinity of the antibody is decided from its objective function value. Select n highest antibodies to be cloned. These antibodies are cloned depending on its affinities. If the affinity is more for the particular antibody it will have more number of clones. It is calculated as

$$N_c = \sum_{i=1}^{n} \text{round}\left(\frac{\beta N}{i}\right) \tag{2.4}$$

where β is the multiplying factor controlling the number of clones and N is the total number of antibodies. These generate repertoire, which undergoes affinity maturation process as shown in Eq. 2.5, which is inversely proportional to its antigenic affinity. If the affinity is high the mutation rate is low.

$$x_{i,m} = x_i + A(\text{rand}[-1, 1])(x_{\max} - x_{\min}) \tag{2.5}$$

where, A is a factor depending on the affinity and decreases as affinity increases. Replace low affinity antibodies with new randomly generated antibodies given by Eq. 2.6

$$x_i = x_{\min} + \text{rand}(0, 1)(x_{\max} - x_{\min}) \tag{2.6}$$

The step-by-step implementation of AIA is explained as follows:

Step 1: Initialize AIA parameters which are necessary for the algorithm. These parameters include population size which indicates the number of individuals, number of generations necessary for the termination criterion, number of antibodies to be cloned, multiplying factor, repertoire rate, number of design variables and respective ranges for the design variables.

Step 2: Generate random population equal to the population size specified. Each population member contains the value of all the design variables. This value of design variable is randomly generated in between the design variable range specified. In AIA, population means the group of antibodies which represents the set of solutions.
Step 3: Obtain the values of the objective function for all the population members. The value of the objective function so obtained indicates antibody affinity. If the problem is a constrained optimization problem, then a specific approach such as static penalty, dynamic penalty and adaptive penalty is used to convert the constrained optimization problem into the unconstrained optimization problem.
Step 4: Select the n highest affinity antibodies from the population which comprises a new set of high affinity antibodies (Eq. 2.4). Clone the n selected antibodies independently and proportional to their affinities. This generates a group of clones. The higher the affinity, the higher the number of clones generated for each of the n selected antibodies.
Step 5: The group of clones undergoes affinity maturation process which is inversely proportional to its affinity (Eq. 2.5). A new set of solutions is generated consisting of matured clones. Determine the affinity of the matured clones. From this set of mature clones, reselect the highest affinity solutions. If the antigenic affinity of this solution is better than the previous iteration solution, then replace the population with the new one.
Step 6: Replace the lowest affinity antibodies from the population depending on the repertoire rate, by new antibodies using Eq. 2.6.
Step 7: Repeat the steps (from step 3) until the specified number of generations or termination criterion is reached.

2.3 Differential Evolution

The algorithm was first proposed by Storn and Price [35]. There are only three real control parameters in the algorithm. These are: (1) differentiation (or mutation) constant F, (2) crossover constant Cr and (3) size of population. The rest of the parameters are (a) dimension of problem S that scales the difficulty of the optimization task; (b) maximal number of generations (or iterations) G, which serves as a stopping condition in our case and (c) high and low boundary constraints, $x_{\max}$ and $x_{\min}$, respectively, that limit the feasible area. DE also starts with a set of random population which consist the initial solution to the problem. Mutant vector $v_{i,m}$ is generated from three different randomly chosen target vectors. This process can be mathematically written as [37],

$$v_{i,m} = x_{i,3} + F(x_{i,1} - x_{i,2}) \tag{2.7}$$

where, $v_{i,m}$ is the obtained mutant vector. In Eq. 2.7 the second term on RHS indicates the weighted difference of two randomly chosen target vectors. The mutant vector is obtained by adding the third target vector to the weighted difference term. New trial vector $u_{i,tar}$ is obtained from the target vector and the mutant vector based on the crossover probability *Cr*. The scaling factor *F* is a user-supplied constant. Trial vector and the current target vector is compared and the best out of them is forwarded to the next generation. The optimal value of *F* for most of the functions lies in the range of 0.4–1.0 [35].

The step-by-step implementation of DE is explained as follows:

Step 1: Initialize DE parameters which are necessary for the algorithm. These parameters include population size which indicates the number of individuals, number of generations necessary for the termination criteria, crossover constant, mutation constant, number of design variables and respective ranges for the design variables.
Step 2: Generate random population equal to the population size specified. Each population member contains the value of all the design variables. This value of design variable is randomly generated in between the design variable range specified. In DE, population means the group of solutions.
Step 3: Obtain the values of the objective function for all the solutions. If the problem is a constrained optimization problem, then a specific approach such as static penalty, dynamic penalty and adaptive penalty is used to convert the constrained optimization problem into the unconstrained optimization problem.
Step 4: Choose three different target vectors. The chosen target vectors should be different from the current target vector. Obtain the mutant vector using Eq. 2.7. In Eq. 2.7, *F* indicates the mutation constant.
Step 5: Obtain trial vector based on the crossover constant. If the crossover constant is greater than the random number between 0 and 1, then the mutant vector becomes the trial vector; otherwise, the current target vector becomes the trial vector.
Step 6: Selection is done between the trial vector and the current target vector. If the objective function value of trial vector is better than the current target vector, then the trial vector enters the new population.
Step 7: Repeat the steps (from step 3) until the specified number of generations or termination criterion is reached.

2.4 Biogeography-Based Optimization

Biogeography-based optimization (BBO) is a population-based optimization algorithm inspired by the natural biogeography distribution of different species [34]. In BBO, each individual is considered as a “habitat” with a habitat suitability index (HSI). A good solution is analogous to an island with a high HSI, and a poor solution indicates an island with a low HSI. High HSI solutions tend to share their features with low HSI solutions. Low HSI solutions accept a lot of new features from high HSI solutions.

In BBO, each individual has its own immigration rate λ and emigration rate μ. A good solution has higher μ and lower λ and vice versa. The immigration rate and the emigration rate are functions of the number of species in the habitat. They can be calculated as follows

$$\lambda_k = I\left(1 - \frac{k}{n}\right) \tag{2.8}$$

$$\mu_k = E\left(\frac{k}{n}\right) \tag{2.9}$$

where, I is the maximum possible immigration rate; E is the maximum possible emigration rate; k is the number of species of the kth individual and n is the maximum number of species. In BBO, there are two main operators, the migration and the mutation.

2.4.1 Migration

Consider a population of candidate which is represented by design variable. Each design variable for particular population member is considered as suitability index variable (SIV) for that population member. Each population member is considered as individual habitat/Island. The objective function value indicates the HSI for the particular population member. Immigration and emigration rates are decided from the curve given in Simon [34]. The nature of the curve is assumed to be same for immigration and emigration but with opposite slopes, which behaves linearly. Value of S represented by the solution depends on its HSI. The emigration and immigration rates of each solution are used to probabilistically share the information between habitats. If a given solution is selected to be modified, then its immigration rate λ is used to probabilistically modify each SIV in that solution. If a given SIV in a given solution S_i is selected to be modified, then its emigration rates μ of the other solutions are used to probabilistically decide which of the solutions should migrate its randomly selected SIV to solution S_i. The above phenomenon is known as migration in BBO.

2.4.2 Mutation

In nature a habitat's HSI can change suddenly due to apparently random events (unusually large flotsam arriving from a neighboring habitat, disease, natural catastrophes, etc.). This phenomenon is termed as SIV mutation, and probabilities of species count are used to determine mutation rates. This probability mutates low HSI as well as high HSI solutions. Mutation of high HSI solutions gives them the chance to further improve. Mutation rate is obtained by using following Eq. 2.10.

$$m(S) = m_{\max}\left(1 - \frac{P_s}{P_{\max}}\right) \tag{2.10}$$

where, $m_{\max}$ is a user-defined parameter called mutation coefficient.

The step-by-step procedure about the implementation of BBO is explained as follows:

Step 1: Initialize BBO parameters which are necessary for the algorithm. These parameters include population size which indicates the number of habitats/islands, number of generations necessary for the termination criterion, maximum immigration and emigration rates, mutation coefficient, number of design variables and respective ranges for the design variables.
Step 2: Generate random population equal to the population size specified. Each population member contains the value of all the design variables. This value of design variable is randomly generated in between the design variable range specified. Every design variable in the population indicates SIVs for that respective population member (Habitat).
Step 3: Obtain the value of objective function for all population members. The value of objective function so obtained indicates the HSI for that Habitat (population member). If the problem is a constrained optimization problem, then a specific approach such as static penalty, dynamic penalty and adaptive penalty is used to convert the constrained optimization problem into the unconstrained optimization problem.
Step 4: Map the value of HSI to obtain the species count. High species count is allotted to the population member having high HSI for maximization optimization problem. If the optimization problem is of minimization type then low HSI member is given high species count.
Step 5: Modify the population using the migration operator considering its immigration and emigration rates. If a given solution is selected to be modified, then its immigration rate λ is used to probabilistically modify each suitability index variable (SIV) in that solution. If a given SIV in a given solution *Si* is selected to be modified, then its emigration rates μ of the other solutions are used to probabilistically decide which of the solutions should migrate the randomly selected SIV to solution S_i. Pseudo code for migration is given as follows.
Select H_i with probability proportional to λ_i (H_i is any solution vector)

```
If H_i is selected
    For j = 1 to n (n is population size)
        Select H_j with probability proportional to μ_i
        If H_j is selected
            Randomly select an SIV σ from H_j
            Replace a random SIV in H_i with σ
        end
    end
end
```

Step 6: Modify population using mutation operator. Calculate probability of existence from the value of immigration and emigration rates as explained earlier. Also calculate the mutation rate considering the user-defined mutation coefficient and probability of existence. The pseudo code for mutation is given as follows:

```
For j = 1 to m (m is number of design variables)
        Use λ_i and μ_i to compute the probability P_i
        Select SIV H_i(j) with probability proportional to P_i and mutation rate
        If H_i(j) is selected
                Replace H_i(j) with a randomly generated SIV
        end
end
```

Step 7: Best obtained results are saved using elitism. All elite members are not modified using migration and mutation operators but can be replaced if better solutions are obtained in any iteration.
Step 8: Repeat the steps (from step 3) until the specified number of generations or termination criterion is reached.

2.5 Particle Swarm Optimization

Particle swarm optimization (PSO) is an evolutionary computation technique developed by Kennedy and Eberhart [20]. It exhibits common evolutionary computation attributes including initialization with a population of random solutions and searching for optima by updating generations. Potential solutions, called particles, are then "flown" through the problem space by following the current optimum particles. The particle swarm concept was originated as a simulation of a simplified social system. The original intent was to graphically simulate the graceful but unpredictable choreography of a bird flock. Each particle keeps track of its coordinates in the problem space, which are associated with the best solution (fitness) it has achieved so far. This value is called '*pBest*'. Another "best" value that is tracked by the global version of the particle swarm optimization is the overall best value and its location obtained so far by any particle in the population. This location is called '*gBest*'. The particle swarm optimization concept consists of, at each step, changing the velocity (i.e. accelerating) of each particle toward its '*pBest*' and '*gBest*' locations (global version of PSO). Acceleration is weighted by a random term with separate random numbers being generated for acceleration toward '*pBest*' and '*gBest*' locations. The updates of the particles are accomplished as per the following Eqs. 2.11 and 2.12.

$$V_{i+1} = w^{*} V_i + c_1^* r_1^* (pBest_i - X_i) + c_2^* r_2^* (gBest_i - X_i) \tag{2.11}$$

$$X_{i+1} = X_i + V_{i+1} \tag{2.12}$$

Equation 2.11 calculates a new velocity (V_{i+1}) for each particle (potential solution) based on its previous velocity, the best location it has achieved ('*pBest*') so far, and the global best location ('*gBest*'), the population has achieved. Equation 2.12 updates individual particle's position (X_i) in solution hyperspace. The two random numbers 'r_1' and 'r_2' in Eq. 2.11 are independently generated in the range [0, 1]. It is observed from Eq. 2.11 that the LHS indicates the velocity term and the RHS has three terms: the first term contains the multiplication of w and V_i, where w is the constant parameter and V_i is the velocity term which indicates the correct dimension as that of LHS, the second and third terms indicate the rate of change of position toward $pBest_i$ and $gBest_i$ from the current position X_i respectively and so both the terms are to be multiplied by 1/Δt, where Δt indicates the time step value. To simplify the algorithm and to reduce the algorithm parameters, the value of Δt is assumed to be unity. Moreover, in Eq. 2.12 the second term on RHS is to be multiplied by Δt, which reduces the term to match the dimension of the position (X_{i+1}) on LHS. So, the Eqs. 2.11 and 2.12 are the final equations after assuming the value of Δt as unity.

The acceleration constants 'c_1' and 'c_2' in Eq. 2.11 represent the weighting of the stochastic acceleration terms that pull each particle toward '*pBest*' and '*gBest*' positions. 'c_1' represents the confidence the particle has in itself (cognitive parameter) and 'c_2' represents the confidence the particle has in swarm (social parameter). Thus, adjustment of these constants changes the amount of tension in the system. Low values of the constants allow particles to roam far from target regions before being tugged back, while high values result in abrupt movement toward, or past through target regions [9]. The inertia weight 'w' plays an important role in the PSO convergence behavior since it is employed to control the exploration abilities of the swarm. The large inertia weights allow wide velocity updates allowing to globally explore the design space while small inertia weights concentrate the velocity updates to nearby regions of the design space. The optimum use of the inertia weight "w" provides improved performance in a number of applications. The effect of w, c_1 and c_2 on convergence for standard numerical benchmark functions was provided by Bergh and Engelbrecht [4].

Particle's velocities on each dimension are confined to a maximum velocity parameter V_{max}, specified by the user. If the sum of accelerations would cause the velocity on that dimension to exceed V_{max}, then the velocity on that dimension is limited to V_{max}.

Unlike genetic algorithm, PSO algorithm does not need complex encoding and decoding process and special genetic operator. PSO takes real number as a particle in the aspect of representation solution and the particles update themselves with internal velocity. In this algorithm, the evolution looks only for the best solution and all particles tend to converge to the best solution.

The step-by-step implementation of PSO is explained as follows:

Step 1: Initialize PSO parameters which are necessary for the algorithm. These parameters include population size which indicates the number of individuals, number of generations necessary for the termination criterion, cognitive constant, social constant, variation of inertia weight, maximum velocity, number of design variables and respective ranges for the design variables.

Step 2: Generate random population equal to the population size specified. Each population member contains the value of all the design variables. This value of design variable is randomly generated in between the design variable range specified. In PSO, population means the group of birds (particles) which represents the set of solutions.
Step 3: Obtain the values of the objective function for all the population members. For the first iteration, value of objective function indicates the *pBest* for the respective particle in the solution. Identify the particle with best objective function value which identifies as *gBest*. If the problem is a constrained optimization problem, then a specific approach such as static penalty, dynamic penalty and adaptive penalty is used to convert the constrained optimization problem into the unconstrained optimization problem.
Step 4: Update the velocity of each particle using Eq. 2.11. Check for the maximum velocity. If the velocity obtained using Eq. 2.11 exceeds the maximum velocity, then reduce the existing velocity to the maximum velocity.
Step 5: Update the position of the particles using Eq. 2.12. Check all the design variables for the upper and lower limits.
Step 6: Obtain the value of objective function for all the particles. The new solution replaces the *pBest* if it has better function value. Identify the *gBest* from the population. Update the value of inertia weight if required.
Step 7: Best obtained results are saved using elitism. All elite members are not modified using crossover and mutation operators but can be replaced if better solutions are obtained in any iteration.
Step 8: Repeat the steps (from step 4) until the specified number of generations or termination criterion is reached.

2.5.1 Modifications in PSO

PSO suggested by Kennedy and Eberhart [20] had no inertia factor term in the algorithm. It was first suggested by Shi and Eberhert [33] and was shown that PSO performs better with introduction of inertia weight factor term. Many research works were reported for the variation of w to increase the performance of PSO. Shi and Eberhart [33] suggested linear variation of weight factor by using following expression:

$$w = ((\max w - \min w)^{*}(\max iter - \mathrm{cur}iter)/\max iter) + \min w \tag{2.13}$$

where, max$w$ and min$w$ are the maximum and minimum value of weight factor (w) respectively; maxiter is the maximum number of generations and curiter is the current iteration of the algorithm. Generally max$w$ is taken as 0.9 and min$w$ as 0.4. Xiaohui et al. [40] suggested random weight factor as:

$$w = 0.5 + 0.5^{*}(\mathrm{rand}) \tag{2.14}$$

where, rand is any random number from 0 to 1. Yong et al. [44] presented Chaotic descending inertia weight. The strategy for the logistic mapping changes inertia weight as:

$$w = \left((\max w - \min w)^{*}(\max iter - \operatorname{cur} iter)/\max iter\right) + \min w^{*}(zr) \qquad (2.15)$$

where, zr = 4 * (rand) * (1−rand). Chaotic descending inertia weight is also applied to the inertia weight suggested by Xiaohui et al. [40] as

$$w = 0.5^{*}(zr) + 0.5^{*}(\text{rand}) \qquad (2.16)$$

where, rand is any random number between 0 and 1.

So, it is observed that there is a significant role of inertia weight for the performance of PSO. Experimentation is carried out in this book to suggest a new inertia weight for the PSO so as to increase its performance. A new variation of inertia weight variation is suggested in this book to increase the success rate for finding the global solution in a few iterations. This saves computation time and less number of function evaluations will be required to arrive at the optimum solution. The procedure to alter the weight factor is explained below.

Set the initial value for the weight (generally 0.9)
Start loop
Set New$w$ = w
Perform PSO operation
w_factor = New$w$/max*iter*
Set w = w–*w_factor*
End loop

The above variation of weight factor follows a nonlinear behavior and it depends on the value of initial weight and maximum number of generations. PSO with the above inertia weight factor is referred to as PSO_M_1 in this book. Moreover, Chaotic descending inertia weight suggested by Yong et al. [44] is also incorporated in the modified inertia weight. Chaotic descending inertia weight changes the value of New$w$=w as New$w$=w*(zr). PSO with modified inertia weight and chaotic descending inertia weight is referred to as PSO_M_2 in this book.

2.6 Artificial Bee Colony Algorithm

Artificial Bee Colony (ABC) algorithm is an optimization algorithm based on the intelligent foraging behavior of honey bee swarm. The colony of artificial bees consists of three groups of bees: employed bees, onlookers and scouts [3, 18]. An employed bee searches the destination where food is available. They collect the food and return back to its origin, where they perform waggle dance depending on the amount of food available at the destination. The onlooker bee watches the dance and follows the employed bee depending on the probability of the available food.

So, more onlooker bees will follow the employed bee associated with the destination having more amount of food. The employed bee whose food source becomes abandoned behaves as a scout bee and it searches for the new food source. This principle of foraging behavior of honey bee is used to solve optimization problems by dividing the population into two parts consisting of employed bees and onlooker bees. An employed bee searches the solution in the search space and the value of objective function associated with the solution is the amount of food associated with that solution. Employed bee updates its position by using Eq. 2.17 and it updates new position if it is better than the previous position, i.e. it follows greedy selection.

$$v_{ij} = x_{ij} + R_{ij}(x_{ij} - x_{kj}) \tag{2.17}$$

where, v_{ij} is the new position of employed bee, x_{ij} is the current position of employed bee, k is a random number between (1, (*population size*)/2) $\neq i$ and $j =$ 1, 2,…, *Number of design variables.* R_{ij} is a random number between (−1, 1).

An onlooker bee chooses a food source depending on the probability value associated with that food source, p_i, calculated by using Eq. 2.18.

$$p_i = \frac{F_i}{\sum_{n=1}^{N/2} F_n} \tag{2.18}$$

where, F_i is the fitness value of the solution i and *N/2* is the number of food sources which is equal to the number of employed bees.

Onlooker bees also update its position by using Eq. 2.17 and also follow greedy selection. The Employed bee whose position of the food source cannot be improved for some predetermined number of cycles than that food source is called abandoned food source. That employed bee becomes scout and searches for the new solution randomly by using Eq. 2.19.

$$x_i^j = x_{\min}^j + \text{rand}\,(0,1)(x_{\max}^j - x_{\min}^j) \tag{2.19}$$

The value of predetermined number of cycles is an important control parameter of the ABC algorithm, which is called "*limit*" for abandonment. The value of limit is generally taken as *Number of employed bees*Number of design variables.*

The step-by-step implementation of ABC is explained as follows:

Step 1: Initialize ABC parameters which are necessary for the algorithm. These parameters include population size which indicates the number of individuals, number of generations necessary for the termination criterion, number of employed bees, number of onlooker bees, limit, number of design variables and respective ranges for the design variables.

Step 2: Generate random population equal to the number of employed bees (generally number of employed bees are half of the population size) specified. Each population member contains the value of all the design variables. This value of design variable is randomly generated in between the design variable range

specified. In ABC, population means the group of honey bees which represents the set of solutions.
Step 3: Obtain the values of the objective function for all the population members. The objective function value in ABC indicates the amount of nectar for the food source. If the problem is a constrained optimization problem, then a specific approach such as static penalty, dynamic penalty and adaptive penalty is used to convert the constrained optimization problem into the unconstrained optimization problem.
Step 4: Update the value of employed bees using Eq. 2.17. Obtain the value of objective function. If the new solution is better than the existing solution, replace the existing solution with the new one. This step indicates the greedy selection procedure for the employed bee phase.
Step 5: Onlooker bees proportionally choose the employed bees depending on the amount of nectar found by the employed bees. Mathematically, for the onlooker bee phase, the solution from the employed bee phase is chosen proportionally based on its objective function value (Eq. 2.18).
Step 6: Update the value of onlooker bees using Eq. 2.17. Obtain the value of objective function. If the new solution is better than the existing solution, replace the existing solution with the new one. Identify the abundant solutions using the limit value. If such solutions exist then these are transformed into the scout bees and the solution is updated using Eq. 2.19.
Step 7: Repeat the steps (from step 4) until the specified number of generations or termination criterion is reached.

2.6.1 Modifications in ABC

As suggested by Karaboga [18], ABC modifies the solution by using the following Eq. 2.20:

$$v_{ij} = x_{ij} + R_{ij}\left(x_{ij} - x_{kj}\right) \tag{2.20}$$

where, R_{ij} is uniformly distributed random number between −1 and 1. Modification in ABC is carried out by changing Eq. 2.20. Uniformly distributed random number is replaced by normally distributed random number with mean equal to zero and svariance equal to one. And also the expression $(x_{ij} - x_{ik})$ is replaced by $(x\text{best}_j - x_{ij})$. Here, $x\text{best}_j$ is the best solution from the population at any particular iteration. The reason for this modification is that, in the Eq. 2.20 the solution tries to move toward any random solution (x_{ik}) and there is no guarantee for the x_{ik} to be better than x_{ij}. So solution can move toward worst solution also, which may require more computational time to reach the optimum solution. By replacing x_{ik} with $x\text{best}_j$, the solution will try to move toward the best solution in every iteration which will lead to optimum solution with less computational effort.

2.7 Harmony Elements Algorithm

According to Chinese philosophy, the five kinds of substances (wood, fire, earth, metal and water) are essential things in the daily life of mankind. Among the five elements, there exist the relations of generation and restriction [24, 46]. The order of generation is: wood generates fire, fire generates earth, earth generates metal, metal generates water and water, in its turn, generates wood. Relationship of restriction for the five elements works in the following order: wood restricts earth, earth water, water fire, fire metal and metal wood. So, they oppose each other and at the same time cooperate with each other, thus a relative balance is maintained between generation and restriction, to ensure normal growth and development of things in nature.

Harmony elements algorithm follows the generation and restriction rules between the elements of the string. It starts the procedure with a random population. Like GA, each individual in the population is made up of string which represents the design variables. Dissimilar to GA, the algorithm initializes the solutions as strings of 0s, 1s, 2s, 3s and 4s to represent 'earth', 'water', 'wood', 'fire' and 'metal', five elements, respectively. Population is modified according to generation and restriction rules to reach its harmonious state.

2.7.1 Modifications in HEA

Harmony Elements Algorithm starts with a random population. Population size, length of each individual string, number of input variables, upper bound and lower bound of input variables are to be initialized at the start of the algorithm. The individual strings will consist of 0s, 1s, 2s, 3s and 4s. Each number corresponds to an element. In this book the initial population matrix is denoted by Q. The basic version of the algorithm reported by Cui and Guo [6] generates random population equal to the population size. Here only one-fifth of the total population size is randomly generated; rest is generated from Q following the generation rule of five elements. This reduces the functional evolutional by 4*population size*number of generations. Four different matrices A, B, C and D are generated from matrix Q by following generation rule (Generation rule: 2 create 3, 3 create 0, 0 create 4, 4 create 1 and 1 create 2.). The above procedure helps to maintain initial harmonious state in the population. The basic algorithm generates one random matrix E equal to the population size to maintain the diversity in the population. Modification is incorporated by introducing the mutation operator to reduce the function evaluations by 1*population size*number of generations. Mutation is incorporated depending on the specified probability. Generally probability for the mutation is very low. The purpose of mutation is to maintain the diversity within the population so that algorithm does not get trapped in local optima. Mutation is carried out by changing the element in the string at randomly selected site. The above procedure for the mutation is same as that in GA. Mutation in HEA is shown as follows:

Before mutation

| 4 3 2 0 2 0 0 4 1 1 1 2 |

⇩ Mutation

After mutation

| 4 3 2 0 2 0 0 4 3 1 1 2 |

There is no bound to follow generation and restriction rule for the mutation. Decode each individual string and evaluate the fitness values of matrix Q, A, B, C and D. Fitness value corresponds to the value of objective function. Rank individual strings in matrix Q, A, B, C and D by fitness value. Check all the matrices for the restriction rule. As all the matrices are ranked and rearranged, the first row of the matrix represents the best string so far. For instance, the best individual string is Q (1, j), for matrix Q. If other individual strings Q (i+1, j) is restricted by Q (1, j), then it is modified by replacing the element following the generation rule. If Q (i, j) and Q (i+1, j) are same, then both of them are modified according to generation rule. (i indicates the population member and j indicated the element number in the string). The restriction rule in HEA is given as follows:

1	0	3	4	3	4	4	1	2	2	2	3
2	3	3	0	0	0	4	1	1	2	2	4
0	0	3	1	4	4	4	2	1	1	2	2

⇩ Generation based on restriction

1	0	3	4	3	4	4	1	2	2	2	3
2	3	0	0	0	0	1	1	2	2	3	1
0	0	0	1	1	4	1	2	2	1	3	2

Merge all the matrices A, B, C, D and Q into one matrix. Decode each individual string and evaluate the fitness values of matrix and then rank individual strings of matrix by fitness value. The so obtained matrix will be the new matrix Q for the next iteration.

The step-by-step procedure about the implementation of HEA is explained as follows:

Step 1: Initialize HEA parameters which are necessary for the algorithm to proceed. These parameters include population size, number of generations necessary for the termination criterion, string length, number of design variables, design variable ranges, and mutation rate.
Step 2: Generate initial population considering all the design variables. All the design variables are initially coded using the string consisting of 0, 1, 2, 3 and 4. Strings for each design variables are combined to form a single string. Initial population is created considering the combined strings which gives matrix Q as

Matrix Q: Randomly generated population

1	0	3	4	3	4	4	1	2	2	2	3	.	.
2	3	3	0	0	0	4	1	1	2	2	4	.	.
0	0	3	1	4	4	4	2	1	1	2	2	.	.
.	.	.	.	.	.	.	.	.	.	.	.		
.	.	.	.	.	.	.	.	.	.	.	.		

Step 3: All other matrices A, B, C and D are created from matrix Q following the generation rule (Generation rule: 2s create 3s, 3s create 0s, 0s create 4s, 4s create 1s, 1s create 2s) as given below.

Matrix Q: Randomly generated population

1	0	3	4	3	4	4	1	2	2	2	3
⋮	⋮	⋮	⋮	⋮	⋮	⋮	⋮	⋮	⋮	⋮	⋮

Generation rule
Corresponding Matrix A: **Created from Matrix Q**

2	4	0	1	0	1	1	2	3	3	3	0
⋮	⋮	⋮	⋮	⋮	⋮	⋮	⋮	⋮	⋮	⋮	⋮

Generation rule
Corresponding Matrix B: **Created from Matrix A**

3	1	4	2	4	2	2	3	0	0	0	4
⋮	⋮	⋮	⋮	⋮	⋮	⋮	⋮	⋮	⋮	⋮	⋮

Generation rule
Corresponding Matrix C: **Created from Matrix B**

0	2	1	3	1	3	3	0	4	4	4	1
⋮	⋮	⋮	⋮	⋮	⋮	⋮	⋮	⋮	⋮	⋮	⋮

Generation rule
Corresponding Matrix D: **Created from Matrix C**

4	3	2	0	2	0	0	4	1	1	1	2
⋮	⋮	⋮	⋮	⋮	⋮	⋮	⋮	⋮	⋮	⋮	⋮

After the creation of all the matrices, all the design variables are decoded and mapped for the considered design variable range.

Step 4: Mutation is carried out considering the mutation probability. Mutation consists of changing one of the bits randomly in the chromosome and so it leads to maintain the diversity in the population and it also does not allow the algorithm to get trapped at local optima. Mutation phenomenon is explained in the earlier section. After performing mutation operation all the matrices are arranged in the ascending order of their objective function value as all the considered problems are for the minimization.

Step 5: Apply generation based on restriction rule for all the matrices and then merge all the matrices to form a single matrix which is the total population size considered. After merging all the matrices they are again arranged in the ascending order. From this arranged matrix one-fifth of the population is selected starting from row one and gives again matrix Q for the next generation. This leads to the best population members obtained in the particular generation. This completes one generation.

Step 6: Repeat the steps (from step 3) until the specified termination criterion is reached.

2.8 Hybrid Algorithms

For the population-based optimization methods, the terms exploration and exploitation have been playing an important role in describing the working of an algorithm. Use of existing information is known as ‘exploitation’. Generation of new solutions in the search space is termed as ‘exploration’. As exploitation and exploration are the opposing forces, its balance is required for the algorithm to search for the global optimum solutions. Any selection procedure in the algorithm is generally characterized as exploitation because the fitness (information) of the individuals is used to determine whether or not an individual should be exploited. So, exploration and exploitation are two important aspects in the population-based optimization algorithms. However, different algorithms employ different operators for exploration and exploitation.

In ABC, a new solution vector is calculated using the current solution and a randomly chosen solution from the population indicates the explorative ability of the algorithm. Moreover, a fitness-based probabilistic selection scheme is used in the ABC algorithm which indicates the exploitation tendency of the algorithm. ABC also has the diversification controlled by the random selection process in the scout bees phase which makes ABC escape from local minima. However, in ABC, a greedy selection scheme is applied between the new solution and the old one and the better one is preferred for inclusion in the population which once again indicates the exploitation tendency of the algorithm. In PSO, a new position vector is calculated using the particle’s current and best solution and the swarm’s best solution. In PSO, the new solution is replaced with the old one without considering which one is better. So, PSO has only explorative tendency and it lacks the exploitation ability. In DE, the existing solution is updated by the difference of the two existing solutions which is weighted by a constant scaling factor, while in ABC it is weighted by a random step size. So, DE also possesses the explorative ability like ABC for updating the solutions. DE also has explicit crossover and also employs greedy selection between the current solution and a new solution. The crossover and greedy selection indicate the exploitation tendency of the DE algorithm. GA also uses both exploration and exploitation of the solutions. The crossover and the mutation operators indicate the exploration ability of the GA algorithm. The selection scheme employed in GA algorithm indicates its exploitation tendency as the information of the individuals is used for the further processes of the algorithm. BBO works by exchanging the design variables from one solution to the other based on the immigration and emigration rate which is similar to the crossover procedure of the GA and so it indicates the explorative ability of the algorithm. But, the mutation process in BBO uses the probability of the solution to decide whether the solution is to be mutated or not. So, mutation process in BBO indicates the exploitation tendency of the algorithm.

It is observed from the above discussion that all the described algorithms have different exploration and exploitation ability. ABC updates the solution in the employed bee phase and generates the new solution by exploration and the greedy selection process is done on the new solution by exploitation. Furthermore, the solutions are exploited by using the proportional selection in the onlooker bee phase and again the new solutions are generated in the onlooker bee phase by exploration. Moreover, the exploration in the employed bee phase and the onlooker bee phase is similar as it is using the similar mathematical expression. Motivated by this point, it is decided to investigate the onlooker bee phase by using some other mathematical expression for the exploration. ABC uses three different exploitation mechanisms (two for the greedy selection and one for the proportional selection) in the onlooker bee phase. The investigation is also carried out to reduce the exploitation in ABC by removing the proportional selection of the onlooker bee. The exploration of the onlooker bee is replaced by the exploration mechanisms of PSO, DE, BBO and GA separately which results in the four different hybrid algorithms.

All hybrid algorithms are developed by keeping ABC as the common algorithm. The four different hybrid algorithms that are developed are HPABC (Hybrid Particle swarm based Artificial Bee Colony), HBABC (Hybrid Biogeography-based Artificial Bee Colony), HDABC (Hybrid Differential evolution based Artificial Bee Colony) and HGABC (Hybrid Genetic algorithm based Artificial Bee Colony). All the developed hybrid algorithms start with the employed bee phase of ABC and then the onlooker bee phase is replaced by the searching mechanism of other algorithms. All the hybrid algorithms are discussed as follows:

2.8.1 HPABC

Both ABC and PSO are good at exploring the search space. HPABC is developed to combine the advantages of both ABC and PSO. HPABC starts with the initial population and updates the solution by following the searching mechanism of the employed bees in ABC. The solutions obtained after the employed bee phase follows the mechanism of particle swarm optimization. The pseudo code for HPABC is given below:

START

Initialize Population size, number of generations, value of w, c_1 and c_2, $V_{\max}$ and range of design variables.

Generate the initial population and evaluate the fitness for each individual

For i =1 to number of generations

For i = 1 to *Population size*

Produce new solutions for the employed bees and evaluate them (Eq. 2.17)

Replace new solution if it is better than the previous one

End

For i = 1 to Population size

Calculate the velocity of each solution (Eq. 2.11)
Check the obtained velocity for the limit ($V_{\max}$)
Produce new solutions (Eq. 2.12)
Replace new solution if it is better than the previous

End

End
STOP

It is observed from the above pseudo code that there is only a little increase in the computational effort of HPABC as compared to basic ABC. However, HPABC eliminates the proportional selection for the onlooker bees and also the scout bees. Solution is updated after the employed bee phase by following the search mechanism of particle swarm optimization and hence it combines the strength of both the algorithms.

2.8.2 *HBABC*

ABC is good at exploring the search space and locating the region of global minimum. On the other hand, BBO has a good exploitation searching tendency for global optimization. Based on these considerations, in order to maximize the exploration and the exploitation a HBABC approach is proposed which combines the strength of ABC and BBO. The pseudo code for HBABC is given below:

START

Initialize Population size, number of generations, immigration rates, emigration rates, mutation rate and range of design variables.

Generate the initial population and evaluate the fitness for each individual

For i = 1 to number of generations

For $i = 1$ to *Population size*

Produce new solutions for the employed bees and evaluate them (Eq. 2.17)

Replace new solution if it is better than the previous one

End

For each individual, map the fitness to the number of species

Calculate the immigration rate λ_i and the emigration rate μ_i for each individual X_i

For $i = 1$ to *Population size*

Select X_i with probability proportional to λ_i

if rand(0, 1) < λ_i

For $j = 1$ to N

Select X_j with probability proportional to μ_j

if rand(0, 1) < μ_j

Randomly select a variable σ from X_j

Replace the corresponding variable in X_i with σ

Endif

Endif

End

Replace new solution if it is better than the previous one

End

End

STOP

It is observed from the above pseudo code that there is only a little increase in the computational effort of HBABC as compared to basic ABC. However, HBABC eliminates the proportional selection for the onlooker bees and also the scout bees. Solution is updated after the employed bee phase by following the search mechanism of Biogeography-based optimization and hence it combines the strength of both the algorithms.

2.8.3 HDABC

ABC and DE have different searching capability and the searching mechanism. Both the algorithms are good at exploring the search space. HDABC is developed to combine the advantages of both ABC and DE. HDABC starts with the initial population and updates the solution by following the searching mechanism of the employed bees in ABC. The solutions obtained after the employed bee phase follows the mechanism of differential evolution. The pseudo code for HDABC is given below.

START

Initialize Population size, number of generations, value of F, and C and range of design variables.

Generate the initial population and evaluate the fitness for each individual

For i = 1 to number of generations

For i = 1 to *Population size*

Produce new solutions for the employed bees and evaluate them (Eq. 2.17)

Replace new solution if it is better than the previous one

End

For i = 1 to *Population size*

Generate mutant vector by using three randomly selected solutions (Eq. 2.7)

Generate trial vector based on crossover probability

If trial vector is better than the current target vector, replace the current solution with the trial solution.

End

End

STOP

It is observed from the above pseudo code that there is only a little increase in the computational effort of HDABC as compared to basic ABC. However, HDABC eliminates the proportional selection for the onlooker bees and also the scout bees. Solution is updated after the employed bee phase by following the search mechanism of differential evolution and hence it combines the strength of both the algorithms.

2.8.4 HGABC

ABC and GA are also having different searching capability and the searching mechanism. ABC is good in the exploration of the search space while GA uses both exploration and exploitation for finding the solution. HGABC is developed to combine the advantages of both ABC and DE. HGABC also starts with the initial population and updates the solution by following the searching mechanism of the employed bees in ABC. The solutions obtained after the employed bee phase follows the mechanism of genetic algorithm to further enhance the solution. The pseudo code for HGABC is given below.

```
START
        Initialize Population size, number of generations, crossover probability,
    mutation probability and range of design variables.
        Generate the initial population and evaluate the fitness for each individual
For i=1 to number of generations
        For i = 1 to Population size
            Produce new solutions for the employed bees and evaluate them
            (Eq. 2.17)

            Replace new solution if it is better than the previous one
        End
        For i = 1 to Population size
            Update solutions by using crossover according to crossover probability
            Update solutions by using mutation according to mutation probability
            Replace solutions if it is better than the existing
        End
End
STOP
```

It is observed from the above pseudo code that there is only a little increase in the computational effort of HGABC as compared to basic ABC. However, HGABC eliminates the proportional selection for the onlooker bees and also the scout bees. Solution is updated after the employed bee phase by following the search mechanism of genetic algorithm and hence it combines the strength of both the algorithms.

2.9 Shuffled Frog Leaping Algorithm

The shuffled frog leaping algorithm is an algorithm based on memetic meta-heuristic. It was brought forward and developed by Eusuff et al. [13]. This algorithm uses the mode of memetic evolvement among frog subgroups in local exploration. The algorithm uses the shuffled strategy and allows the message changing in local exploration. The shuffled frog leaping algorithm combines the advantages of memetic evolvement algorithm and particle swarm optimization (PSO). The algorithm changes message not only in the local exploration but also in the global exploration. So, the local and the global are combined well in the SFLA. The local search makes memetic to transfer among the individuals and the shuffled strategy makes memetic to transfer among the global. As genetic algorithm (GA) and particle swarm optimization (PSO), the shuffled frog leaping algorithm (SFLA) is an optimization algorithm based on colony.

The SFLA is a combination of determinacy method and random method. The determinacy strategy allows the algorithm to exchange messages effectively. The randomicity ensures the algorithm's flexibility and robustness. The SFLA progresses by transforming frogs (solutions) in a memetic evolution. In this algorithm, individual frogs are not so important; rather, they are seen as hosts for memes and described as a memetic vector. In the SFLA, the population consists of a set of frogs (solutions) that is partitioned into subsets referred to as memeplexes. The different memeplexes are considered as different cultures of frogs, each performing a local search. Within each memeplex, the individual frogs hold ideas, which can be influenced by the ideas of other frogs, and evolve through a process of memetic evolution. After a defined number of memetic evolution steps, ideas are passed among memeplexes in a shuffling process. The local search and the shuffling processes continue until defined convergence criteria are satisfied.

The algorithm begins the random selection of frog groups. The frog groups are divided to some subgroups. These subgroups can accomplish the local exploration independently and in different directions. A frog of each subgroup can affect others in the subgroup. So, they undergo the memetic evolution. The memetic evolution improves the individual memetics quality and strengthens the executive ability to goal. It is possible to increase the good frog's weight and to decrease the bad frog's weight for a good goal. When some memetics accomplish the evolution, the frog subgroups are shuffled. The memetics are optimized in global scope and produce some new frog subgroups by mixture mechanism.

The shuffling enhances the quality of memetics which are affected by the different subgroups. The local exploration and the global exploration are shuffled until the end of the constringency condition. The balance strategy between the global message exchange and local deep search makes the algorithm to jump out the local extremum easily (Yue et al., [45]). The flowchart of SFLA algorithm is shown in Fig. 2.1.

2.10 Grenade Explosion Algorithm

The idea of the presented algorithm is based on observation of a grenade explosion, in which the thrown pieces of shrapnel destruct the objects near the explosion location. Le is the length of explosion along each coordinate, in which the thrown piece of shrapnel may destruct the objects. The loss caused by each piece of shrapnel is calculated. A high value for loss per piece of shrapnel in an area indicates there are valuable objects in that area. To make more loss, the next grenade is thrown where the greatest loss occurs. Although the objects near grenade's location are more likely to be damaged, the probability of destruction is still kept for farther objects by choosing a high value for Le. This process would result in finding the best place for throwing the grenades, even though shrapnel cannot discover this area in early iterations. The loss caused by destruction of an object is considered as the fitness of the objective function at the object's location. Suppose that X is the current location of a grenade and $X = \{X_m\}$, $m = 1, 2, \ldots, n$. n is the

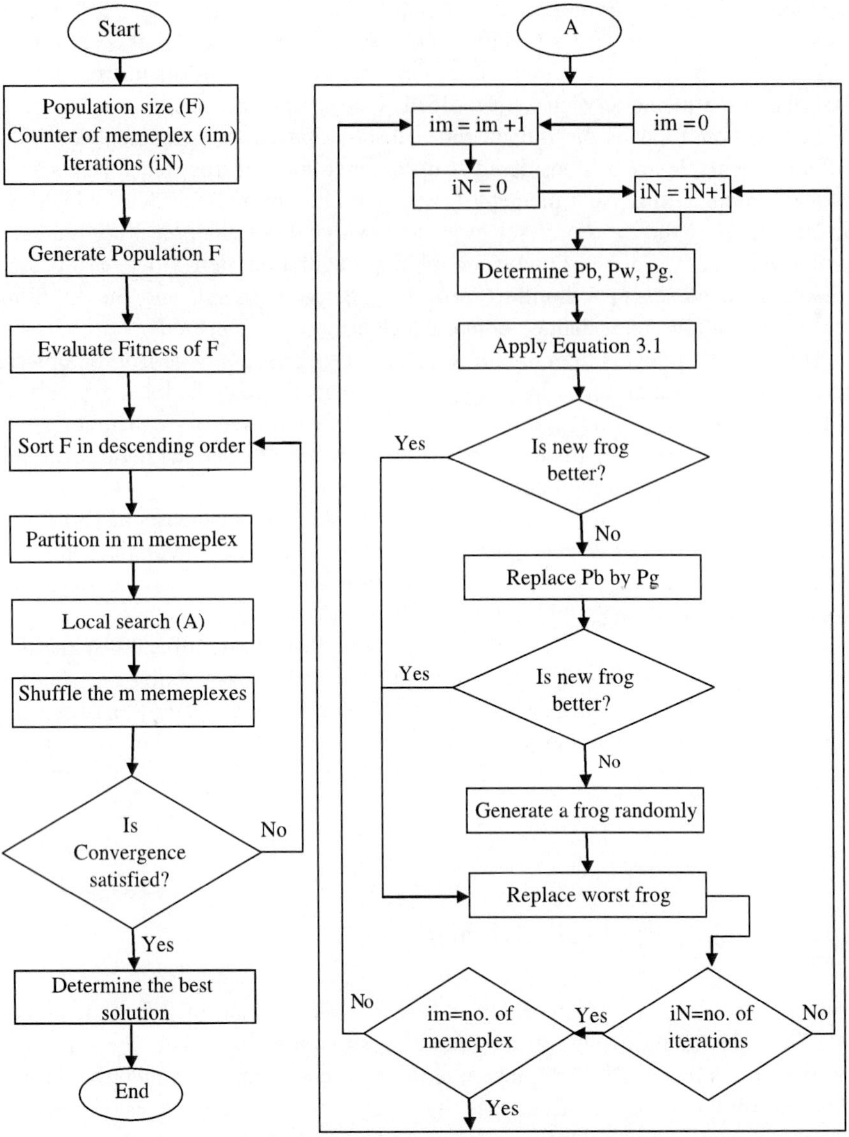

Fig. 2.1 Flowchart of SFLA (from [2]; reprinted with permission from Springer Science+ Business Media)

search space dimension and is equal to the number of independent variables. Now N_q pieces of shrapnel are produced by the grenade explosion and destruct objects that are in X'_j location (from [1]; reprinted with permission from Elsevier):

$$X'_j = \{X_m + \text{sign}(r_m) * p|\, r_m| * Le\}, j = 1, 2.\ldots.N_q \tag{2.21}$$

where r_m is a uniformly distributed random number in [−1, 1] and p is a constant. A high value for p lets pieces of shrapnel search the region near the exploded grenade more accurately, while a low one lets them to explore farther regions better.

Considering Eq. 2.21, it is obvious that exploration for more valuable items performs in an n-dimensional cubic space extended 2Le units along each coordinate and the grenade is located at the center of this cube. To use this algorithm, an independent variable range is scaled to [−1, 1]. Using Eq. 2.21, some produced shrapnel may collide to objects outside the feasible space. To increase the convergence rate and exploration of near-boundary regions more accurately, such a collision location is transported to a new location inside the feasible region according to the following scheme:

$$\text{If } X'_j \text{ doesn't belong to } [-1,1]^n,\ \left(B'_j = X'_j/\text{Largest component of } X'_j \text{ in value}\right) \tag{2.22}$$

$$B''_j = r'_j * \left(B'_j - X\right) + X \tag{2.23}$$

$$j = 1 \text{ to Nq (shrapnel number) and } 0 < r'_j < 1 \text{(random number)}$$

where, X'_j is the collision location outside the feasible space and B''_j is the new location inside the feasible space. One of the special concepts of this algorithm is the agent's territory radius (R_t), which means an agent (in this algorithm agents are grenades) does not let other agents come closer than a specific distance, which is specified by R_t. When several agents are exploring the feasible space, a high value for this parameter makes sure that grenades are spread quite uniformly in the feasible space and the whole space is being explored. While a low value for this parameter lets the grenades get closer to search the local region all together, a higher value for the explosion range makes it possible to explore farther regions (better exploration), while a lower one lets the grenades focus on the region nearby (better exploitation). The value of exponent p determines the intensity of exploration. This parameter is updated based on the value of Tw:

$$P = \max\{1/n, \log(R_t/\text{Le})/\log(\text{Tw})\} \tag{2.24}$$

where Tw is the probability that a produced piece of shrapnel collides an object in an n-dimension hyper-box which circumscribes the grenade's territory.

To increase the global search ability, a high value for R_t should be chosen at the beginning ($R_{t-\text{initial}}$) and reduced gradually to let the grenades search the probable global minimum location found altogether for a precise answer. A simple method to reduce R_t is given by Eq. 2.25.

$$R_t = \left(R_{t-\text{initial}}/\ R_{rd}^{\text{iteration number/total iterations}}\right) \tag{2.25}$$

The value of R_{rd} is set before the algorithm starts. This parameter represents the ratio of the value of R_t in the first iteration to its value in the final iteration. Furthermore, Le is reduced according to the following equation:

$$\text{Le} = (\text{Le initial})^{\text{m}}(\text{Rt})^{1-m}, \quad 0 \leq m \leq 1 \tag{2.26}$$

which indicates that Le is reduced more slowly than R_t during the iterations in order to save the global search ability. m can be constant during the algorithm, or reduced from a higher value to a lower one.

The next chapter presents the applications of many existing optimization algorithms to the design optimization of mechanical elements.

References

1. Ahrari A, Atai A (2010) Grenade Explosion Method-A novel tool for optimization of multimodal functions. Appl Soft Comput 10(4):1132–1140
2. Amiri B, Fathian M, Maroosi A (2009) Application of shuffled frog leaping algorithm on clustering. Int J Adv Manuf Technol 45:199–209
3. Basturk B, Karaboga D (2006) An artificial bee colony (ABC) algorithm for numeric function optimization. IEEE Swarm Intelligence Symposium, 12–14 May ,Indianapolis
4. Bergh F, Engelbrecht AP (2006) A study of particle swarm optimization particle trajectories. Inf Sci 176:937–971
5. Cai X, Cui Y, Tan Y (2009) Predicted modified PSO with time-varying accelerator coefficients. Int J Bio-Inspired Comput 1:50–60
6. Cui YH, Guo R (2008) Harmony elements algorithmHarmony elements algorithm. http://www.mathworks.com/matlabcentral/fileexchange/21963-harmony-element-algorithm
7. Cui H, Turan O (2010) Application of a new multi-agent hybrid co-evolution based particle swarm optimisation methodology in ship design. Comput-Aided Des 2:1013–1027
8. Dong HK, Ajith A, Jae HC (2007) A hybrid genetic algorithm and bacterial foraging approach for global optimization. Inf Sci 177:3918–3937
9. Dong Y, Tang J, Xu B, Wang D (2005) An application of swarm optimization to nonlinear programming. Comput Math Appl 49:1655–1668
10. Dorigo M (1992) Optimization, learning and natural algorithms. PhD Dissertation, Politecnico di Milano, Italy
11. Emma H, Jon T (2008) Application areas of AIS: the past, the present and the future. Appl Soft Comput 8:191–201
12. Eusuff M, Lansey E (2003) Optimization of water distribution network design using the shuffled frog leaping algorithm. J Water Resour Plan Manag ASCE 129:210–225
13. Eusuff M, Lansey K, Pasha F (2006) Shuffled frog-leaping algorithm: a memetic meta-heuristic for discrete optimization. Eng Optim 38(2):129–154
14. Farmer JD, Packard N, Perelson A (1986) The immune system, adaptation and machine learning. Physica 22:187–204
15. Geem ZW, Kim JH, Loganathan GV (2001) A new heuristic optimization algorithm: Harmony Search. Simul, the Soc for Model and Simul Int 76(2):60–68
16. Holland J (1975) Adaptation in natural and artificial systems. University of Michigan Press, Ann Arbor
17. Hui L, Zixing C, Yong W (2010) Hybridizing particle swarm optimization with differential evolution for constrained numerical and engineering optimization. Appl Soft Comput 10:629–640
18. Karaboga D (2005) An idea based on honey bee swarm for numerical optimization. Technical Report-TR06, Erciyes University, Engineering Faculty, Computer Engineering Department, Turkey

19. Karaboga D, Akay B (2010) A modified artificial bee colony (ABC) algorithm for constrained optimization problems. Appl Soft Comput doi:10.1016/j.asoc.2010.12.001
20. Kennedy J, Eberhart RC (1995) Particle swarm optimization. Proceedings IEEE International Conference on Neural Networks, Piscataway, 1942–1948
21. Leandro NC, Fernando JVZ (2002) Learning and optimization using the clonal selection principle. IEEE Trans Evol Comput Spec Issue Artif Immune Sys 6(3):239–251
22. Li R, Chang X (2006) A modified genetic algorithm with multiple subpopulations and dynamic parameters applied in CVAR model. Comput Intell for Model, Control and Autom, Sydney, p 151
23. Liu J, Tang LA (1999) Modified genetic algorithm for single machine scheduling. Comput Ind Eng 37:43–46
24. Maciocia G (2005) The foundations of chinese medicine. Elsevier, London
25. Montalvo I, Izquierdo J, Perez-Garcia R, Herrera M (2010) Improved performance of PSO with self-adaptive parameters for computing the optimal design of water supply systems. Eng Appl Artif Intell 23:727–735
26. Mouti FSA, Hawary MEE (2009) Modified artificial bee colony algorithm for optimal distributed generation sizing and allocation in distribution systems. IEEE Electr Power and Energy Conf (EPEC), Montreal, pp 1–9
27. Passino KM (2002) Biomimicry of bacterial foraging for distributed optimization and control. IEEE Control Syst Mag 22:52–67
28. Preechakul C, Kheawhom S (2009) Modified genetic algorithm with sampling techniques for chemical engineering optimization. J Ind and Eng Chem 15:101–107
29. Rashedi E, Nezamabadi-pour H, Saryazdi S (2009) GSA: a gravitational search algorithm. Inf Sci 179:2232–2248
30. Rodin V, Benzinou A, Guillaud A, Ballet P, Harrouet F, Tisseau J, Le Bihan J (2004) An immune oriented multi-agent system for biological image processing. Pattern Recogn 37:631–645
31. Shahla N, Mohammad EB, Nasser G, Mehdi HA (2009) A novel ACO–GA hybrid algorithm for feature selection in protein function prediction. Expert Sys Appl 36:12086–12094
32. Shen Q, Jiang J, Tao J, Shen G, Yu R (2005) Modified ant colony optimization algorithm for variable selection in QSAR modeling: QSAR Studies of Cyclooxygenase Inhibitors. J Chem Inf Model 45:1024–1029
33. Shi Y, Eberhart RC (1998) A modified particle swarm optimization. Proceedings the International Conference on Evolutionary Computer, Anchorage, pp 69–73
34. Simon D (2008) Biogeography-based optimization. IEEE Trans Evol Comput 12:702–713
35. Storn R, Price K (1997) Differential evolution–a simple and efficient heuristic for global optimization over continuous spaces. J Glob Optim 11:341–359
36. Tung Y, Erwie Z (2008) A hybrid genetic algorithm and particle swarm optimization for multimodal functions. Appl Soft Comput 8:849–857
37. Vitaliy F (2006) Differential evolution–in search of solutions. Springer, New York
38. Wang X, Gao XZ, Ovaska SJ (2004) Artificial immune optimization methods and applications–a survey. IEEE Int Conf Sys Man Cybern 4:3415–3420
39. Wen YL (2010) A GA–DE hybrid evolutionary algorithm for path synbook of four-bar linkage. Mech Mach Theory 45:1096–1107
40. Xiaohui H, Eberhart RC, Shi Y (2003) Engineering optimization with particle swarm. Proceedings of swarm intelligence symposium, West Lafayette, pp 53–57
41. Yannis M, Magdalene M (2010) Hybrid multi-swarm particle swarm optimization algorithm for the probabilistic traveling salesman problem. Comput Oper Res 37:432–442
42. Yildiz AR (2009) A novel particle swarm optimization approach for product design and manufacturing. Int J Adv Manuf Technol 40:617–628
43. Ying PC (2010) An ant direction hybrid differential evolution algorithm in determining the tilt angle for photovoltaic modules. Expert Sys Appl 37:5415–5422

44. Yong F, Yong MY, Wang AX (2007) Comparing with chaotic inertia weights in particle swarm optimization. Proceedings the Sixth International Conference on Machine Learning and Cybernetics, Hong Kong, pp 19–22
45. Yue H, Gu G, Liu H, Shen J, Zhao J (2009) A modified ant colony optimization algorithm for tumor marker gene selection. Genomics Proteomics Bioinforma 7:200–208
46. Zhang EQ (1992) Basic theory of traditional chinese medicine. Shanghai University of Traditional Medicine, Shanghai

Chapter 3
Mechanical Design Optimization Using the Existing Optimization Techniques

In this chapter, an effort is made to verify if any improvement in the solution is possible by employing advanced optimization techniques to some of the mechanical element design optimization problems available in the research literature. Seven different mechanical element design problems, like optimization of a gear train, radial ball bearing, Belleville spring, multi-plate disc clutch brake, robot gripper, hydrostatic thrust bearing and four stage gear train, are considered in this chapter for the analysis. The considered problems were solved by using either GA or PSO or by both the methods by the other researchers. These problems are tested by using other advanced optimization techniques such as ABC, DE, BBO and AIA in this book. PSO is applied to the problems to which only GA was applied. The descriptions of different mechanical elements considered in this work are given in the next section.

3.1 Description of Different Mechanical Design Optimization Problems

3.1.1 Example 1: Optimization of a Gear Train

This problem is taken from Yokota et al. [1]. The single stage spur gear considered by Yokota et al. [1] is shown in Fig. 3.1 with all its geometries. The optimal gear design problem defined by Yokota et al. [1] consists of a nonlinear objective function with five nonlinear constraints involving five design variables. Design variable considered are width of the pinion and gear (b), shaft diameter of pinion and gear (d_1, d_2), number of teeth on pinion (Z_1) and module of pinion and gear (m). The design variables with their ranges, objective function and the constraints are given below:

R. V. Rao and V. J. Savsani, *Mechanical Design Optimization Using Advanced Optimization Techniques*, Springer Series in Advanced Manufacturing,
DOI: 10.1007/978-1-4471-2748-2_3,

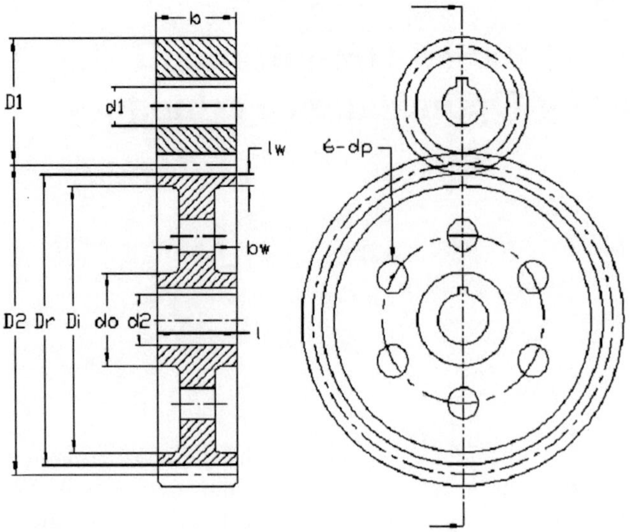

Fig. 3.1 Single stage spur gear (from Yokota et al. [1] reprinted with permission from Elsevier)

- Design variables: $X = (b, d_1, d_2, Z_1, m)$, where, $20 \leq b \leq 32$, $10 \leq d_1 \leq 30$, $30 \leq d_2 \leq 40$, $18 \leq Z_1 \leq 25$ and $m = (2.75, 3, 3.5, 4)$
- Objective function to be minimized:

$$\begin{aligned} \text{Weight} &= \text{F}(x) \\ &= (\pi\rho/4{,}000)\left[bm^2Z_1^2\left(1+a^2\right) - \left(D_i^2 - d_0^2\right)(l - b_w) - nd_p^2 b_w - \left(d_1^2 + d_2^2\right)b\right] \end{aligned} \tag{3.1}$$

where, ρ is the density of material, a is the gear ratio, D_i is the inside diameter of rim, d_o is the outside diameter of boss, l is the length of boss, b_w is the thickness of web, d_p is the drill hole diameter, n is the number of holes, d_1 and d_2 is the diameter of pinion and gear shaft, respectively.

- Constraints:
  - For bending strength of tooth:

$$g_1(x) = F_s \geq b_1 \tag{3.2}$$

  - For surface durability:

$$g_2(x) = \left(F_s/F_p\right) \geq b_2 \tag{3.3}$$

where, F_s is the bending strength of gear and F_p is the wear load.

- For torsional strength of pinion shaft:

$$g_3(x) = d_1^3 \geq b_3 \tag{3.4}$$

- For torsional strength of gear shaft:

$$g_4(x) = d_2^3 \geq b_4 \tag{3.5}$$

- For the center distance:

$$g_5(x) = (1 + a)mZ_1/2 \leq b_5 \tag{3.6}$$

where, $D_r = m(aZ_1 - 2.5)$, $l_w = 2.5$ m, $D_i = D_r - 2l_w$, $b_w = 3.5$ m, $d_0 = d_2 + 25$, $d_p = 0.25(D_i - d_0)$, $D_1 = mZ_1$, $D_2 = amZ_1$, $N_2 = N_1/a$, $Z_2 = Z_1D_2/D_1$, $v = \pi D_1N_1/60{,}000$, $b_1 = 1{,}000P/v$, P is the power to be transmitted, v *is the* pitch line velocity, $b_3 = 48.68e6P/(N_1\tau)$, τ is the shear strength of shaft, $b_4 = 48.68e6P/(N_2\tau)$, $F_s = \pi K_vK_w\sigma bmy$, y is the tooth form factor, σ is the allowable stress of gear, $F_p = 2K_vK_wD_1bZ_2/(Z_1 + Z_2)$, Z_2 is the number of teeth on gear, $a = 4$, $\rho = 8$, $P = 7.5$, $n = 6$, $\sigma = 294.3$, $y = 0.102$, $b_2 = 0.193$, $\tau = 19.62$, $K_w = 0.8$, $K_v = 0.389$.

Yokota et al. [1] solved this constrained optimization problem by considering all the design variables as discrete. The model of Yokota et al. [1] is modified by using American Gear Manufacturers Association (AGMA) standard Equations [2, 3] which include many detailed design factors such as; K_v, which cannot be constant as it depends on pitch line velocity which is again the function of pitch diameters of pinion/gear; Form factor y, which depends on the number of teeth and cannot be taken as constant. Moreover, in the above design considered by Yokota et al. [1] there is no mention of hardness, which plays a very crucial role for the surface fatigue strength. So design is modified considering many additional factors which are practically required for the optimal gear design. Refined design includes six design variables including hardness as an additional design variable and 8 constraints which are given below:

- Design variables: $x = (b, d_1, d_2, Z_1, m, H)$, where, $200 \leq H \leq 400$
- Constraints:
  - For the bending fatigue strength:

$$g_1(x) = S_nC_sK_rK_{ms}bJm/(K_vK_oK_m) \geq b_1 \tag{3.7}$$

  where, J is the Lewis gear geometry factor, K_v is the coefficient of pitch line velocity, K_o is the coefficient of degree of non-uniformity, K_m is the coefficient of accuracy of gear alignment, S_n is the endurance limit, K_r is the reliability factor, K_{ms} is the mean stress factor, C_s is the surface factor.
  - for the surface fatigue strength:

$$g_2(x) = S_{fe}^2C_l^2C_r^2bD_1I/(C_p^2K_vK_oK_m) \geq b_1 \tag{3.8}$$

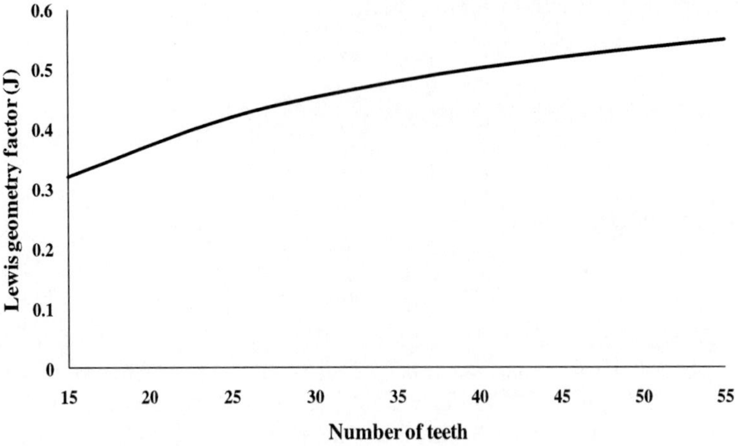

Fig. 3.2 Plot of third order polynomial fit for Lewis geometry factor (J)

where, S_{fe} represents the reference value of surface strength for a specific material with a level of 99% reliability at 10^7 cycles, C_l is the surface fatigue life factor, C_r is the reliability adjustment factor, I is the geometry factor and C_p is the elastic coefficient of material.

- For avoiding the interference:

$$g_3(x) = \sin^2\phi D_1(2D_2 + D_1)/(4m) - -D_2 - 1 \geq 0 \tag{3.9}$$

where, $\emptyset$ *is the* pressure angle

- For uniform load distribution:

$$g_4(x) = b/m \geq 8 \tag{3.10}$$

$$g_5(x) = b/m \leq 16 \tag{3.11}$$

- For torsional strength of shafts:

$$g_6(x) = d_1^3 \geq b_3 \tag{3.12}$$

$$g_7(x) = d_2^3 \geq b_4 \tag{3.13}$$

- For center distance:

$$g_8(x) = (1 + a)mZ_1/2 \leq b_5 \tag{3.14}$$

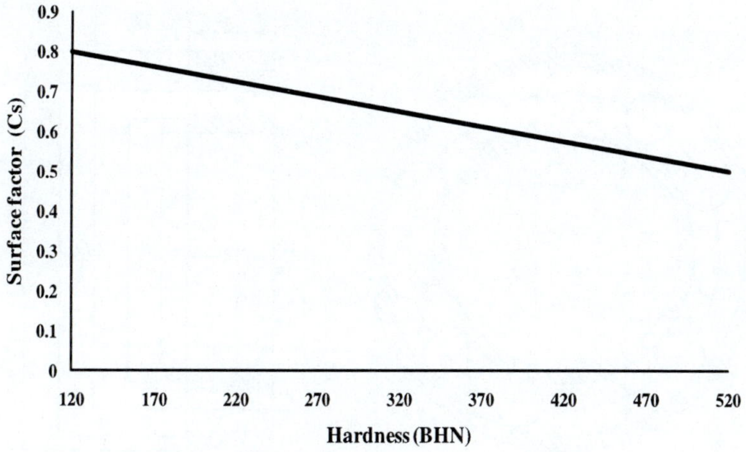

Fig. 3.3 Variation of surface factor with hardness

where, values of C_s and J are determined from Figs. 3.2 and 3.3, respectively, $S_n = 1.7236H$, $K_v = (78 + \sqrt{(196.85v)})/78$, $S_{fe} = 2.8H{-}69$, $I = a\sin\text{Ø}\cos\text{Ø}/(2(a + 1))$, $K_r = 0.814$, $K_{ms} = 1.4$, $K_o = 1$, $K_m = 1.3$, $C_p = 191$, $C_l = 1$, $C_r = 1$ and $\text{Ø} = 25$. The objective function is same as that given by Eq. 3.1.

The design proposed by Yokota et al. [1] will be referred to as Example 1A in this book. The same design is attempted by using mixed discrete–continuous design variables by considering b, d_1 and d_2 as continuous variables and Z_1 and m as the discrete variables. The design with mixed discrete–continuous design variables will be referred to as Example 1B in this book. The modified design according to AGMA with all the discrete design variables will be referred to as Example 1C in this book and the modified design with mixed discrete–continuous design variables will be referred to as Example 1D in this book.

3.1.2 Example 2: Optimization of a Radial Ball Bearing

Rolling element bearings appear to have a simple outer geometry, but their internal geometry can have varying effects on the amount of stresses, deflections and load distributions it can handle. Therefore, the internal geometry plays a vital role. Deflection of the bearing accounts for the stiffness of the bearing, which also depends on bearings internal geometry. The internal geometry has a direct effect on the performance and the life of a bearing.

Figure 3.4 shows the geometries of a radial ball bearing. Generally bearing is specified by three standardized boundary dimensions, namely, bore diameter (d), the outer diameter (D) and the bearing width (B_w). Keeping these boundary dimensions fixed, internal parameters can be altered to have the maximum performance of the bearing. Internal parameters include ball diameter (D_b), pitch

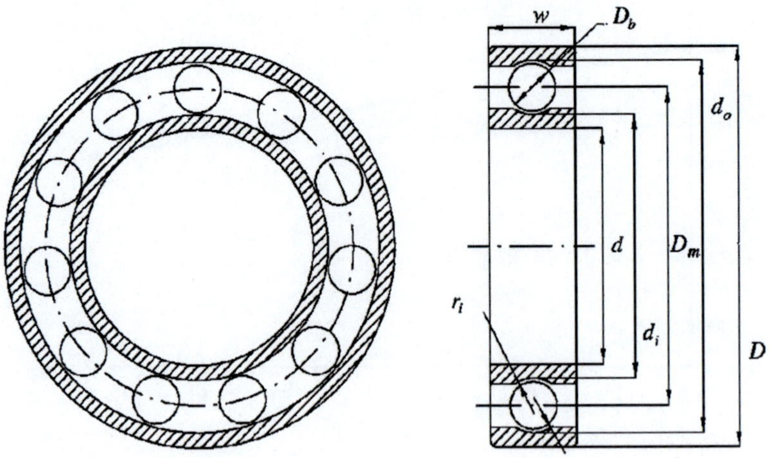

Fig. 3.4 Radial ball bearing (from Gupta et al. [4] reprinted with permission from Elsevier)

diameter (D_m), inner and outer raceway curvature coefficients (f_i and f_o) and number of balls (Z). The purpose of optimization is to evaluate the above mentioned internal geometry to maximize the performance of the bearing.

Gupta et al. [4] presented the optimization aspects of rolling element bearings by using a non-dominated sorting genetic algorithm (NSGA-II). The design parameters, objective functions and constraints for defining feasible design parameter space considered by the authors are shown below.

- **Design variables:**

As discussed above, internal geometries are very important for the performance of a bearing and all these parameters are to be considered as design variables. There are many parameters such as $K_{D\min}$, $K_{D\max}$, ε, e and ζ which only appear in constraints and indirectly affect the internal geometry. These parameters were considered constant by Changsen [5] but were taken as design variables with some studied range by Gupta et al. [4]. So, a total of ten design variables are considered for the optimization problem and these are:

$$X = [D_m, D_b, Z, f_i, f_o, K_{D\min}, K_{D\max}, \varepsilon, e, \zeta]$$

All the design variables are continuous in nature, except Z (number of balls) which varies as integer during the optimization process. Ranges of all these ten design variables were given as: $0.5(D + d) \leq D_m \leq 0.6(D + d), 0.15(D - d) \leq D_\mathrm{b} \leq 0.45(D - d)$, $4 \leq Z \leq 50$, $0.515 \leq f_i \leq 0.6$, $0.515 \leq f_o \leq 0.6$, $0.4 \leq K_{D\min} \leq 0.5$, $0.6 \leq K_{D\max} \leq 0.7, 0.3 \leq \varepsilon \leq 0.4, 0.02 \leq e \leq 0.1, 0.6 \leq \zeta \leq 0.85$

where, ε is the parameter for outer ring strength consideration, ζ is the bearing width limiter, e is the parameter for mobility condition, $K_{D\min}$ is the minimum ball diameter limiter and $K_{D\max}$ is the maximum ball diameter limiter.

- **Objective functions:**

Three different objective functions were considered by Gupta et al. [4] which deal with the performance of the bearing. These were maximization of dynamic capacity, minimum film thickness and static capacity. All these objectives were expressed in mathematical form as described in the following sub-sections.

- **Dynamic Capacity (C_d)**

Bearing is a rotating element and experiences continuous reversal of stresses induced in balls and rings. These stresses affect the fatigue life of the bearing. Dynamic capacity or dynamic load rating for the bearing is a direct measure of the fatigue life. Dynamic capacity for the bearing for outer ring fixed and inner ring rotating is expressed as:

$$C_d = f_c Z^{2/3} D_b^{1.8} \quad \text{If } D_b \leq 25.4 \text{ mm} \tag{3.15}$$

and

$$C_d = 3.647 f_c Z^{2/3} D_b^{1.4} \quad \text{If } D_b > 25.4 \text{ mm} \tag{3.16}$$

where,

$$f_c = 37.91 \left[1 + \left\{ 1.04 \left(\frac{1-\gamma}{1+\gamma} \right)^{1.72} \left(\frac{f_i(2f_o - 1)}{f_o(2f_i - 1)} \right)^{0.41} \right\}^{10/3} \right]^{-0.3}$$

$$\gamma = \frac{D_b \cos\alpha}{D_m}, \; f_i = \frac{r_i}{D_b}, \; f_o = \frac{r_o}{D_b} \tag{3.17}$$

r_o, r_i is the outer and inner raceway groove curvature radius, respectively, α is the radial contact angle

Deep groove ball bearing was considered by Gupta et al. [4] and hence the value of α was taken as zero. The aim of optimization is the maximization of dynamic capacity.

- **Elastohydrodynamic minimum film thickness ($H_{\min}$)**

Long wear life is also required for the bearing along with fatigue life. Minimum film thickness is important to increase the wear life as it avoids the metal-to-metal contact for the rotating bearing. Minimum film thickness was predicted by elastohydrodynamic lubrication theory, which gives minimum film thickness as:

$$\begin{aligned} H_{\min,\text{ring}} &= 3.63 \alpha_1 R_{x,\text{ring}}^{0.466} \, E_o^{-0.117} Q^{-0.073} \left\{ \frac{\pi n_i D_m \eta_o (1-\gamma^2)}{120} \right\}^{0.68} \\ &\times \left[1 - \exp\left\{ -0.703 \left(\frac{R_{y,\text{ring}}}{R_{x,\text{ring}}} \right)^{0.636} \right\} \right] \end{aligned} \tag{3.18}$$

Here i represents the number of rows and it was considered as 1 for single row deep groove rolling bearing. Other expressions in the objective function are given below.

$$Q = \frac{5F_r}{iZ\cos\alpha} \tag{3.19}$$

$$R_{x,\text{inner}} = \frac{D_b}{2(1-\gamma)}, \quad R_{x,\text{outer}} = \frac{D_b}{2(1+\gamma)}$$

$$R_{y,\text{inner}} = \frac{f_i D_b}{2f_i - 1}, \quad R_{y,\text{outer}} = \frac{f_o D_b}{2f_o - 1}$$

The expression for $H_{\min}$ is applicable for the inner and outer ring separately. So for the optimization process the minimum of the two is maximized. So $H_{\min}$ can be expressed as:

$$H_{\min} = \min\left(H_{\min,\text{inner}}, H_{\min,\text{outer}}\right)$$

The subscript 'ring' that appears in the objective function can take value as the inner or outer ring. Operating conditions for the bearing given by Gupta et al. [4] are:

$\alpha_1 = 1\text{e} - 8$, $\eta_o = 0.02$, $n_i = 5000$, $E_o = 2.25\text{e}11$, and $F_r = 15000$

where, E_o is the equivalent modulus of elasticity, F_r is the radial load, n_i is the rotational speed of inner ring, η_o is the dynamic viscosity at atmospheric pressure, α_1 is the pressure coefficient of viscosity.

- **Static capacity (C_s)**

Static capacity is the load which a bearing can sustain in the stationary position. The static capacity is also defined for the inner and outer rings separately. It was expressed as:

$$C_{s,\text{inner}} = \frac{23.8ZiD_b^2\left(a_i^* b_i^*\right)^3 \cos\alpha}{\left(4 - \frac{1}{f_i} + \frac{2\gamma}{1-\gamma}\right)^2} \tag{3.20}$$

$$C_{s,\text{outer}} = \frac{23.8ZiD_b^2\left(a_o^* b_o^*\right)^3 \cos\alpha}{\left(4 - \frac{1}{f_o} - \frac{2\gamma}{1+\gamma}\right)^2} \tag{3.21}$$

For the calculation of a^* and b^*, it is required to calculate $F(\rho)$ separately for inner and outer race as:

$$F(\rho)_i = \frac{\frac{1}{f_i} + \frac{2\gamma}{1-\gamma}}{\left(4 - \frac{1}{f_i} + \frac{2\gamma}{1-\gamma}\right)} \tag{3.22}$$

$$F(\rho)_o = \frac{\frac{1}{f_o} - \frac{2\gamma}{1+\gamma}}{\left(4 - \frac{1}{f_o} - \frac{2\gamma}{1+\gamma}\right)} \tag{3.23}$$

The values of a^* and b^* can be taken from Harris [6] once the value of $F(\rho)$ is found.

- **Constraints:**

For the assembly of a bearing, assembly angle and the number of balls should satisfy the following condition:

$$g_1(X) = \frac{\phi_o}{2\sin^{-1}(D_b/D_m)} - Z + 1 \geq 0 \tag{3.24}$$

In the above expression ϕ_o is the assembly angle expressed in radian and can be formulated as:

$$\phi_o = 2\pi - 2\cos^{-1}\frac{\left[\{(D-d)/2 - 3(T/4)\}^2 + \{D/2 - (T/4) - D_b\}^2 - \{d/2 + (T/4)\}^2\right]}{2\{(D-d)/2 - 3(T/4)\}\{D/2 - (T/4) - D_b\}} \tag{3.25}$$

where,

$$T = D - d - 2D_b$$

Upper and lower bounds for the balls can be expressed by the following constraints:

$$g_2(X) = 2D_b - K_{D\min}(D-d) \geq 0 \tag{3.26}$$

$$g_3(X) = K_{D\max}(D-d) - 2D_b \geq 0 \tag{3.27}$$

Additional constraint on the size of ball was decided by the width of the bearing and thus it also formed a constraint as,

$$g_4(X) = \zeta B_w - D_b \leq 0 \tag{3.28}$$

To ensure the running mobility of bearings, there should be difference between the pitch diameter and the average diameter and also inner ring thickness must be more than outer ring thickness, and these lead to the following two constraints.

$$g_5(X) = D_m - 0.5(D + d) \geq 0 \tag{3.29}$$

$$g_6(X) = (0.5 + e)(D + d) - D_m \geq 0 \tag{3.30}$$

The thickness of a bearing ring at the outer raceway bottom should not be less than εD, where ε is a parameter obtained from the simple strength consideration of the outer ring and thus leads to the following constraint:

$$g_7(X) = 0.5\ (D - D_m - D_b) - \varepsilon D_b \geq 0 \tag{3.31}$$

The groove curvature radii of the inner and outer raceways of a bearing should not be less than $0.515D_b$. These lead to the following two constraints.

$$g_8(X) = f_i \geq 0.515 \tag{3.32}$$

$$g_9(X) = f_o \geq 0.515 \tag{3.33}$$

Designers are always interested in the design where all the objectives are maximized simultaneously because all the objective functions are of importance. This concept leads to the multi-objective optimization design. In this book, weight method is implemented to convert multi-objective functions into a single objective function. A combined objective function F is formulated considering three objective functions C_d, $H_{\min}$ and C_s as

$$F = w_1(C_d/C_{d\max}) + w_2(H_{\min}/H_{\min\max}) + w_3(C_s/C_{s\max}) \tag{3.34}$$

where, w_1, w_2 and w_3 are different weight factors and $C_{d\max}$, $H_{\min\max}$and $C_{s\max}$ are the maximum values of the objective functions C_d, $H_{\min}$ and C_s, respectively, when these objectives are considered independently under the given constraints. As all of the three objective functions are considered of equal importance for the design of rolling element bearing, the weight factors are taken as 0.33333 each. In this book optimization of dynamic capacity, static capacity and elastohydrodynamic minimum film thickness will be referred to as Example 2A, Example 2B and Example 2C, respectively.

3.1.3 Example 3: Optimization of a Belleville Spring

The objective is to design a Belleville spring having minimum weight and satisfying a number of constraints. The problem has 4 design variables: external diameter of the spring ($D_e(x_3)$,) internal diameter of the spring ($D_i(x_4)$,), thickness of the spring ($t(x_1)$,) and the height ($h(x_2)$,) of the spring as shown in Fig. 3.5. Of these design variables, t is a discrete variable and the rest are continuous variables. Constraints are for compressive stress, deflection, height to deflection, height to maximum height, outer diameter, inner diameter and slope.

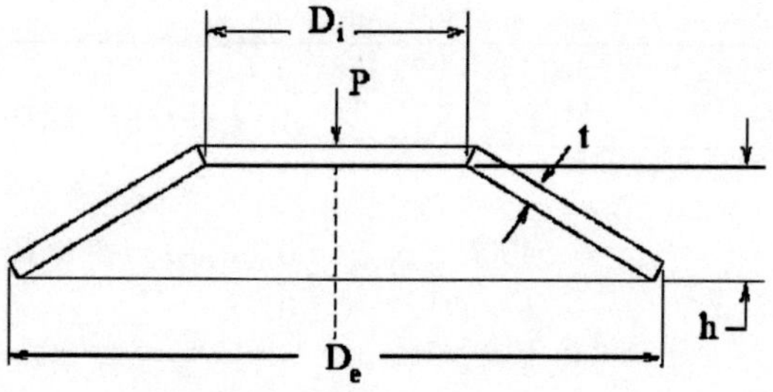

Fig. 3.5 Belleville spring (from [7] reprinted with permission from Elsevier)

Minimize:

$$f(x) = 0.07075\pi(D_e^2 - D_i^2)t \tag{3.35}$$

Subject to:

$$g_1(x) = S - \frac{4E\delta_{\max}}{(1-\mu^2)\alpha D_e^2}\left[\beta\left(h - \frac{\delta_{\max}}{2}\right) + \gamma t\right] \geq 0 \tag{3.36}$$

where, S is the allowable strength by the spring material, E is the modulus of elasticity for the spring material, μ is the Poisson's ratio, $\delta_{\max}$ is the maximum deflection of the spring

$$g_2(x) = \left(\frac{4E\delta}{(1-\mu^2)\alpha D_e^2}\left[\left(h - \frac{\delta}{2}\right)(h-\delta)t + t^3\right]\right)_{\delta=\delta_{\max}} - P_{\max} \geq 0 \tag{3.37}$$

$$g_3(x) = \delta_l - \delta_{\max} \geq 0 \tag{3.38}$$

$$g_4(x) = H - h - t \geq 0 \tag{3.39}$$

where, $P_{\max}$ is the maximum load acting on the spring, H is the overall height of the spring.

$$g_5(x) = D_{\max} - D_e \geq 0 \tag{3.40}$$

$$g_6(x) = D_e - D_i \geq 0 \tag{3.41}$$

$$g_7(x) = 0.3 - \frac{h}{D_e - D_i} \geq 0 \tag{3.42}$$

where,

$$\alpha = \frac{6}{\pi \ln K}\left(\frac{K-1}{K}\right)^2, \quad \beta = \frac{6}{\pi \ln K}\left(\frac{K-1}{\ln K} - 1\right), \quad \gamma = \frac{6}{\pi \ln K}\left(\frac{K-1}{2}\right),$$

Table 3.1 Variation of *f(a)* with *a*

a	≤1.4	1.5	1.6	1.7	1.8	1.9	2	2.1	2.2	2.3	2.4	2.5	2.6	2.7	≥2.8
$f(a)$	1	0.85	0.77	0.71	0.66	0.63	0.6	0.58	0.56	0.55	0.53	0.52	0.51	0.51	0.5

$$P_{\max} = 5400lb, \;\; \delta_{\max} = 0.2in, \;\; S = 200kPsi, \;\; E = 30e6psi, \mu = 0.3,$$
$$H = 2in, \;\; D_{\max} = 12.01in,$$
$$K = \frac{D_e}{D_i}, \quad \delta_l = f(a)a, \quad a = h/t$$
$$0.01 \le x_1 \le 0.6, \;\; 0.05 \le x_2 \le 0.5, \;\; 5 \le x_3, \; x_4 \le 15.$$

Variation of *f(a)* with *a* is given in Table 3.1.

3.1.4 Example 4: Optimization of a Multiple Disc Clutch Brake

This problem is taken from Osyczka [8]. Figure 3.6 shows a multiple disc clutch brake. The objective is to minimize the mass of the multiple disc clutch brake with five discrete variables: inner radius(r_i = *60, 61 ,62 …80*), outer radius (r_o = *90, 91, 92 …110*), thickness of discs (t = *1, 1.5, 2,2.5, 3*), actuating force (F = *600, 610, 620, …1,000*) and number of friction surfaces(Z = *2, 3, 4, 5, 6, 7, 8, 9*). The problem can be stated as:

Minimize $f(x) = \pi(r_0^2 - r_i^2)t(Z+1)\rho$

Subject to:

$$g_1(x) = r_o - r_i - \Delta r \ge 0 \tag{3.43}$$

$$g_2(x) = l_{\max} - (Z+1)(t+\delta) \ge 0 \tag{3.44}$$

$$g_3(x) = p_{\max} - p_{rz} \ge 0 \tag{3.45}$$

$$g_4(x) = p_{\max} v_{sr\max} - p_{rz} v_{sr} \ge 0 \tag{3.46}$$

$$g_5(x) = v_{sr\max} - v_{sr} \ge 0 \tag{3.47}$$

$$g_6(x) = T_{\max} - T \ge 0 \tag{3.48}$$

$$g_7(x) = M_h - sM_s \ge 0 \tag{3.49}$$

$$g_8(x) = T \ge 0 \tag{3.50}$$

where,

$$M_h = \frac{2}{3}\mu FZ\frac{r_o^3 - r_i^3}{r_o^2 - r_i^2}, \;\; p_{rz} = \frac{F}{\pi(r_o^2 - r_i^2)}, \;\; v_{sr} = \frac{2\pi n(r_o^3 - r_i^3)}{90(r_o^2 - r_i^2)}, \;\; T = \frac{I_z \pi n}{30(M_h + M_f)},$$

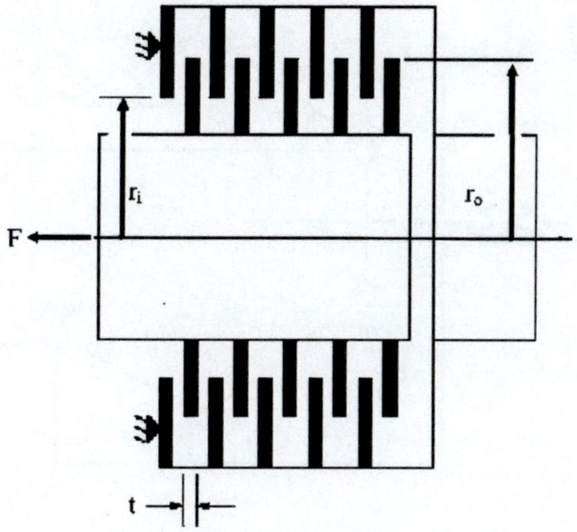

Fig. 3.6 Multiple disc clutch brake (from Rao et al. [9] reprinted with permission from Elsevier)

minimum difference between radii (Δr) = 20 mm, maximum disc thickness ($t_{\max}$) = 3 mm, minimum disc thickness ($t_{\min}$) = 1.5 mm, maximum length ($l_{\max}$) = 30 mm, maximum number of disc ($Z_{\max}$) = 10, maximum velocity of slip stick ($v_{sr\max}$) = 10 m/s, coefficient of friction (μ) = 0.5, distance between disc when unloaded (δ) = 1.5, static input torque (M_s) = 40 Nm, frictional resistance torque (M_f) = 3 Nm, input speed (n) = 250 rpm, maximum allowable pressure on disc ($p_{\max}$) = 1 MPa, disc mass moment of inertia(I_z) = 55 kgmm^2, maximum stopping time ($T_{\max}$) = 15 s, maximum actuating force ($F_{\max}$) = 1,000 N, minimum inner diameter ($r_{i\min}$) = 55 mm, maximum outer diameter ($r_{o\max}$) = 110 mm and density of material (ρ) = 0.0000078 kg/m^3.

3.1.5 Example 5: Optimization of a Robot Gripper

The objective is to minimize the difference between maximum and minimum force applied by the gripper for the range of gripper end displacements. There are seven continuous design variables (a, b, c, e, f, l, δ) as shown in Figs. 3.7 and 3.8. All the design variables are associated with the geometric dimensions of the robot gripper. There are six different geometric constraints associated with the robot gripper problems. The problem is taken from Osyczka et al. [10] which is stated as:

$$\text{Minimize} \quad f(x) = \max_z F_k(x,z) - \min_z F_k(x,z) \tag{3.51}$$

Subject to:

$$g_1(x) = Y_{\min} - y(x, Z_{\max}) \geq 0 \tag{3.52}$$

$$g_2(x) = y(x, Z_{\max}) \geq 0 \tag{3.53}$$

$$g_3(x) = y(x, 0) - Y_{\max} \geq 0 \tag{3.54}$$

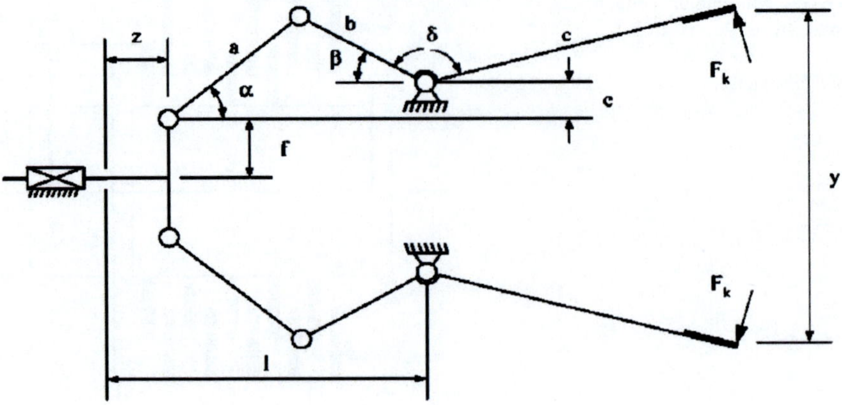

Fig. 3.7 Robot gripper (from Rao et al. [9] reproduced with permission from Elsevier)

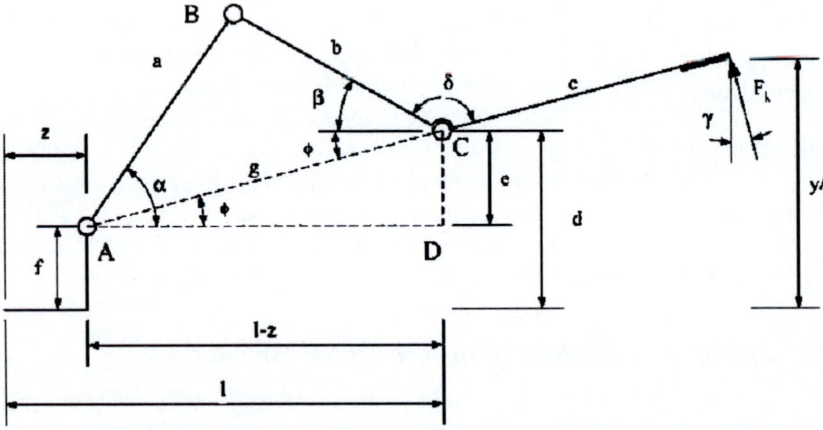

Fig. 3.8 Different geometries of robot gripper (from Rao et al. [9] reproduced with permission from Elsevier)

$$g_4(x) = (a+b)^2 - l^2 - e^2 \geq 0 \tag{3.55}$$

$$g_5(x) = (l - Z_{\max})^2 + (a-e)^2 - b^2 \geq 0 \tag{3.56}$$

$$g_6(x) = l - Z_{\max} \geq 0 \tag{3.57}$$

Following parameters are derived from Fig. 3.8:

$$g = \sqrt{(l-z)^2 + e^2} \tag{3.58}$$

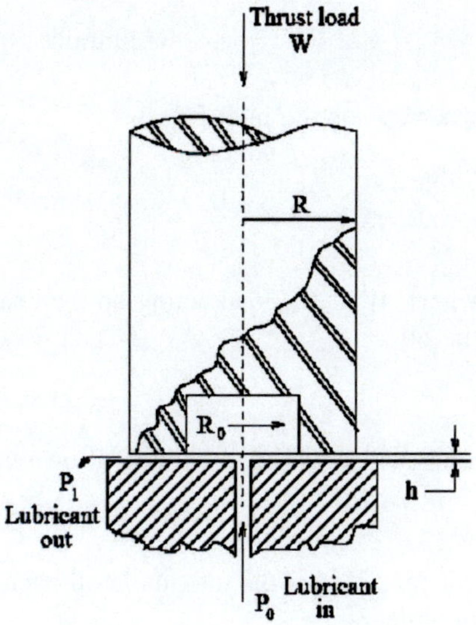

Fig. 3.9 Hydrodynamic thrust bearing (from [7] reprinted with permission from Elsevier)

$$\alpha = \arccos\left(\frac{a^2 + g^2 - b^2}{2ag}\right) + \phi \tag{3.59}$$

$$\beta = \arccos\left(\frac{b^2 + g^2 - a^2}{2bg}\right) - \phi \tag{3.60}$$

$$\phi = \arctan\left(\frac{e}{l - z}\right) + \phi \tag{3.61}$$

$$F_k = \left(\frac{Pb\sin(\alpha + \beta)}{2c\cos(\alpha)}\right) \tag{3.62}$$

$$y(x, z) = 2(e + f + c\sin(\beta + \delta)) \tag{3.63}$$

where, $Y_{\min} = 50$, $Y_{\max} = 100$, $Z_{\max} = 100$, $P = 100$, $10 \le a,b,f \le 150$, $100 \le c \le 200$, $0 \le e \le 50$, $100 \le l \le 300$, $1 \le \delta \le 3.14$

3.1.6 Example 6: Optimization of a Hydrodynamic Thrust Bearing

The objective is to minimize the power loss in a hydrostatic thrust bearing. There are four design variables bearing step radius (R), recess radius (R_o), oil viscosity (μ) and flow rate (Q). Figure 3.9 shows a hydrodynamic thrust bearing. Seven different constraints are associated with the problem based on load carrying capacity, inlet oil pressure, oil temperature rise, oil film thickness and physical constraints. The problem can be stated as:

$$\text{Minimize}: f(x) = \frac{QP_o}{0.7} + E_f \tag{3.64}$$

where, P_o is the inlet pressure and E_f is the power loss due to friction

Subject to:

$$g_1(x) = W - W_s \geq 0 \tag{3.65}$$

where, W_s is the load acting on the bearing and W is the load carrying capacity of the oil.

$$g_2(x) = P_{\max} - P_o \geq 0 \tag{3.66}$$

where, $P_{\max}$ is the maximum allowable pressure.

$$g_3(x) = \Delta T_{\max} - \Delta T \geq 0 \tag{3.67}$$

where, $\Delta T_{\max}$ is the maximum allowable temperature difference and ΔT is the rise in temperature.

$$g_4(x) = h - h_{\min} \geq 0 \tag{3.68}$$

where, h-oil film thickness and $h_{\min}$-minimum required oil film.

$$g_5(x) = R - R_o \geq 0$$

$$g_6(x) = 0.001 - \frac{\gamma}{gP_o}\left(\frac{Q}{2\pi Rh}\right) \geq 0 \tag{3.69}$$

$$g_7(x) = 5,000 - \frac{W}{\pi(R^2 - R_o^2)} \geq 0 \tag{3.70}$$

where,

$$W = \frac{\pi P_o}{2}\frac{R^2 - R_o^2}{\ln\frac{R}{R_o}}$$

$$P_o = \frac{6\mu Q}{\pi h^3}\ln\frac{R}{R_o}$$

$$E_f = 9,336\, Q\gamma C\Delta T$$

$$\Delta T = 2(10^P - 560)$$

$$P = \frac{\log_{10}\log_{10}(8.122e6\mu + 0.8) - C_1}{n}$$

$$h = \left(\frac{2\pi N}{60}\right)^2 \frac{2\pi\mu}{E_f}\left(\frac{R^4}{4} - \frac{R_o^4}{4}\right)$$

$\gamma = 0.0307$, $C = 0.5$, $n = -3.55$, $C_1 = 10.04$, $W_s = 101{,}000$, $P_{\max} = 1{,}000$, $\Delta T_{\max} = 50$, $h_{\min} = 0.001$, $g = 386.4$, $N = 750$, γ is the density of oil, C is the specific heat of oil, C_1 and n is the constant for the given oil, $1 \le R, Ro, Q \le 16$, $1e-6 \le \mu \le 16e-6$.

3.1.7 Example 7: Discrete Optimization of a Four Stage Gear Train

The four stage gear train design problem considered in this study was originally introduced by Pomrehn and Papalambros [11]. A gear train is to be designed by using eight spur gears arranged into four-stages. The gear train has to allow an input power of 55.9 W at 5000 rpm and the output speed of the system must be 250 ± 5 rpm in the same rotational direction as the input. Power loss through the gear train can be assumed negligible. The gearbox housing provides discrete locations for the gear/pinion shafts and also four gear blank thicknesses are available.

The objective is to minimize the total weight of the given gear train. With the material specified (aluminum–bronze), the objective is minimizing the total gear volume. The problem can be stated as:

Minimize

$$f = \pi \sum_{i=1}^{4} \frac{b_i c_i^2 (N_{pi}^2 + N_{gi}^2)}{(N_{pi} + N_{gi})^2} \tag{3.71}$$

where, b_i is the pinion and gear thickness, c_i is the distance between pinion and gear center, N_{gi} is the number of teeth on gear, N_{pi} is the number of teeth on pinion, i is the gear stage index and its values are 1, 2, 3, 4

Subject to:

- Gear-Tooth Bending Fatigue Strength Constraints:

$$g_1 = \left[\frac{366{,}000}{\pi\omega_1} + \frac{2c_1 N_{p1}}{(N_{p1} + N_{g1})}\right]\left[\frac{(N_{p1} + N_{g1})^2}{4b_1 c_1^2 N_{p1}}\right] \le \frac{\sigma_N J_R}{0.0167 W K_o K_m} \tag{3.72}$$

$$g_2 = \left[\frac{366{,}000 N_{g1}}{\pi\omega_1 N_{p1}} + \frac{2c_2 N_{p2}}{(N_{p2} + N_{g2})}\right]\left[\frac{(N_{p2} + N_{g2})^2}{4b_2 c_2^2 N_{p2}}\right] \le \frac{\sigma_N J_R}{0.0167 W K_o K_m} \tag{3.73}$$

$$g_3 = \left[\frac{366{,}000 N_{g1} N_{g2}}{\pi\omega_1 N_{p1} N_{p2}} + \frac{2c_3 N_{p3}}{(N_{p3} + N_{g3})}\right]\left[\frac{(N_{p3} + N_{g3})^2}{4b_3 c_3^2 N_{p3}}\right] \le \frac{\sigma_N J_R}{0.0167 W K_o K_m} \tag{3.74}$$

$$g_4 = \left[\frac{366,000N_{g1}N_{g2}N_{g3}}{\pi\omega_1 N_{p1}N_{p2}N_{p3}} + \frac{2c_4N_{p4}}{(N_{p4}+N_{g4})}\right]\left[\frac{(N_{p4}+N_{g4})^2}{4b_4c_4^2N_{p4}}\right] \le \frac{\sigma_N J_R}{0.0167WK_oK_m} \tag{3.75}$$

where, W is the Input power, J_R *is the* Geometry factor, K_M is the Mounting factor, K_o is the Overload factor, σ_N is the allowable bending stress, ω_1 is the Input speed,

- Gear-Tooth Contact Strength Constraints:

$$g_5 = \left[\frac{366,000}{\pi\omega_1} + \frac{2c_1N_{p1}}{(N_{p1}+N_{g1})}\right]\left[\frac{(N_{p1}+N_{g1})^3}{4b_1c_1^2N_{g1}N_{p1}^2}\right] \le \left(\frac{\sigma_H}{C_p}\right)^2\left(\frac{\sin\phi\cos\phi}{0.0334WK_oK_m}\right) \tag{3.76}$$

$$g_6 = \left[\frac{366,000N_{g1}}{\pi\omega_1 N_{p1}} + \frac{2c_2N_{p2}}{(N_{p2}+N_{g2})}\right]\left[\frac{(N_{p2}+N_{g2})^3}{4b_2c_2^2N_{g2}N_{p2}^2}\right] \le \left(\frac{\sigma_H}{C_p}\right)^2\left(\frac{\sin\phi\cos\phi}{0.0334WK_oK_m}\right) \tag{3.77}$$

$$g_7 = \left[\frac{366,000N_{g1}N_{g2}}{\pi\omega_1 N_{p1}N_{p2}} + \frac{2c_3N_{p3}}{(N_{p3}+N_{g3})}\right]\left[\frac{(N_{p3}+N_{g3})^3}{4b_3c_3^2N_{g3}N_{p3}^2}\right] \le \left(\frac{\sigma_H}{C_p}\right)^2\left(\frac{\sin\phi\cos\phi}{0.0334WK_oK_m}\right) \tag{3.78}$$

$$g_8 = \left[\frac{366,000N_{g1}N_{g2}N_{g3}}{\pi\omega_1 N_{p1}N_{p2}N_{p3}} + \frac{2c_4N_{p4}}{(N_{p4}+N_{g4})}\right]\left[\frac{(N_{p4}+N_{g4})^3}{4b_4c_4^2N_{g4}N_{p4}^2}\right] \le \left(\frac{\sigma_H}{C_p}\right)^2\left(\frac{\sin\phi\cos\phi}{0.0334WK_oK_m}\right) \tag{3.79}$$

where, σ_H is the allowable fatigue stress and C_p is the elastic coefficient.

- Gear-Tooth Contact Ratio Constraints:

$$g_{9-12} = N_{pi}\sqrt{\frac{\sin^2\phi}{4} + \frac{1}{N_{pi}} + \left(\frac{1}{N_{pi}}\right)^2} + N_{gi}\sqrt{\frac{\sin^2\phi}{4} + \frac{1}{N_{gi}} + \left(\frac{1}{N_{gi}}\right)^2} - \frac{\sin\phi(N_{pi}+N_{gi})}{2} \ge CR_{\min}\pi\cos\phi \tag{3.80}$$

where, ϕ is the pressure angle and $CR_{\min}$ is the allowable contact ratio.

- Minimum Pinion Size Constraints:

$$g_{13-16} = d_{\min} \leq \frac{2c_i N_{pi}}{N_{pi} + N_{gi}} \tag{3.81}$$

- Minimum Gear Size Constraints:

$$g_{17-20} = d_{\min} \leq \frac{2c_i N_{gi}}{N_{pi} + N_{gi}} \tag{3.82}$$

- Gear Housing Constraints for Pinions:

$$g_{21} = x_{p1} + \frac{(N_{p1} + 2)c_1}{N_{p1} + N_{g1}} \leq L_{\max} \tag{3.83}$$

$$g_{22-24} = \left(x_{g(i-1)} + \frac{(N_{pi} + 2)c_i}{N_{pi} + N_{gi}} \right)_{i=2,3,4} \leq L_{\max} \tag{3.84}$$

$$g_{25} = -x_{p1} + \frac{(N_{p1} + 2)c_1}{N_{p1} + N_{g1}} \leq 0 \tag{3.85}$$

$$g_{26-28} = \left(-x_{g(i-1)} + \frac{(N_{pi} + 2)c_i}{N_{pi} + N_{gi}} \right)_{i=2,3,4} \leq 0 \tag{3.86}$$

$$g_{29} = y_{p1} + \frac{(N_{p1} + 2)c_1}{N_{p1} + N_{g1}} \leq L_{\max} \tag{3.87}$$

$$g_{30-32} = \left(y_{g(i-1)} + \frac{(N_{pi} + 2)c_i}{N_{pi} + N_{gi}} \right)_{i=2,3,4} \leq L_{\max} \tag{3.88}$$

$$g_{33} = -y_{p1} + \frac{(N_{p1} + 2)c_1}{N_{p1} + N_{g1}} \leq 0 \tag{3.89}$$

$$g_{34-36} = \left(-y_{g(i-1)} + \frac{(N_{pi} + 2)c_i}{N_{pi} + N_{gi}} \right)_{i=2,3,4} \leq 0 \tag{3.90}$$

where, x_{gi} is the x-coordinate of gear shaft, x_{pi} is the x-coordinate of pinion shaft, y_{gi} is the y-coordinate of gear shaft, y_{pi} is the y-coordinate of pinion shaft.

- Gear Housing Constraints for Gears:

$$g_{37-40} = x_{gi} + \frac{(N_{gi} + 2)c_i}{N_{pi} + N_{gi}} \leq L_{\max} \tag{3.91}$$

$$g_{41-44} = -x_{gi} + \frac{(N_{gi} + 2)c_i}{N_{pi} + N_{gi}} \leq 0 \tag{3.92}$$

$$g_{45-48} = y_{gi} + \frac{(N_{gi} + 2)c_i}{N_{pi} + N_{gi}} \leq L_{\max} \tag{3.93}$$

$$g_{49-52} = -y_{gi} + \frac{(N_{gi} + 2)c_i}{N_{pi} + N_{gi}} \leq 0 \tag{3.94}$$

- Gear Pitch Constraints:

$$g_{53-56} = (0.945c_i - N_{pi} - N_{gi})(b_i - 5.715)(b_i - 8.255)(b_i - 12.70)(-1) \leq 0 \tag{3.95}$$

$$g_{57-60} = (0.646c_i - N_{pi} - N_{gi})(b_i - 3.175)(b_i - 8.255)(b_i - 12.70)(+1) \leq 0 \tag{3.96}$$

$$g_{61-64} = (0.504c_i - N_{pi} - N_{gi})(b_i - 3.175)(b_i - 5.715)(b_i - 12.70)(-1) \leq 0 \tag{3.97}$$

$$g_{65-68} = (0.0c_i - N_{pi} - N_{gi})(b_i - 3.175)(b_i - 5.715)(b_i - 8.255)(+1) \leq 0 \tag{3.98}$$

$$g_{69-72} = (N_{pi} + N_{gi} - 1.812c_i)(b_i - 5.715)(b_i - 8.255)(b_i - 12.70)(-1) \leq 0 \tag{3.99}$$

$$g_{73-76} = (N_{pi} + N_{gi} - 0.945c_i)(b_i - 3.175)(b_i - 8.255)(b_i - 12.70)(+1) \leq 0 \tag{3.100}$$

$$g_{77-80} = (N_{pi} + N_{gi} - 0.646c_i)(b_i - 3.175)(b_i - 5.715)(b_i - 12.70)(-1) \leq 0 \tag{3.101}$$

$$g_{81-84} = (N_{pi} + N_{gi} - 0.504c_i)(b_i - 3.175)(b_i - 5.715)(b_i - 8.255)(+1) \leq 0 \tag{3.102}$$

- Kinematic Constraints:

$$g_{85} = \omega_{\min} \leq \omega_1 \frac{N_{p1}N_{p2}N_{p3}N_{p4}}{N_{g1}N_{g2}N_{g3}N_{g4}} \tag{3.103}$$

$$g_{86} = \omega_1 \frac{N_{p1}N_{p2}N_{p3}N_{p4}}{N_{g1}N_{g2}N_{g3}N_{g4}} \leq \omega_{\max} \tag{3.104}$$

where,

$$x_{p1}, y_{p1}, x_{gi}, y_{gi} \in (12.7,\ 25.4,\ 38.1,\ 50.8,\ 63.5, 76.2, 88.9, 101.6, 114.3),$$

$b_i \in (3.175, 5.715, 8.255, 12.7)$

$N_{pi}, N_{gi} \in (7, 8, 9.........76)$, $c_i = \sqrt{(x_{gi} - x_{pi})^2 + (y_{gi} - y_{pi})^2}$,
$CR_{\min} = 1.4$, $d_{\min} = 25.4$ mm, $\phi = 20^{\circ}$, $W = 55.9$ W, $J_R = 0.2$, $K_M = 1.6$, $K_O = 1.5$, maximum housing dimension ($L_{\max}$) = 127 mm, $\sigma_H = 3{,}290$ kgf/cm^2, $\sigma_N = 2{,}090$ kgf/cm^2, $\omega_I = 5{,}000$ rpm, minimum output speed ($\omega_{\min}$) = 245 rpm, maximum output speed ($\omega_{\max}$) = 255 rpm, $C_p = 464$.

3.2 Applications of Advanced Optimization Techniques to Different Design Optimization Problems of Mechanical Elements

3.2.1 Example 1: Optimization of Gear Train

This example is taken from Yokota et al. [1]. It was solved by Yokota et al. [1] by using improved Genetic Algorithm by considering population size = 20 and number of generation = 1,000 resulting in 20,000 function evaluations.

Table 3.2 Comparison of results for Example 1

Optimization method	A*	B*	C*	D*	FE
IGA [1]	3,512.6	–	–	–	20,000
PSO	3,142.712756	3,135.535546	2,994.844118	2,993.674547	2,000
ABC	3,142.712756	3,135.515852	2,994.844118	2,993.665605	2,000
BBO	3,142.712756	3,136.1597	2,994.844118	2,995.0367	2,000
DE	3,142.712756	3,135.515852	2,994.844118	2,993.6657	2,000
AIA	3,142.712756	3,136.106724	2,994.844118	2,994.160357	2,000

*A** Design considered by Yokota et al. [1] with discrete design variables (Example 1A)
*B** Design considered by Yokota et al. [1] with mixed continuous-discrete design variables (Example 1B)
*C** Modified design with discrete design variables (Example 1C)
*D** Modified design with mixed continuous-discrete design variables (Example 1D), *FE*-Function evaluations, *IGA* Improved GA

The crossover probability and mutation probability was taken as 0.4 and 0.1, respectively. The best value reported by Yokota et al. [1] is $f(X) = 3512.6$ with $b = 24$, $d_1 = 30$, $d_2 = 30$, $z_1 = 18$, $m = 2.75$. In this book the same problem is attempted by using PSO, ABC, BBO, DE and AIA. Following parameters are taken for the analysis:

- Population size = 20
- Number of generations = 100
- For PSO: w varies linealy from 0.9 to 0.4, $c_1 = 2$, $c_2 = 2$, $V_{\max} = 4$
- For ABC: number of employed bees = number of onlooker bees = Population size/2, limit = number of generations.
- For DE: $F = 0.5$, $C = 0.5$
- For BBO: Immigration rate = emigration rate = 1, mutation factor = 0.01,
- For AIA: Clone size = population size, $\beta = 1$, repertoire rate = 0.25

Results obtained by different algorithms for different cases are given in Table 3.2. Moreover, values for the objective function, design variables and constraints for all the cases are given in Table 3.3.

It is observed from Table 3.2 that application of PSO, ABC, BBO, DE and AIA for the weight optimization of gear train has produced better results than the results reported by Yokota et al. [1]. Solutions obtained by using ABC, PSO, DE, BBO and AIA have resulted in 10.53% reduction in weight than that of reported by Yokota et al. [1] by using 90% less function evaluations. All the considered algorithms have produced similar best solutions.

The same problem presented by Yokota et al. [1] is solved by considering the mixed continuous-discrete design variables. Value of Z_1 and m are considered as discrete and rest of the design variables are considered as continuous. It is observed from the results that by considering mixed continuous-discrete design variables, ABC and DE have produced better results than the rest of the algorithms. Results obtained by using PSO is near to the results obtained by using ABC and DE, but results of BBO and AIA is inferior to the results of ABC and DE.

Table 3.3 Values of objective function, constraints and design variables for Example 2

	A*	B*	C*	D*
Design variables				
x_1	24	23.9192958690	22	22.000000000
x_2	30	30.0000000000	30	30.0000000000
x_3	37	36.7664865516	37	36.7474482862
x_4	18	18.0000000000	18	18.0000000000
x_5	–	–	337	391.6830127930
Constraints				
$g_1(X)$	0.0034	0.0000000001	2.0156251441	2.2793804992
$g_2(X)$	0.7295	0.7294527202	0.0297001031	0.4217886196
$g_3(X)$	1.1730	1.1730382294	0.7993003505	0.7993003505
$g_4(X)$	0.0192	0.0000000040	0.0000000000	0.0000000000
$g_5(X)$	0.6162	0.6161616162	6.0000000000	6.0000000000
$g_6(X)$	–	–	1.1764174199	1.1764174199
$g_7(X)$	–	–	0.0207599219	0.0000000001
$g_8(X)$	–	–	0.6161616162	0.6161616162
Objective function				
$f(X)$	3,142.712756	3,135.515852	2,994.844118	2,993.665605

*A** Design considered by Yokota et al. [1] with discrete design variables (Example 1A)
*B** Design considered by Yokota et al. [1].with mixed continuous-discrete design variables (Example 1B)
*C** Modified design with discrete design variables (Example 1C)
*D** Modified design with mixed continuous-discrete design variables (Example 1D)

The consideration of mixed continuous-discrete design variables have resulted in 0.2% reduction in weight than that of considering discrete design variables.

For the modified design also, all the considered algorithms have produced similar best solutions for the discrete design variables. The same modified design is solved by considering mixed continuous-discrete design variables. It is observed from the results that for the modified design ABC have outperformed PSO, DE, BBO and AIA. The results of PSO and DE are near to the results of ABC, but results of BBO and AIA is inferior to the results of ABC, DE and PSO. Modifications in the design have produced 14.73% reduction in weight than the design considered by Yokota et al. [1] by using GA. Moreover, application of advanced optimization techniques has produced 4.7% reduction in weight than improved GA for the design of Yokota et al. [1]. Considerations of mixed continuous-discrete design variables have produced 0.04% reduction in weight.

3.2.2 Example 2: Optimization of Radial Ball Bearing

The problem for rolling element bearing was presented by Gupta et al. [4]. The best reported function value by Gupta et al. [4] for the dynamic capacity and static capacity are 6,029.54 and 3,672.966, respectively with design variables

Table 3.4 Comparison of results for Example 2

Optimization method	A*	B*	C*	FE
NSGA-II [4]	6,029.54	3,672.966	0.2193	225,000
PSO	6,032.249216	3,792.419941	0.223973	10,000
ABC	6,032.315099	3,792.420036	0.223974	10,000
BBO	5,790.02657	3,665.5299	0.22369	10,000
DE	6,032.314158	3,792.4186	0.22397	10,000
AIA	5,728.578202	3,632.841191	0.223503	10,000

*A** optimization of dynamic capacity
*B** optimization of static capacity
*C** optimization of elastohydrodynamic minimum film thickness, *FE* Function evaluations

$X = (20.05977, 6.2111, 7, 0.51499, 0.51504, 0.4298, 0.6342, 0.300063, 0.0345, 0.7143)$, and for the elastohydrodynamic minimum film thickness is 0.2193 with design variables $X = (22.546, 4.6579, 9, 0.51499, 0.5167, 0.4068, 0.6275, 0.300024, 0.0794, 0.6386)$. Multi-objective optimization was also carried out by considering all the objective functions simultaneously using NSGA-II. The values reported by Gupta et al. [4] for multi-objective optimization are $X = (20.702, 5.81, 7\ 0.5149, 0.5159, 0.4046, 0.6057, 0.300011, 0.057, 0.693)$ which give dynamic capacity, static capacity and elastohydodynamic minimum film thickness as 5,511.5, 3,401.91 and 0.2096, respectively. This problem was solved by using NSGA-II by Gupta et al. [4] by considering population size of 4,500 and maximum number of generations of 50, resulting in 225,000 function evaluations. Now the same problem is attempted in this book by using PSO, ABC, BBO, DE and AIA. Following parameters are taken for the analysis:

- Population size = 50
- Number of generations = 200

All other algorithm parameters for ABC, PSO, BBO, DE and AIA are kept same as in Example 1. Results for the best solutions obtained by different algorithms by considering all the objective function separately are given in Table 3.4. Moreover, values for the objective function, design variables and constraints by considering single and multi-objective optimization are given in Table 3.5.

It is observed from Table 3.4 that the applications of PSO, ABC and DE have produced better results than the results reported by Gupta et al. [4] for the optimization of dynamic capacity, static capacity and elastohydrodynamic minimum film thickness. Results obtained by using BBO and AIA are inferior to the results of other optimization methods. Moreover, results of ABC are better than that of PSO and DE. Application of ABC has produced 0.046, 3.25 and 2.13% better results for the optimization of dynamic capacity, static capacity and elastohydrodynamic minimum film thickness if all the objective functions are considered separately. For the multi-objective optimization application of ABC has produced 9.45, 11.48 and 2.5% better results than NSGA-II for the dynamic capacity, static capacity and elastohydodynamic minimum film thickness, respectively. Moreover,

Table 3.5 Values of objective function, constraints and design variables for Example 2

	A*	B*	C*	D*
Design variables				
x_1	6.2129710190	6.2129710190	4.7531163709	6.2129710186
x_2	20.0592463694	20.0592463695	22.3950136213	20.0592463700
x_3	7.0000000000	7.0000000000	7.0000000000	7.0000000000
x_4	0.5150000000	0.5150000000	0.5822163979	0.5150000000
x_5	0.5150000000	0.5366358675	0.5150000001	0.5150000000
x_6	0.4632363940	0.4467854117	0.4000000000	0.4000000000
x_7	0.6350400888	0.6933689170	0.7000000000	0.6921816040
x_8	0.3000000000	0.3000000000	0.3000000025	0.3000000000
x_9	0.0359216980	0.0690762845	0.0701615425	0.0582934447
x_{10}	0.8487454906	0.7964001636	0.7000000000	0.8031896022
Constraints				
$g_1(X)$	0.0000000000	0.0000000000	0.0000068524	0.0000000004
$g_2(X)$	3.1612141580	3.4902338040	1.5062327418	4.4259420372
$g_3(X)$	0.2748597380	1.4414363020	4.4937672582	1.4176900428
$g_4(X)$	1.4257383964	0.9546304534	1.5468836291	1.0157354012
$g_5(X)$	0.0592463694	0.0592463695	2.3950136213	0.0592463700
$g_6(X)$	1.3776215506	2.7038050105	0.4114480787	2.2724914180
$g_7(X)$	0.0000000001	0.0000000001	0.0000000807	0.0000000001
$g_8(X)$	0.0000000000	0.0000000000	0.0672163979	0.0000000000
$g_9(X)$	0.0000000000	0.0216358675	0.0000000001	0.0000000000
Objective function				
$f(X)$	6,032.315099	3,792.420036	0.223974	$C_d = 6032.315098$ $C_s = 3792.420036$ $H_{\min} = 0.214859$

*A** optimization of dynamic capacity
*B** optimization of static capacity
*C** optimization of elastohydrodynamic minimum film thickness
*D** multi-objective optimization

optimization by using ABC requires 95.5% less function evaluations than NSGA-II for the optimization of radial ball bearing.

3.2.3 Example 3: Optimization of Belleville Spring

The design problem of Belleville spring was attempted by Coello [7] by using a new constraint handling technique (NCHT) and Deb and Goyal [12] by using GeneAS. The best function value reported by Coello [7] is 2.121964 with design variables as $X = (0.208, 0.2, 8.751, 11.067)$ by considering population size = 160 and number of generations = 150 resulting in 24,000 function evaluations. Now the same problem is attempted in this book by using PSO, ABC, BBO, DE and AIA. Following parameters are taken for the analysis:

Table 3.6 Comparison of results for Example 3

Optimization method	Function value	Function evaluations
NCHT [7]	2.121964	24,000
PSO	2.12232	15,000
ABC	1.987799	15,000
BBO	2.119003	15,000
DE	1.984374	15,000
AIA	2.108494	15,000

- Population size = 50
- Number of generations = 300

All other algorithm parameters for ABC, PSO, BBO, DE and AIA are kept same as in Example 1. Results for the best solutions obtained by different algorithms are given in Table 3.6. Moreover, values for the objective function, design variables and constraints are given in Table 3.7.

It is observed from Table 3.6 that the applications of ABC, BBO, DE and AIA to the Belleville spring design problem have produced better results than that reported by Coello [7]. PSO has produced nearly the same results as that reported by Coello [7]. Result produced by DE is better than all the results produced by other algorithms. Results obtained by using DE have shown 6.48% weight reduction than that given by Coello [7] by requiring 37.5% less function evaluations.

3.2.4 Example 4: Optimization of Multiple Disc Clutch Brake

The problem for multiple clutch brake was also attempted by Deb and Srinivasan [13] by using NSGA-II. The value of minimum mass reported by Deb and Srinivasan [13] is 0.4704 kg with $r_i = 70$ mm, $r_o = 90$ mm, $t = 1.5$ mm, $F = 1{,}000$ N and $Z = 3$. Now the same problem is attempted in this book by using PSO, ABC, BBO, DE and AIA. Following parameters are taken for the analysis:

- Population size = 20
- Number of generations = 30

All other algorithm parameters for ABC, PSO, BBO, DE and AIA are kept same as in Example 1. Results for the best solutions obtained by different algorithms are given in Table 3.8. Moreover, values for the objective function, design variables and constraints are given in Table 3.9.

It is observed from Table 3.8 that the applications of PSO, ABC, BBO, DE and AIA to the multiplate clutch disc brake design problem have produced better results than that reported by Deb and Srinivasan [13]. All the algorithms have produced similar results except AIA. Application of advanced optimization techniques have shown 23% weight reduction than that given in Deb and Srinivasan [13].

Table 3.7 Values of objective function, constraints and design variables for Example 3

Design variables	
x_1	0.204262
x_2	0.200021
x_3	10.003783
x_4	11.990978
Objective function	
$f(X)$	1.984374
Constraints	
$g_1(X)$	159.9270601084
$g_2(X)$	0.1924019464
$g_3(X)$	0.0000210000
$g_4(X)$	1.5957170000
$g_5(X)$	0.0190220000
$g_6(X)$	1.9871950000
$g_7(X)$	0.1993450567

Table 3.8 Comparison of results for Example 4

Optimization method	Function value
NSGA-II [13]	0.4074
PSO	0.313657
ABC	0.313657
BBO	0.313657
DE	0.313657
AIA	0.321498

3.2.5 Example 5: Optimization of a Robotic Gripper

Robot gripper problem was attempted by Osyczka et al. [10] by using GA with population size of 400 and number of generations as 400 i.e. requiring 160,000 function evaluations. The value of the objective function reported by Osyczka et al. [10] is $f(X) = 5.02$ N with $a = 150$, $b = 131.1$, $c = 196.5$, $e = 12.94$, $f = 133.80$, $l = 175$ and $\delta = 2.60$. Now the same problem is attempted in this book by using PSO, ABC, BBO, DE and AIA. Following parameters are taken for the analysis:

- Population size = 50
- Number of generations = 500

All other algorithm parameters for ABC, PSO, BBO, DE and AIA are kept same as in Example 1. Results for the best solutions obtained by different algorithms are given in Table 3.10. Moreover, values for the objective function, design variables and constraints are given in Table 3.11. It is observed from the results

Table 3.9 Values of objective function, constraints and design variables for Example 4

Design variable	
x_1	70
x_2	90
x_3	1
x_4	860
x_5	3
Objective function	
$f(X)$	0.313657
Constraints	
$g_1(X)$	0.0000000000
$g_2(X)$	24.0000000000
$g_3(X)$	0.9144542181
$g_4(X)$	9,819.9001736111
$g_5(X)$	7,894.6965897818
$g_6(X)$	1.5099273180
$g_7(X)$	3,737.5000000000
$g_8(X)$	13.4900726820

that applications of PSO, ABC and BBO to the robot gripper design problem have produced better results than that reported by Osyczka et al. [10]. DE fails to find the feasible solution and AIA has shown inferior results than that reported by Osyczka et al. [10]. Application of ABC has shown 15.3% improvement in the result than that given in Osyczka et al. [10].

3.2.6 Example 6: Optimization of a Hydrostatic Thrust Bearing

The problem for hydrostatic bearing was attempted by He et al. [14] by using improved PSO, by Coello [7] by using new constrained handling techniques and by Deb and Goyal [12] by using GeneAS. The best reported results are by He et al. [14] with the function value of 1,632.2149 and $R_o = 5.956868685$, $R_i = 5.389175395$, $\mu = 5.40213310$ and $Q = 2.30154678$ by using 90,000 function evaluations. Now the same problem is attempted in this book by using ABC, BBO, DE and AIA. Following parameters are taken for the analysis:

- Population size = 50
- Number of generations = 1,500

All other algorithm parameters for ABC, BBO, DE and AIA are kept same as in Example 1. Results for the best solutions obtained by different algorithms are given in Table 3.12. Moreover, values for the objective function, design variables and constraints are given in Table 3.13.

Table 3.10 Comparison of results for Example 5

Optimization method	Objective function value	Function evaluations
GA [10]	5.02	160,000
PSO	4.496555	25,000
ABC	4.2476	25,000
BBO	4.6753	25,000
DE	–	25,000
AIA	5.3421	2,500

Table 3.11 Values of objective function, constraints and design variables for Example 5

Design variables	
x_1	150
x_2	150
x_3	200
x_4	0
x_5	150
x_6	100
x_7	2.339539113
–	–
Constraints	
$g_1(X)$	28.0928485273
$g_2(X)$	21.9071514727
$g_3(X)$	33.6495870810
$g_4(X)$	79,999.9999999800
$g_5(X)$	0.0000000000
$g_6(X)$	0.0000000001
Objective function	
$f(X)$	4.2476436

It is observed from Table 3.12 that the applications of DE to the hydrostatic thrust bearing design has produced good results than that reported by He et al. [14]. All other algorithms fail to find better results than DE by considering 75,000 function evaluations. Application of DE has shown 0.37% improvement in the result than that given in He et al. [14].

3.2.7 Example 7: Discrete Optimization of a Four Stage Gear Train

Example 7 is taken from Pomrehn and Papalambros [11]. This problem was also solved by Khorshid and Seireg [15]. The best reported results are by Khorshid and Seireg [15] with the function value $f(X) = 38.13$ and design variables $X = (20, 23,$

Table 3.12 Comparison of results for Example 6

Optimization method	Objective function value	Function evaluations
PSO (He et al. [14])	1,632.215	90,000
ABC	1,721.136	75,000
BBO	2,248.463	75,000
DE	1,626.164	75,000
AIA	2,476.342	75,000

Table 3.13 Values of objective function, constraints and design variables for Example 6

Design variables	
x_1	5.9579210698323877
x_2	5.3908549781783925
x_3	0.000005359109319111625I
x_4	2.2722386008326123
Objective function	
$f(X)$	1626.96165
Constraints	
$g_1(X)$	13.8754180063
$g_2(X)$	0.5641388623
$g_3(X)$	0.0065630723
$g_4(X)$	0.0003252256
$g_5(X)$	0.5670660917
$g_6(X)$	0.0009963589
$g_7(X)$	3.7005956376

13, 13 44, 48, 26,28, 12.7 × {7, 4, 7, 4, 3, 5, 7, 6, 6, 3}, 3.175, 3.175, 3.175, 3.175, 3.175). Now the same problem is attempted in this book by using PSO, ABC, BBO, DE and AIA. Following parameters are taken for the analysis:

- Population size = 100
- Number of generations = 1,000

All other algorithm parameters for ABC, PSO, BBO, DE and AIA are kept same as in Example 1. For this example all the algorithms are slightly changed to handle discrete design variables by rounding the continuous value to its nearer integer.

It is observed from Table 3.14 that only BBO is successful in giving feasible solution. All other algorithms fail to find the feasible solution by using 100,000 function evaluations. Application of BBO has shown 4.1% reduction in volume of a four stage gear box than that given by Khorshid and Seireg [15].

The next chapter presents the applications of modified PSO, modified ABC and modified HEA to the unconstrained and constrained benchmark functions and also to the design optimization problems of few mechanical elements.

Table 3.14 Comparison of results for Example 7

Design variables	A*	B*	C*	D*	E*	F*	G*	H*
x_1	21	20	No feasible solution obtained	19	No feasible solution obtained	No feasible solution obtained	No feasible solution obtained	No feasible solution obtained
x_2	19	23		19				
x_3	9	13		20				
x_4	9	13		22				
x_5	37	44		38				
x_6	49	48		49				
x_7	14	26		42				
x_8	25	28		40				
x_9 (× 12.7)	4	7		7				
x_{10} (× 12.7)	2	4		4				
x_{11} (× 12.7)	6	7		6				
x_{12} (× 12.7)	4	4		3				
x_{13} (× 12.7)	6	3		5				
x_{14} (× 12.7)	3	5		7				
x_{15} (× 12.7)	5	7		7				
x_{16} (× 12.7)	4	6		4				
x_{17} (× 12.7)	2	6		5				
x_{18} (× 12.7)	6	3		7				

(continued)

Table 3.14 (continued)

	A*	B*	C*	D*	E*	F*	G*	H*
Design variables								
x_{19}	3.175	3.175		3.175				
x_{20}	3.175	3.175	3.175					
x_{21}	8.255	3.175	3.175					
x_{22}	8.255	3.175	3.175					
Objective function								
$f(X)$	91.88	38.13	36.57					

*A** Pomrehn and Papalambros [11]
*B** Khorshid and Seireg [15]
*C** Dolen et al. [16]
*D** BBO
*E** PSO
*F** ABC
*G** DE
*H** AIA

References

1. Yokota T, Taguchi T, Gen M (1998) A solution method for optimal weight design problem of the gear using genetic algorithms. Comput Ind Eng 35:523–526
2. Norton RL (2001) Machine design: an integrated approach. Person Education, Asia New Delhi
3. Juvinall RC, Marshek KM (2000) Fundamentals of machine components design. Wiley, New York
4. Gupta S, Tiwari R, Shivashankar BN (2007) Multi-objective design optimization of rolling bearings using genetic algorithm. Mech and Mach Theory 42:1418–1443
5. Changsen W (1991) Analysis of rolling element bearings. Mechanical Engineering Publications Ltd, London
6. Harris TA (2000) Rolling bearing analysis. Wiley, New York
7. Coello CAC (2000) Use of a self-adaptive penalty approach for engineering optimization problems. Comput in Ind 41:113–127
8. Osyczka A (2002) Evolutionary algorithms for single and multicriteria design optimization. Stud Fuzzyness Soft Comput, Physica-Verlag, Heidelberg
9. Rao RV, Savsani VJ, Vakharia DP (2011) Teaching–learning-based optimization: a novel method for constrained mechanical design optimization problems. Comput Aided Des 43:303–315
10. Osyczka A, Krenich S, Karas K (1999) Optimum design of robot grippers using genetic algorithms. In: Proceedings of the 3rd world congress of structural and multidisciplinary optimization, New York
11. Pomrehn LP, Papalambros PY (1995) Discrete optimal formulation with application to gear train design. ASME J Mech Des 117(3):419–424
12. Deb K, Goyal M (1996) A combined genetic adaptive search (geneAS) for engineering design. Comput sci and informatics 26:30–35
13. Deb K, Srinivasan A (2005) Innovization: innovation of design principles through optimization. KanGAL Report No. 2005007. Kanpur Genetic Algorithms Laboratory, Department of Mechanical Engineering, Indian Institute of Technology Kanpur, India
14. He S, Wu QH, Wen JY, Saunders JR, Paton RC (2004) A particle swarm optimizer with passive congregation. Biosystems 78(1–3):135–147
15. Khorshid E, Seireg A (1999) Discrete nonlinear optimisation by constraint decomposition and designer interaction. Int J Comput Appl in Technol 12:76–82
16. Dolen M, Kaplan H, Seireg A (2005) Discrete parameter-nonlinear constrained optimisation of a gear train using genetic algorithms. Int J Comput Appl Tech 24(2):110–121

Chapter 4
Applications of Modified Optimization Algorithms to the Unconstrained and Constrained Problems

It is observed from the literature that modification in a particular optimization method suits well to a specific problem [1–18]. However, the same modification may not work for the other applications. So, if any modification is done in any optimization algorithm, it is required to check that algorithm for a wide variety of problems before drawing any general conclusion for the modification incorporated.

To check the performance of the modified algorithms, thirteen unconstrained and twenty-four constrained benchmark problems are considered in this book. Moreover, twenty different mechanical element design optimization problems are considered to check the effectiveness of any modifications in optimization algorithms for suitability to such mechanical design optimization problems. Details of different benchmark problems and mechanical design problems considered in this book are given in the following sections.

4.1 Unconstrained Benchmark Functions (BM-UC)

Thirteen different benchmark problems are considered in this book. All the considered unconstrained benchmark functions possess different characteristics like separability, multimodality and regularity [19]. A function is multimodal if it has two or more local optima. A function is separable if it can be written as a sum of functions of variable separately. Function is regular if it is differentiable at each point of their domain. Non-seperable functions are more difficult to optimize and difficulty increases if the function is multimodel. Complexity increases when the local optima are randomly distributed. Moreover, complexity increases with the increase in dimensionality. Details of different unconstrained benchmark functions along with their characteristics are given in Table 4.1.

R. V. Rao and V. J. Savsani, *Mechanical Design Optimization Using Advanced Optimization Techniques*, Springer Series in Advanced Manufacturing,
DOI: 10.1007/978-1-4471-2748-2_4,

Table 4.1 Details of unconstrained benchmark functions

S.No.	Name	Unconstrained benchmark functions	M*	S*	R*
1	Sphere	$f(x) = \sum_{i=1}^{n} x_i^2$ $-100 \le x_i \le 100, \quad \min(f) = f(0, 0\ldots0) = 0$	n	y	y
2	Schwefel 2.22	$f(x) = \sum_{i=1}^{n} \lvert x_i \rvert + \prod_{i=1}^{n} \lvert x_i \rvert - 100 \le x_i \le 100, \quad \min(f) = f(0, 0\ldots0) = 0$	n	n	n
3	Schwefel 1.2	$nf(x) = \sum_{i=1}^{n} \left(\sum_{j=1}^{i} x_j \right)^2 -100 \le x_i \le 100, \quad \min(f) = f(0, 0\ldots0) = 0$	n	n	y
4	Schwefel 2.21	$gnf(x) = \max_i \{\lvert x_i \rvert, \ 1 \le i \le n\} - 100 \le x_i \le 100, \quad \min(f) = f(0, 0\ldots0) = 0$	n	n	n
5	Rosenbrock	$f(x) = \sum_{i=1}^{n-1} \left[100(x_{i+1} - x_i^2)^2 + (x_i - 1)^2 \right]$ $-30 \le x_i \le 30, \quad \min(f) = f(1, 1\ldots1) = 0$	n	n	y
6	Step	$f(x) = \sum_{i=1}^{n} [x_i + 0.5]^2$ $-100 \le x_i \le 100, \quad \min(f) = f(0, 0\ldots0) = 0$	n	y	n
7	Quartic	$nf(x) = \sum_{i=1}^{n} [ix_i^4] - 1.28 \le x_i \le 1.28, \quad \min(f) = f(0, 0\ldots0) = 0$	n	y	y
8	Schwefel 2.26	$f(x) = -\sum_{i=1}^{30} \left(x_i \sin\left(\sqrt{\lvert x_i \rvert} \right) \right)$ $-500 \le x_i \le 500,$ $\min(f) = f(420.9687, 420.9687\ldots420.9687) = -12569.5$	y	y	n
9	Rastrigin	$f(x) = \sum_{i=1}^{n} [x_i^2 - 10\cos(2\pi x_i) + 10]$ $-5.12 \le x_i \le 5.12, \quad \min(f) = f(0, 0\ldots0) = 0$	y	y	y

(continued)

Table 4.1 (continued)

S.No.	Name	Unconstrained benchmark functions	M*	S*	R*
10	Ackley	$f(x) = \sum_{i=1}^{n} -20\exp\left(-0.2\sqrt{\frac{1}{30}\sum_{i=1}^{30} x_i^2}\right) - \exp\left(\frac{1}{30}\sum_{i=1}^{30}\cos 2\pi x_i\right)$ $-32 \le x_i \le 32, \quad \min(f) = f(0,0\ldots0) = 0$	y	n	y
11	Griewank	$f(x) = \frac{1}{4,000}\sum_{i=1}^{n} x_i^2 - \prod_{i=1}^{n}\cos\left(\frac{x_i}{\sqrt{i}}\right) + 1$ $-600 \le x_i \le 600, \quad \min(f) = f(0,0\ldots0) = 0$	y	n	y
12	Penalty #1	$f(x) = \frac{\pi}{30}\left[10\sin^2(\pi y_1) + \sum_{i=1}^{n-1}(y_i - 1)^2\{1 + 10\sin^2(\pi y_{i+1})\} + (y_n - 1)^2\right]$ $+\sum_{i=1}^{n} u(x_i, 10, 100, 4)$ $-50 \le x_i \le 50, \quad \min(f) = f(1,1\ldots1) = 0$	y	n	y
13	Penalty #2	$f(x) = 0.1\left[\sin^2(\pi 3x_1) + \sum_{i=1}^{n-1}(x_i - 1)^2\{1 + \sin^2(3\pi x_{i+1})\} + (x_n - 1)^2(1 + \sin^2(2\pi x_{30})\right]$ $+\sum_{i=1}^{n} u(x_i, 5, 100, 4)$ $-50 \le x_i \le 50, \quad \min(f) = f(1,1\ldots1) = 0$ $u(x_i, a, k, m) = \begin{cases} k(x_i - a)^m & x_i > a, \\ 0, & -a \le x_i \le a, \\ k(-x_i - a)^m, & x_i < -a \end{cases}$ $y_i = 1 + 1/4(x_i + 1)$	y	n	y

M* Multi-modality, S* Separability, R* Regularity, n no, y yes

4.2 Constrained Benchmark Functions (BM-C)

Twenty-four constrained benchmark functions are considered in this book with different characteristics [20]. Objective functions and constraints are either linear, nonlinear or quadratic in nature. Constraints are in the form of inequality, equality or either of the both. Moreover, the ratio of the feasible to the total search space is also different for the considered problems. Number of active constraints also varies with the problems. Different characteristic of the considered constrained benchmark functions are given in Table 4.2.

n is the number of decision variables, ρ is the estimated ratio between the feasible region and the search space, LI is the number of linear inequality constraints, NI is the number of nonlinear inequality constraints, LE is the number of linear equality constraints, NE is the number of nonlinear equality constraints and a is the number of active constraints at the optimum solution, O is the optimum result.

Details of constrained benchmark functions are given as follows [20]:

G01:

Minimize:

$$f(X) = 5\sum_{i=1}^{4} x_i - 5\sum_{i=1}^{4} x_i^2 - \sum_{i=5}^{13} x_i \tag{4.1}$$

Subject to:

$$g_1(X) = 2x_1 + 2x_2 + x_{10} + x_{11} - 10 \le 0 \tag{4.2}$$

$$g_2(X) = 2x_1 + 2x_3 + x_{10} + x_{12} - 10 \le 0 \tag{4.3}$$

$$g_3(X) = 2x_2 + 2x_3 + x_{11} + x_{12} - 10 \le 0 \tag{4.4}$$

$$g_4(X) = -8x_1 + x_{10} \le 0 \tag{4.5}$$

$$g_5(X) = -8x_2 + x_{11} \le 0 \tag{4.6}$$

$$g_6(X) = -8x_3 + x_{12} \le 0 \tag{4.7}$$

$$g_7(X) = -2x_4 - x_5 + x_{10} \le 0 \tag{4.8}$$

$$g_8(X) = -2x_6 - x_7 + x_{11} \le 0 \tag{4.9}$$

$$g_9(X) = -2x_8 - x_9 + x_{12} \le 0 \tag{4.10}$$

G02:

Minimize:

Table 4.2 Characteristics of constrained benchmark functions

Functions	n	Type of function	ρ (%)	LI	NI	LE	NE	a	O
G01	13	Quadratic	0.0111	9	0	0	0	6	−15
G02	20	Nonlinear	99.9971	0	2	0	0	1	−0.8036
G03	10	Polynomial	0.0000	0	0	0	1	1	−1.0005
G04	5	Quadratic	52.1230	0	6	0	0	2	−3,0665.5
G05	4	Cubic	0.0000	2	0	0	3	3	5,126.496
G06	2	Cubic	0.0066	0	2	0	0	2	−6,961.81
G07	10	Quadratic	0.0003	3	5	0	0	6	24.3062
G08	2	Nonlinear	0.8560	0	2	0	0	0	−0.09582
G09	7	Polynomial	0.5121	0	4	0	0	2	680.63
G10	8	Linear	0.0010	3	3	0	0	6	7,049.248
G11	2	Quadratic	0.0000	0	0	0	1	1	0.7499
G12	3	Quadratic	4.7713	0	1	0	0	0	−1
G13	5	Nonlinear	0.0000	0	0	0	3	3	0.0539
G14	10	Nonlinear	0.0000	0	0	3	0	3	−47.7649
G15	3	Quadratic	0.0000	0	0	1	1	2	961.715
G16	5	Nonlinear	0.0204	4	34	0	0	4	−1.9051
G17	6	Nonlinear	0.0000	0	0	0	4	4	8,853.539
G18	9	Quadratic	0.0000	0	13	0	0	6	−0.86602
G19	15	Nonlinear	33.4761	0	5	0	0	0	32.6556
G20	24	Linear	0.0000	0	6	2	12	16	0.20498
G21	7	Linear	0.0000	0	1	0	5	6	193.7245
G22	22	Linear	0.0000	0	1	8	11	19	236.4309
G23	9	Linear	0.0000	0	2	3	1	6	−400.055
G24	2	Linear	79.6556	0	2	0	0	2	−5.50801

$$f(X) = -\left|\frac{\sum_{i=1}^{n}\cos^4(x_i) - 2\prod_{i=1}^{n}\cos^2(x_i)}{\sqrt{\sum_{i=1}^{n} ix_i^2}}\right| \tag{4.11}$$

Subject to:

$$g_1(X) = 0.75 - \prod_{i=1}^{n} x_i \le 0 \tag{4.12}$$

$$g_2(X) = \sum_{i=1}^{n} x_i - 7.5n \le 0 \tag{4.13}$$

G03:

Minimize:

$$f(X) = -(\sqrt{n})^n \prod_{i=1}^{n} x_i \tag{4.14}$$

Subject to:

$$h_1(X) = \sum_{i=1}^{n} x_i^2 - 1 = 0 \tag{4.15}$$

G04:

Minimize:

$$f(X) = 5.3578547x_3^2 + 0.8356891x_1x_5 + 37.293239x_1 - 4,0792.141 \tag{4.16}$$

Subject to:

$$g_1(X) = 85.334407 + 0.005685x_2x_5 + 0.0006262x_1x_4 - 0.0022053x_3x_5 - 92 \le 0 \tag{4.17}$$

$$g_2(X) = -85.334407 - 0.005685x_2x_5 - 0.0006262x_1x_4 + 0.0022053x_3x_5 \le 0 \tag{4.18}$$

$$g_3(X) = 80.51249 + 0.0071317x_2x_5 + 0.002995x_1x_2 + 0.0021813x_3^2 - 110 \le 0 \tag{4.19}$$

$$g_4(X) = -80.51249 - 0.0071317x_2x_5 - 0.002995x_1x_2 - 0.0021813x_3^2 + 90 \le 0 \tag{4.20}$$

$$g_5(X) = 9.300961 + 0.0047026x_3x_5 + 0.0012547x_1x_3 + 0.0019085x_3x_4 - 25 \le 0 \tag{4.21}$$

$$g_6(X) = -9.300961 - 0.0047026x_3x_5 - 0.0012547x_1x_3 - 0.0019085x_3x_4 + 20 \le 0 \tag{4.22}$$

G05:

Minimize:

$$f(X) = 3x_1 + 0.000001x_1^3 + 2x_2 + (0.000002/3)x_2^3 \tag{4.23}$$

Subject to:

$$g_1(X) = -x_4 + x_3 - 0.55 \le 0 \tag{4.24}$$

$$g_2(X) = -x_3 + x_4 - 0.55 \le 0 \tag{4.25}$$

$$h_3(X) = 1,000\sin(-x_3 - 0.25) + 1,000\sin(-x_4 - 0.25) + 894.8 - x_1 = 0 \tag{4.26}$$

$$h_4(X) = 1,000\sin(x_3 - 0.25) + 1,000\sin(x_3 - x_4 - 0.25) + 894.8 - x_2 = 0 \tag{4.27}$$

$$h_5(X) = 1,000\sin(x_4 - 0.25) + 1,000\sin(x_4 - x_3 - 0.25) + 1,294.8 = 0 \tag{4.28}$$

G06:

Minimize:

$$f(X) = (x_1 - 10)^3 + (x_2 - 20)^3 \tag{4.29}$$

Subject to:

$$g_1(X) = -(x_1 - 5)^2 - (x_2 - 5)^2 + 100 \le 0 \tag{4.30}$$

$$g_2(X) = (x_1 - 6)^2 + (x_2 - 5)^2 - 82.81 \le 0 \tag{4.31}$$

where $13 \le x_1 \le 100$ and $0 \le x_2 \le 100$

G07:

Minimize:

$$f(X) = x_1^2 + x_2^2 - x_1x_2 - 14x_1 - 16x_2 + (x_3 - 10)^2 + 4(x_4 - 5)^2 + (x_5 - 3)^2 + 2(x_6 - 1)^2 + 5x_7^2 + 7(x_8 - 11)^2 + 2(x_9 - 10)^2 + (x_{10} - 7)^2 + 45 \tag{4.32}$$

Subject to:

$$g_1(X) = -105 + 4x_1 + 5x_2 - 3x_7 + 9x_8 \le 0 \tag{4.33}$$

$$g_2(X) = 10x_1 - 8x_2 - 17x_7 + 2x_8 \le 0 \tag{4.34}$$

$$g_3(X) = -8x_1 + 2x_2 + 5x_9 - 2x_{10} - 12 \le 0 \tag{4.35}$$

$$g_4(X) = 3(x_1 - 2)^2 + 4(x_2 - 3)^2 + 2x_3^2 - 7x_4 - 120 \le 0 \tag{4.36}$$

$$g_5(X) = 5x_1^2 + 8x_2 + (x_3 - 6)^2 - 2x_4 - 40 \le 0 \tag{4.37}$$

$$g_6(X) = x_1^2 + 2(x_2 - 2)^2 - 2x_1x_2 + 14x_5 - 6x_6 \le 0 \tag{4.38}$$

$$g_7(X) = 0.5(x_1 - 8)^2 + 2(x_2 - 4)^2 + 3x_5^2 - x_6 - 30 \le 0 \tag{4.39}$$

$$g_8(X) = -3x_1 + 6x_2 + 12(x_9 - 8)^2 - 7x_{10} \le 0 \tag{4.40}$$

G08:

Minimize:

$$f(X) = -\frac{\sin^3(2\pi x_1)\sin(2\pi x_2)}{x_1^3(x_1 + x_2)} \tag{4.41}$$

Subject to:

$$g_1(X) = x_1^2 - x_2 + 1 \le 0 \tag{4.42}$$

$$g_2(X) = 1 - x_1 + (x_2 - 4)^2 \le 0 \tag{4.43}$$

G09:

Minimize:

$$f(X) = (x_1 - 10)^2 + 5(x_2 - 12)^2 + x_3^4 + 3(x_4 - 11)^2 + 10x_5^6 + 7x_6^2 + x_7^4 - 4x_6x_7 - 10x_6 - 8x_7 \tag{4.44}$$

Subject to:

$$g_1(X) = -127 + 2x_1^2 + 3x_2^4 + x_3 + 4x_4^2 + 5x_5 \leq 0 \quad (4.45)$$

$$g_2(X) = -282 + 7x_1 + 3x_2 + 10x_3^2 + x_4 - x_5 \leq 0 \quad (4.46)$$

$$g_3(X) = -196 + 23x_1 + x_2^2 + 6x_6^2 - 8x_7 \leq 0 \quad (4.47)$$

$$g_4(X) = 4x_1^2 + x_2^2 - 3x_1x_2 + 2x_3^2 + 5x_6 - 11x_7 \leq 0 \quad (4.48)$$

where, $-10 \leq x_i \leq 10$ $(i = 1, \ldots, 7)$

G10:

Minimize:

$$f(X) = x_1 + x_2 + x_3 \quad (4.49)$$

Subject to:

$$g_1(X) = -1 + 0.0025(x_4 + x_6) \leq 0 \quad (4.50)$$

$$g_2(X) = -1 + 0.0025(x_5 + x_7 - x_4) \leq 0 \quad (4.51)$$

$$g_3(X) = -1 + 0.01(x_8 - x_5) \leq 0 \quad (4.52)$$

$$g_4(X) = -x_1x_6 + 833.3325x_4 + 100x_1 - 83{,}333.333 \leq 0 \quad (4.53)$$

$$g_5(X) = -x_2x_7 + 1{,}250x_5 + x_2x_4 - 1{,}250x_4 \leq 0 \quad (4.54)$$

$$g_6(X) = -x_3x_8 + 125{,}000 + x_3x_5 - 2{,}500x_5 \leq 0 \quad (4.55)$$

G11:

Minimize:

$$f(X) = x_1^2 + (x_2 - 1)^2 \quad (4.56)$$

Subject to:

$$h(X) = x_2 - x_1^2 = 0 \quad (4.57)$$

G12:

Minimize:

$$f(X) = -(100 - (x_1 - 5)^2 - (x_2 - 5)^2 - (x_3 - 5)^2)/100 \quad (4.58)$$

Subject to:

$$g(X) = (x_1 - p)^2 + (x_2 - q)^2 + (x_3 - r)^2 - 0.0625 \leq 0 \quad (4.59)$$

where $0 \leq x_i \leq 10$ $(i = 1, 2, 3)$ and $p, q, r = 1, 2, \ldots, 9$.

G13:

Minimize:

$$f(X) = e^{x_1 x_2 x_3 x_4 x_5} \quad (4.60)$$

Subject to:

$$h_1(X) = x_1^2 + x_2^2 + x_3^2 + x_4^2 + x_5^2 - 10 = 0 \quad (4.61)$$

$$h_2(X) = x_2 x_3 - 5 x_4 x_5 = 0 \quad (4.62)$$

$$h_3(X) = x_1^3 + x_2^3 + 1 = 0 \quad (4.63)$$

G14:

Minimize:

$$f(X) = \sum_{i=1}^{10} x_i \left(c_i + \ln \frac{x_i}{\sum_{j=i}^{10} x_j} \right) \quad (4.64)$$

Subject to:

$$h_1(X) = x_1 + 2x_2 + 2x_3 + x_6 + x_{10} - 2 = 0 \quad (4.65)$$

$$h_2(X) = x_4 + 2x_5 + x_6 + x_7 - 1 = 0 \quad (4.66)$$

$$h_3(X) = x_3 + x_7 + x_8 + 2x_9 + x_{10} - 1 = 0 \quad (4.67)$$

where, $0 \leq x_i \leq 10$ $(i = 1, \ldots, 10)$ and $c_1 = -6.089$, $c_2 = -17.164$, $c_3 = -34.054$, $c_4 = -5.914, c_5 = -24.721$, $c_6 = -14.986$, $c_7 = -24.1$, $c_8 = -10.708$, $c_9 = -26.662$, $c_{10} = -22.179$

G15:

Minimize:

$$f(X) = 1000 - x_1^2 - 2x_2^2 - x_3^2 - x_1 x_2 - x_1 x_3 \quad (4.68)$$

Subject to:

$$h_1(X) = x_1^2 + x_2^2 + x_3^2 - 25 = 0 \quad (4.69)$$

$$h_2(X) = 8x_1 + 14x_2 + 7x_3 - 56 = 0 \quad (4.70)$$

where $0 \le x_i \le 10$ $(i = 1, 2, 3)$

G16:

Minimize:

$$f(X) = 0.000117y_{14} + 0.1365 + 0.00002358y_{13} + 0.000001502y_{16} + 0.0321y_{12} + 0.004324y_5 + 0.001\frac{c_{15}}{c_{16}} + 37.48\frac{y_2}{c_{12}} - 0.0000005843y_{17} \quad (4.71)$$

Subject to:

$$g_1(X) = \frac{0.28}{0.72}y_5 - y_4 \le 0 \quad (4.72)$$

$$g_2(X) = x_3 - 1.5x_2 \le 0 \quad (4.73)$$

$$g_3(X) = 3,496\frac{y_2}{c_{12}} - 21 \le 0 \quad (4.74)$$

$$g_4(X) = 110.6 + y_1 - \frac{62,212}{c_{17}} \le 0 \quad (4.75)$$

$$g_5(X) = 213.1 - y_1 \le 0 \quad (4.76)$$

$$g_6(X) = y_1 - 405.23 \le 0 \quad (4.77)$$

$$g_7(X) = 17.505 - y_2 \le 0 \quad (4.78)$$

$$g_8(X) = y_2 - 1,053.6667 \le 0 \quad (4.79)$$

$$g_9(X) = 11.275 - y_3 \le 0 \quad (4.80)$$

$$g_{10}(X) = y_3 - 35.03 \le 0 \quad (4.81)$$

$$g_{11}(X) = 214.228 - y_4 \le 0 \quad (4.82)$$

$$g_{12}(X) = y_4 - 665.585 \le 0 \quad (4.83)$$

$$g_{13}(X) = 7.458 - y_5 \le 0 \quad (4.84)$$

$$g_{14}(X) = y_5 - 584.463 \le 0 \quad (4.85)$$

$$g_{15}(X) = 0.961 - y_6 \le 0 \quad (4.86)$$

$$g_{16}(X) = y_6 - 265.961 \le 0 \quad (4.87)$$

$$g_{17}(X) = 1.612 - y_7 \le 0 \quad (4.88)$$

$$g_{18}(X) = y_7 - 7.046 \le 0 \quad (4.89)$$

$$g_{19}(X) = 0.146 - y_8 \leq 0 \tag{4.90}$$

$$g_{20}(X) = y_8 - 0.222 \leq 0 \tag{4.91}$$

$$g_{21}(X) = 107.99 - y_9 \leq 0 \tag{4.92}$$

$$g_{22}(X) = y_9 - 273.366 \leq 0 \tag{4.93}$$

$$g_{23}(X) = 922.693 - y_{10} \leq 0 \tag{4.94}$$

$$g_{24}(X) = y_{10} - 1,286.105 \leq 0 \tag{4.95}$$

$$g_{25}(X) = 926.832 - y_{11} \leq 0 \tag{4.96}$$

$$g_{26}(X) = y_{11} - 1,444.046 \leq 0 \tag{4.97}$$

$$g_{27}(X) = 18.766 - y_{12} \leq 0 \tag{4.98}$$

$$g_{28}(X) = y_{12} - 537.141 \leq 0 \tag{4.99}$$

$$g_{29}(X) = 1072.163 - y_{13} \leq 0 \tag{4.100}$$

$$g_{30}(X) = y_{13} - 3,247.039 \leq 0 \tag{4.101}$$

$$g_{31}(X) = 8,961.448 - y_{14} \leq 0 \tag{4.102}$$

$$g_{32}(X) = y_{14} - 26,844.086 \leq 0 \tag{4.103}$$

$$g_{33}(X) = 0.063 - y_{15} \leq 0 \tag{4.104}$$

$$g_{34}(X) = y_{15} - 0.386 \leq 0 \tag{4.105}$$

$$g_{35}(X) = 71,084.33 - y_{16} \leq 0 \tag{4.106}$$

$$g_{36}(X) = -140,000 + y_{16} \leq 0 \tag{4.107}$$

$$g_{37}(X) = 2,802,713 - y_{17} \leq 0 \tag{4.108}$$

$$g_{38}(X) = y_{17} - 12,146,108 \leq 0 \tag{4.109}$$

Where, $y_1 = x_2 + x_3 + 41.6$, $c_1 = 0.024x_4 - 4.62$, $y_2 = \dfrac{12.5}{c_1} + 12$, $c_2 = 0.0003535x_1^2 + 0.5311x_1 + 0.08705y_2x_1$, $c_3 = 0.052x_1 + 78 + 0.002377y_2x_1$, $y_3 = \dfrac{c_2}{c_3}$, $y_4 = 19y_3$, $c_4 = 0.04782(x_1 - y_3) + \dfrac{0.1956(x_1 - y_3)^2}{x_2} + 0.6376y_4$ $+1.594y_3$, $c_5 = 100x_2$, $c_6 = x_1 - y_3 - y_4$, $c_7 = 0.950-$ $\frac{c_4}{c_5}$, $y_5 = c_6c_7$, $y_6 = x_1 - y_5 - y_4 - y_3$, $c_8 = (y_5 + y_4)0.995$, $y_7 = \frac{c_8}{y_1}$, $y_8 = \frac{c_8}{3,798}$, $c_9 = y_7 - \frac{0.0663y_7}{y_8} - 0.3153$, $y_9 = \frac{96.82}{c_9} + 0.321y_1$, $y_{10} = 1.29y_5 + 1.258y_4 + 2.29y_3 + 1.71y_6$, $y_{11} = 1.71x_1 - 0.452y_4 + 0.580y_3$, $c_{10} = \frac{12.3}{752.3}$, $c_{11} = (1.75y_2)(0.995x_1)$, $c_{12} = 0.995y_{10}$

$+1,998$, $y_{12} = c_{10}x_1 + \frac{c_{11}}{c_{12}}$, $y_{13} = c_{12} - 1.75y_2, g22y_{14} = 3,623 + 64.4x_2$ $+58.4x_3 + \frac{146{,}312}{y_6+x_5}$, $c_{13} = 0.995y_{10} + 60.8x_2 + 48x_4 - 0.1121y_{14} - 5,095$, $y_{15} = \frac{y_{13}}{c_{13}}$, $y_{16} = 148,000 - 331,000y_{15} + 40y_{13} - 61y_{15}y_{13}$, $c_{14} = 2,324y_{10} - 28,740,000y_2, y_{17} = 14,130,000 - 1,328y_{10} - 531y_{11} + \frac{c_{14}}{c_{12}}$, $c_{15} = \frac{y_{13}}{y_{15}} - \frac{y_{13}}{0.52}$, $c_{16} = 1.104 - 0.72y_{15}, c_{17} = y_9 + x_5$, $704.4148 \le x_1 \le 906.3855$, $68.6 \le x_2 \le 288.88$, $0 \le x_3 \le 134.75$, $193 \le x_4 \le 287.0966$ and $25 \le x_5 \le 84.1988$

G17:

Minimize:

$$f(X) = f(x_1) + f(x_2) \tag{4.110}$$

where,

$$f_1(x_1) = \begin{cases} 30x_1 & 0 \le x_1 < 300 \\ 31x_1 & 300 \le x_1 < 400 \end{cases} \tag{4.111}$$

$$f_2(x_2) = \begin{cases} 28x_2 & 0 \le x_2 < 100 \\ 29x_2 & 100 \le x_2 < 200 \\ 30x_2 & 200 \le x_2 < 1,000 \end{cases} \tag{4.112}$$

Subject to:

$$h_1(X) = -x_1 + 300 - \frac{x_3x_4}{131.078}\cos(1.48477 - x_6) + \frac{0.90798x_3^2}{131.078}\cos(1.47588) \tag{4.113}$$

$$h_2(X) = -x_2 - \frac{x_3x_4}{131.078}\cos(1.48477 + x_6) + \frac{0.90798x_4^2}{131.078}\cos(1.47588) \tag{4.114}$$

$$h_3(X) = -x_5 - \frac{x_3x_4}{131.078}\sin(1.48477 + x_6) + \frac{0.90798x_4^2}{131.078}\sin(1.47588) \tag{4.115}$$

$$h_4(X) = 200 - \frac{x_3x_4}{131.078}\sin(1.48477 - x_6) + \frac{0.90798x_3^2}{131.078}\sin(1.47588) \tag{4.116}$$

where the bounds are $0 \le x_1 \le 400$, $0 \le x_2 \le 1000$, $340 \le x_3 \le 420$, $340 \le x_4 \le 420$, $-1000 \le x_5 \le 1000$ and $0 \le x_6 \le 0.5236$.

G18:

Minimize:

$$f(X) = -0.5(x_1x_4 - x_2x_3 + x_3x_9 - x_5x_9 + x_5x_8 - x_6x_7) \quad (4.117)$$

Subject to:

$$g_1(X) = x_3^2 + x_4^2 - 1 \le 0 \quad (4.118)$$

$$g_2(X) = x_9^2 - 1 \le 0 \quad (4.119)$$

$$g_3(X) = x_5^2 + x_6^2 - 1 \le 0 \quad (4.120)$$

$$g_4(X) = x_1^2 + (x_2 - x_9)^2 - 1 \le 0 \quad (4.121)$$

$$g_5(X) = (x_2 - x_5)^2 + (x_2 - x_6)^2 - 1 \le 0 \quad (4.122)$$

$$g_6(X) = (x_2 - x_7)^2 + (x_2 - x_8)^2 - 1 \le 0 \quad (4.123)$$

$$g_7(X) = (x_3 - x_5)^2 + (x_4 - x_6)^2 - 1 \le 0 \quad (4.124)$$

$$g_8(X) = (x_3 - x_7)^2 + (x_4 - x_8)^2 - 1 \le 0 \quad (4.125)$$

$$g_9(X) = x_7^2 + (x_8 - x_9)^2 - 1 \le 0 \quad (4.126)$$

$$g_{10}(X) = x_2x_3 - x_1x_4 - 1 \le 0 \quad (4.127)$$

$$g_{11}(X) = -x_3x_9 \le 0 \quad (4.128)$$

$$g_{12}(X) = x_5x_9 \le 0 \quad (4.129)$$

$$g_{13}(X) = x_6x_7 - x_5x_8 \le 0 \quad (4.130)$$

where the bounds are $-10 \le x_i \le 10$ $(i = 1, \ldots, 8)$ and $0 \le x_9 \le 20$

G19:

Minimize:

$$f(X) = \sum_{j=1}^{5}\sum_{i=1}^{5} c_{ij}x_{(10+i)}x_{(10+j)} + 2\sum_{j=1}^{5} d_j x_{(10+j)}^3 - \sum_{i=1}^{10} b_i x_i \quad (4.131)$$

Subject to:

$$g_j(X) = -2\sum_{i=1}^{5} c_{ij}x_{(10+i)} - 3d_j x_{(10+j)}^2 - e_j + \sum_{i=1}^{10} a_{ij}x_i \le 0 \quad j = 1, \ldots, 5 \quad (4.132)$$

where $b = [-40, -2, -0.25, -4, -4, -1, -40, -60, 5, 1]$ and the remaining data is given in the Table 4.3

The bounds are $0 \le x_i \le 10$ $(i = 1, \ldots, 8)$

G20:

Minimize:

$$f(X) = \sum_{i=1}^{24} a_i x_i \tag{4.133}$$

Subject to:

$$g_i(X) = \frac{\left(x_i + x_{(i+12)}\right)}{\sum_{j=1}^{24} x_j + e_i} \le 0 \quad i = 1, 2, 3 \tag{4.134}$$

$$g_i(X) = \frac{\left(x_{(i+3)} + x_{(i+15)}\right)}{\sum_{j=1}^{24} x_j + e_i} \le 0 \quad i = 4, 5, 6 \tag{4.135}$$

$$h_i(X) = \frac{x_{(i+12)}}{b_{(i+12)} \sum_{j=13}^{24} \frac{x_j}{b_j}} - \frac{c_i x_i}{40 b_i \sum_{j=1}^{12} \frac{x_j}{b_j}} = 0 \quad i = 1, \ldots, 12 \tag{4.136}$$

$$h_{13}(X) = \sum_{i=1}^{24} x_i - 1 = 0 \tag{4.137}$$

$$h_{14}(X) = \sum_{i=1}^{12} \frac{x_i}{d_i} + k \sum_{i=13}^{24} \frac{x_i}{b_i} - 1.671 = 0 \tag{4.138}$$

where $k = (0.7302)(530)\left(\frac{14.7}{40}\right)$ and data set is detailed in Table 4.4. The bounds are $0 \le x_i \le 10$ $(i = 1, \ldots, 24)$

G21:

Minimize:

$$f(X) = x_1 \tag{4.139}$$

Subject to:

$$g_1(X) = -x_1 + 35x_2^{0.6} + 35x_3^{0.6} \le 0 \tag{4.140}$$

$$h_1(X) = -300x_3 + 7{,}500x_5 - 7{,}500x_6 - 25x_4x_5 + 25x_4x_6 + x_3x_4 = 0 \tag{4.141}$$

$$h_2(X) = 100x_2 + 155.365x_4 + 2{,}500x_7 - x_2x_4 - 25x_4x_7 - 15{,}536.5 = 0 \tag{4.142}$$

$$h_3(X) = -x_5 + \ln(-x_4 + 900) = 0 \tag{4.143}$$

Table 4.3 Data set for benchmark function G19

j	1	2	3	4	5
e_j	−15	−27	−36	−18	−12
c_{1j}	30	−20	−10	32	−10
c_{2j}	−20	39	−6	−31	32
c_{3j}	−10	−6	10	−6	−10
c_{4j}	32	−31	−6	39	−20
c_{5j}	−10	32	−10	−20	30
d_j	4	8	10	6	2
a_{1j}	−16	2	0	1	0
a_{2j}	0	−2	0	0.4	2
a_{3j}	−3.5	0	2	0	0
a_{4j}	0	−2	0	−4	−1
a_{5j}	0	−9	−2	1	−2.8
a_{6j}	2	0	−4	0	0
a_{7j}	−1	−1	−1	−1	−1
a_{8j}	−1	−2	−3	−2	−1
a_{9j}	1	2	3	4	5
a_{10j}	1	1	1	1	1

Table 4.4 Data set for benchmark function G20

i	a_i	b_i	c_i	d_i	e_i
1	0.0693	44.094	123.7	31.244	0.1
2	0.0577	58.12	31.7	36.12	0.3
3	0.05	58.12	45.7	34.784	0.4
4	0.2	137.4	14.7	92.7	0.3
5	0.26	120.9	84.7	82.7	0.6
6	0.55	170.9	27.7	91.6	0.3
7	0.06	62.501	49.7	56.708	-
8	0.1	84.94	7.1	82.7	-
9	0.12	133.425	2.1	80.8	-
10	0.18	82.507	17.7	64.517	-
11	0.1	46.07	0.85	49.4	-
12	0.09	60.097	0.64	49.1	-
13	0.0693	44.094	-	-	-
14	0.0577	58.12	-	-	-
15	0.05	58.12	-	-	-
16	0.2	137.4	-	-	-
17	0.26	120.9	-	-	-
18	0.55	170.9	-	-	-
19	0.06	62.501	-	-	-
20	0.1	84.94	-	-	-
21	0.12	133.425	-	-	-
22	0.18	82.507	-	-	-
23	0.1	46.07	-	-	-
24	0.09	60.097	-	-	-

$$h_4(X) = -x_6 + \ln(x_4 + 300) = 0 \quad (4.144)$$

$$h_5(X) = -x_7 + \ln(-2x_4 + 700) = 0 \quad (4.145)$$

where, the bounds are $0 \le x_1 \le 1,000$, $0 \le x_2, x_3 \le 1,000$, $100 \le x_4 \le 300$, $6.3 \le x_5 \le 6.7, 5.9 \le x_6 \le 6.4$ and $4.5 \le x_7 \le 6.25$

G22:

Minimize:

$$f(X) = x_1 \quad (4.146)$$

Subject to:

$$g_1(X) = -x_1 + x_2^{0.6} + x_3^{0.6} + x_4^{0.6} \le 0 \quad (4.147)$$

$$h_1(X) = x_5 - 100,000x_8 + 1 \times 10^7 = 0 \quad (4.148)$$

$$h_2(X) = x_6 + 100,000x_8 - 100,000x_9 = 0 \quad (4.149)$$

$$h_3(X) = x_7 + 100,000x_9 - 5 \times 10^7 = 0 \quad (4.150)$$

$$h_4(X) = x_5 + 100,000x_{10} - 3.3 \times 10^7 = 0 \quad (4.151)$$

$$h_5(X) = x_6 + 100,000x_{11} - 4.4 \times 10^7 = 0 \quad (4.152)$$

$$h_6(X) = x_7 + 100,000x_{12} - 6.6 \times 10^7 = 0 \quad (4.153)$$

$$h_7(X) = x_5 - 120x_2x_{13} = 0 \quad (4.154)$$

$$h_8(X) = x_6 - 80x_3x_{14} = 0 \quad (4.155)$$

$$h_9(X) = x_7 - 40x_4x_{15} = 0 \quad (4.156)$$

$$h_{10}(X) = x_8 - x_{11} + x_{16} = 0 \quad (4.157)$$

$$h_{11}(X) = x_9 - x_{12} + x_{17} = 0 \quad (4.158)$$

$$h_{12}(X) = -x_{18} + \ln(x_{10} - 100) = 0 \quad (4.159)$$

$$h_{13}(X) = -x_{19} + \ln(-x_8 + 300) = 0 \quad (4.160)$$

$$h_{14}(X) = -x_{20} + \ln(x_{16}) = 0 \quad (4.161)$$

$$h_{15}(X) = -x_{21} + \ln(-x_9 + 400) = 0 \quad (4.162)$$

$$h_{16}(X) = -x_{22} + \ln(x_{17}) = 0 \quad (4.163)$$

$$h_{17}(X) = -x_8 - x_{10} + x_{13}x_{18} - x_{13}x_{19} + 400 = 0 \quad (4.164)$$

$$h_{18}(X) = x_8 - x_{10} - x_{11} + x_{14}x_{20} - x_{14}x_{21} + 400 = 0 \quad (4.165)$$

$$h_{19}(X) = x_9 - x_{12} - 4.6051x_{15} + x_{15}x_{22} + 100 = 0 \quad (4.166)$$

Where, the bounds are $0 \leq x_1 \leq 20,000$, $0 \leq x_2, x_3, x_4 \leq 1 \times 10^6$, $0 \leq x_5, x_6, x_7 \leq 4 \times 10^7$, $100 \leq x_8 \leq 299.99$, $100 \leq x_9 \leq 399.99$, $100.01 \leq x_{10} \leq 300$, $100 \leq x_{11} \leq 400$, $100 \leq x_{12} \leq 6000 \leq x_{13}, x_{14}, x_{15} \leq 500$, $0.01 \leq x_{17} \leq 400$, $-4.7 \leq x_{18}, x_{19}, x_{20}, x_{21}, x_{22} \leq 6.25$

G23:

Minimize:

$$f(X) = -9x_5 - 15x_8 + 6x_1 + 16x_2 + 10(x_6 + x_7) \quad (4.167)$$

Subject to:

$$g_1(X) = x_9x_3 + 0.02x_6 - 0.025x_5 \leq 0 \quad (4.168)$$

$$g_2(X) = x_9x_4 + 0.02x_7 - 0.015x_8 \leq 0 \quad (4.169)$$

$$h_1(X) = x_1 + x_2 - x_3 - x_4 = 0 \quad (4.170)$$

$$h_2(X) = 0.03x_1 + 0.01x_2 - x_9(x_3 + x_4) = 0 \quad (4.171)$$

$$h_3(X) = x_3 + x_6 - x_5 = 0 \quad (4.172)$$

$$h_4(X) = x_4 + x_7 - x_8 = 0 \quad (4.173)$$

where the bounds are $0 \leq x_1, x_2, x_6 \leq 300$, $0 \leq x_3, x_4, x_5 \leq 100$, $0 \leq x_4, x_8 \leq 200$ and $0.01 \leq x_9 \leq 0.03$

G24:

Minimize:

$$f(X) = -x_1 - x_2 \quad (4.174)$$

Subject to:

$$g_1(X) = -2x_1^4 + 8x_1^3 - 8x_1^2 + x_2 - 2 \leq 0 \quad (4.175)$$

$$g_2(X) = -4x_1^4 + 32x_1^3 - 88x_1^2 + 96x_1 + x_2 - 36 \leq 0 \quad (4.176)$$

where the bounds are $0 \leq x_1 \leq 3$ and $0 \leq x_2 \leq 4$

4.3 Additional Mechanical Element Design Optimization Problems (MD)

Seven different mechanical element design optimization problems are presented in Chap. 2. In this chapter, thirteen more mechanical design element design optimization problems are presented. Details of these additional mechanical element design optimization problems are given as follows:

4.3.1 *Example 8: Design of Pressure Vessel*

A cylindrical vessel is capped at both ends by hemispherical heads as shown in Fig. 4.1. The objective is to minimize the total cost, including the cost of the material, forming and welding. There are four design variables: thickness of the shell (T_s), thickness of the head (T_h), inner radius (R) and length of the cylindrical section of the vessel, not including the head (L) so design vector $X = (x_1, x_2, x_3, x_4) = (T_s, T_h, R, L)$. T_s and T_h are integer multiples of 0.0625 inch, which are the available thicknesses of rolled steel plates, and R and L are continuous. The above problem was solved by many researchers by using different optimization methods like branch and bound approach [21], an augmented Lagrangian Multiplier approach [22], Genetic Adaptive Search method (GeneAS) [23], self adaptive penalty approach [24], society and civilization algorithm [25], Ant colony algorithm [34] ($\mu + \lambda$)-Evolutionary Strategy(ES) [26], Unified Particle Swarm Optimization (UPSO) [27], Co-evolutionary Particle Swarm Optimization (CPSO) [28], Co-evolutionary Differential Evolution (CoDE) [29], modified particle swarm optimization [7], Hybrid PSO-DE [30], Artificial Bee Colony (ABC) [31], etc. the best solution reported is $f(X) = 6059.714339$ with $X = (0.8125, 0.4375, 42.098446, 176.636596)$. The problem can be stated as follows:

Minimize:

$$f(x) = 0.6224x_1x_3x_4 + 1.7781x_2x_3^2 + 3.1661x_1^2x_4 + 19.84x_1^2x_3 \tag{4.177}$$

Subject to:

$$g_1(x) = -x_1 + 0.0193x_3 \le 0 \tag{4.178}$$

$$g_2(x) = -x_2 + 0.00954x_3 \le 0 \tag{4.179}$$

$$g_3(x) = -\pi x_3^2x_4 - \frac{4}{3}\pi x_3^3 + 1,296,000 \le 0 \tag{4.180}$$

$$g_4(x) = x_4 - 240 \le 0 \tag{4.181}$$

where,

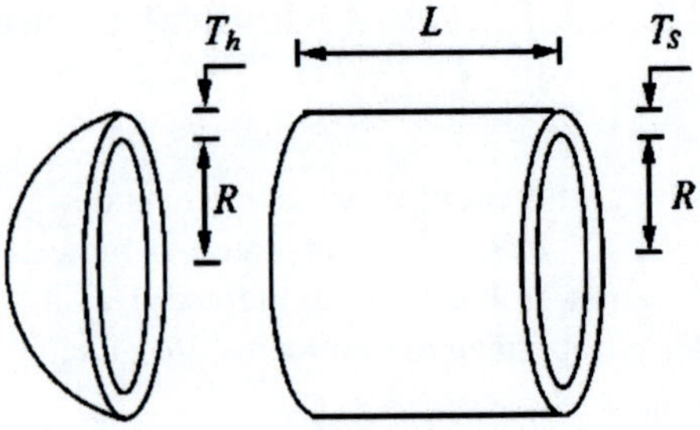

Fig. 4.1 Pressure vessel design (from [24] reprinted with permission from Elsevier)

$$0.1 \leq x_1 \leq 99,\ 0.1 \leq x_2 \leq 99,\ 10 \leq x_3 \leq 200,\ 10 \leq x_4 \leq 200$$

4.3.2 Example 9: Design of Welded Beam

The objective is to design a welded beam for minimum cost. There are four design variables height of weld (h), length of weld (L), height of beam (t) and width of beam (b) as shown in Fig. 4.2. Design vector can be defined as $X = (x_1, x_2, x_3, x_4) = (h, L, t, b)$. Design is subjected to the constraints on shear stress (τ), bending stress in the beam (σ), buckling load on the bar (P_c), end deflection of the beam (δ) and side constraints. This problem is solved by many researchers by using different optimization methods such as geometric programming [32], Genetic Adaptive Search method (GeneAS) [23], self adaptive penalty approach [24], society and civilization algorithm [25], Ant colony algorithm [34] ($\mu + \lambda$)-Evolutionary Strategy(ES) [26], Unified Particle Swarm Optimization (UPSO) [27], Co-evolutionary Particle Swarm Optimization (CPSO) [28], Co-evolutionary Differential Evolution (CoDE) [29], modified particle swarm optimization [7], Hybrid PSO-DE [30], Artificial Bee Colony (ABC) [31], etc. the best value reported in the literature is $f(X) = 1.724852$ with $X = (0.205730,\ 3.470489,\ 9.036624,\ 0.205730)$. The problem can be stated as follows:

Minimize:

$$f(x) = 1.10471 x_1^2 x_2 + 0.04811 x_3 x_4 (14.0 + x_2) \tag{4.182}$$

Subject to:

$$g_1(x) = \tau(x) - \tau_{max} \leq 0 \tag{4.183}$$

$$g_2(x) = \sigma(x) - \sigma_{max} \leq 0 \tag{4.184}$$

$$g_3(x) = x_1 - x_4 \leq 0 \tag{4.185}$$

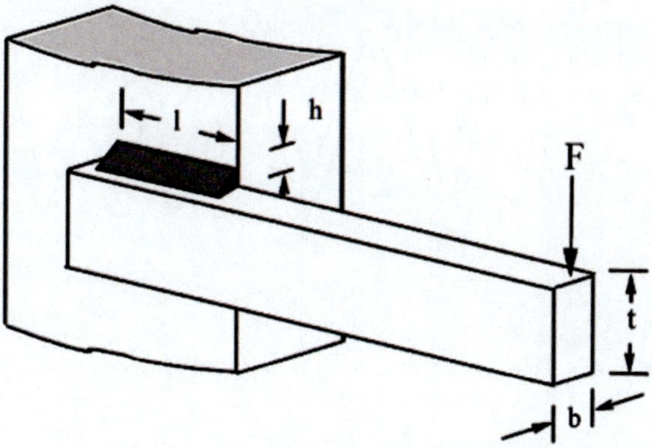

Fig. 4.2 Welded beam (from [24] reprinted with permission from Elsevier)

$$g_4(x) = 0.10471x_1^2 + 0.04811x_3x_4(14.0 + x_2) - 5.0 \le 0 \tag{4.186}$$

$$g_5(x) = 0.125 - x_1 \le 0 \tag{4.187}$$

$$g_6(x) = \delta(x) - \delta_{max} \le 0 \tag{4.188}$$

$$g_7(x) = P - P_c(x) \le 0 \tag{4.189}$$

where,

$$\tau(x) = \sqrt{(\tau')^2 + 2\tau'\tau''\frac{x_2}{2R} + (\tau'')^2},\ \tau' = \frac{P}{\sqrt{2}x_1x_2},\ \tau'' = \frac{MR}{J},\ M = P(L + \frac{x_2}{2})$$

$$R = \sqrt{\frac{x_2^2}{4} + \left(\frac{x_1 + x_3}{2}\right)^2},\ J = 2\left[\sqrt{2}x_1x_2\left\{\frac{x_2^2}{12} + \left(\frac{x_1 + x_3}{2}\right)^2\right\}\right],\ \sigma(x) = \frac{6PL}{x_4x_3^2}$$

$$\delta(x) = \frac{4PL^3}{Ex_3^3x_4},\ P_c(x) = \frac{4.013E\sqrt{\frac{x_3^2x_4^6}{36}}}{L^2}\left(1 - \frac{x_3}{2L}\sqrt{\frac{E}{4G}}\right),$$

$$P = 6{,}000\ lb, L = 14\ in, E = 30e6\ psi, G = 12e6\ psi,$$
$$\tau_{max} = 13{,}600\ psi, \sigma_{max} = 30{,}000\ psi,$$
$$\delta_{max} = 0.25\ in$$

τ_{max} is the design shear stress of weld, τ is the weld shear stress, σ_{max} is the design normal stress for beam material, σ is the maximum beam bending stress, P_c is the bar buckling load, δ is the beam end deflection, E is the modulus of elasticity for the beam material, and G is the modulus of rigidity for the beam material.

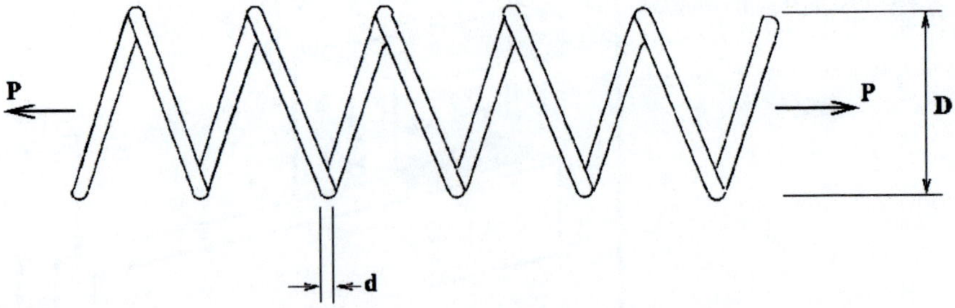

Fig. 4.3 Tension compression spring (from [24] reprinted with permission from Elsevier)

4.3.3 Example 10: Design of Tension/Compression Spring

This problem is taken from Belegundu [33] which consists of minimizing the weight of a tension/compression spring as shown in Fig. 4.3) subject to constraints on minimum deflection, shear stress, surge frequency, limits on outside diameter and on design variables. The design variables are the wire diameter (d), the mean coil diameter (D) and the number of active coils (N). Design vector can be defined as $X = (x_1, x_2, x_3) = (d, D, N)$. This problem is solved by many researchers by using different optimization techniques such as self adaptive penalty approach [24], society and civilization algorithm [25], Ant colony algorithm [34] ($\mu + \lambda$)-Evolutionary Strategy(ES) [26], Unified Particle Swarm Optimization (UPSO) [27], Co-evolutionary Particle Swarm Optimization (CPSO) [28], Co-evolutionary Differential Evolution (CoDE) [29], modified particle swarm optimization [7], Hybrid PSO-DE [30], Artificial Bee Colony (ABC) [31], etc. the best result reported is $f(X) = 0.012665$ with $X = (0.051749, 0.358179, 11.203763)$. The problem can be expressed as:

Minimize:

$$f(x) = (N+2)Dd^2 \tag{4.190}$$

Subject to:

$$g_1(x) = 1 - \frac{D^3N}{71785d^4} \le 0 \tag{4.191}$$

$$g_2(x) = \frac{4D^2 - dD}{12,566(Dd^3 - d^4)} + \frac{1}{5,108d^2} - 1 \le 0 \tag{4.192}$$

$$g_3(x) = 1 - \frac{140.45d}{D^2N} \le 0 \tag{4.193}$$

$$g_4(x) = \frac{D+d}{1.5} - 1 \le 0 \tag{4.194}$$

where

$$0.05 \le x_1 \le 2, 0.25 \le x_2 \le 1.3, 2 \le x_3 \le 15$$

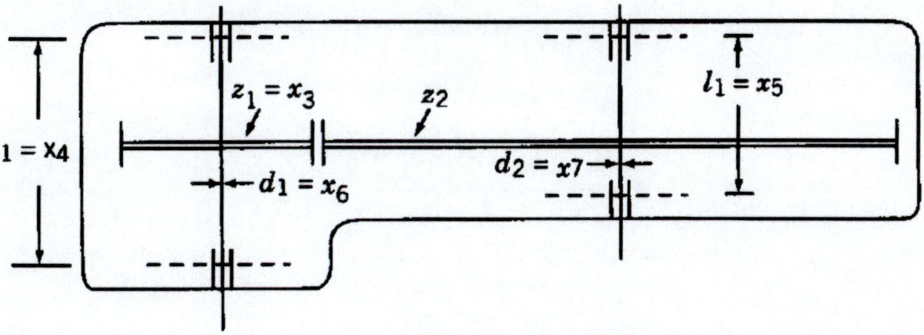

Fig. 4.4 Speed reducer

4.3.4 Example 11: Design of a Speed Reducer

The weight of the speed reducer as shown in Fig. 4.4 is to be minimized subject to constraints on bending stress of the gear teeth, surfaces stress, transverse deflections of the shafts and stresses in the shafts. The variables x_1, x_2, x_3, x_4, x_5, x_6 and x_7 are the face width, module of teeth, number of teeth in the pinion, length of the first shaft between bearings, length of the second shaft between bearings and the diameter of the first and second shafts, respectively. The third variable is integer, the rest of them are continuous. This problem is solved by many researchers by using different optimization techniques such as $(\mu + \lambda)$-Evolutionary Strategy (ES) [26], Unified Particle Swarm Optimization (UPSO) [27], Hybrid PSO-DE [30], ABC [31], etc. the best result reported by Montes and Coello [26] is $f(X) = 2996.3481$ with $X = (3.49999, 0.6999, 17, 7.3, 7.8, 3.3502, 5.2866)$. The problem can be stated as:

Minimize:

$$f(x) = 0.7854x_1x_2^2(3.3333x_3^2 + 14.9334x_3 - 43.0934) - 1.508x_1(x_6^2 + x_7^2) + 7.4777(x_6^3 + x_7^3) + 0.7854(x_4x_6^2 + x_5x_7^2) \tag{4.195}$$

Subject to:

$$g_1(x) = \frac{27}{x_1x_2^2x_3} - 1 \leq 0 \tag{4.196}$$

$$g_2(x) = \frac{397.5}{x_1x_2^2x_3^2} - 1 \leq 0 \tag{4.197}$$

$$g_3(x) = \frac{1.93x_4^3}{x_2x_3x_6^4} - 1 \leq 0 \tag{4.198}$$

$$g_4(x) = \frac{1.93x_5^3}{x_2x_3x_7^4} - 1 \leq 0 \quad (4.199)$$

$$g_5(x) = \frac{\sqrt{\left(\frac{745x_4}{x_2x_3}\right)^2 + 16.9e6}}{110x_6^3} - 1 \leq 0 \quad (4.200)$$

$$g_6(x) = \frac{\sqrt{\left(\frac{745x_5}{x_2x_3}\right)^2 + 157.5e6}}{85x_7^3} - 1 \leq 0 \quad (4.201)$$

$$g_7(x) = \frac{x_2x_3}{40} - 1 \leq 0 \quad (4.202)$$

$$g_8(x) = \frac{5x_2}{x_1} - 1 \leq 0 \quad (4.203)$$

$$g_9(x) = \frac{x_1}{12x_2} - 1 \leq 0 \quad (4.204)$$

$$g_{10}(x) = \frac{1.5x_6 + 1.9}{x_4} - 1 \leq 0 \quad (4.205)$$

$$g_{11}(x) = \frac{1.1x_7 + 1.9}{x_5} - 1 \leq 0 \quad (4.206)$$

where,

$$2.6 \leq x_1 \leq 3.6, 0.7 \leq x_2 \leq 0.8, 17 \leq x_3 \leq 28, 7.3 \leq x_4 \leq 8.3, 7.8 \leq x_5 \leq 8.3,$$
$$2.9 \leq x_6 \leq 3.9, 5.0 \leq x_7 \leq 5.5$$

4.3.5 Example 12: Design of Stiffened Cylindrical Shell

This problem is taken from Jarmai et al. [35]. The objective is to optimize cost of a cylindrical shell member that is orthogonally stiffened by using ring stiffeners of box cross section and stringers of halved I-section Figs. 4.5 and 4.6.

The objective function for the stiffened shell is given as

$$f(X) = K_M + \sum_i K_{Fi} + K_P \quad (4.207)$$

where, K_m is the cost of material, K_{Fi} is the cost of fabrication and K_P is the cost of painting.

where,

$$K_M = k_{M1}5\rho V_1 + k_{M1}\rho n_r V_R + k_{M2}\rho n_s A_s L$$

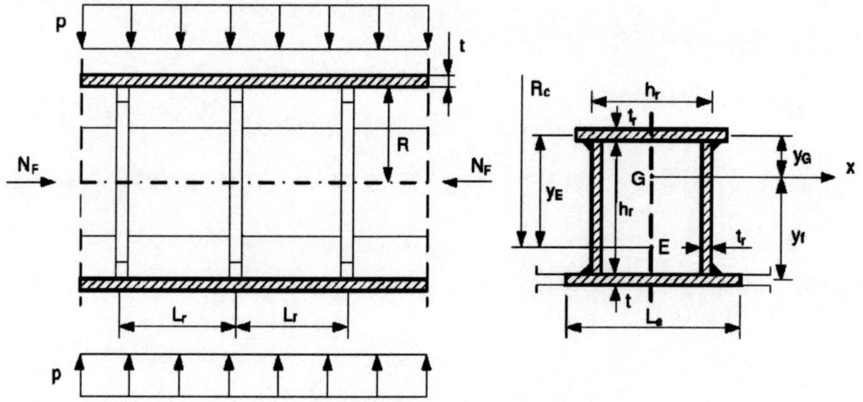

Fig. 4.5 Orthogonally stiffened cylindrical shell with stringer and ring stiffener acted by compression and external pressure (from Jarmai et al. [35] reprinted with permission from Elsevier)

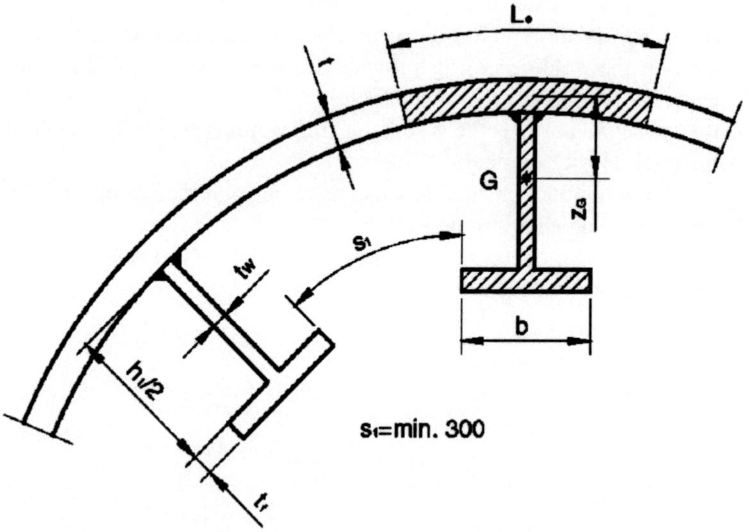

Fig. 4.6 Halved rolled I—section as a stringer (from Jarmai et al. [35] reprinted with permission from Elsevier)

$$K_{F0} = 5k_F\Theta e^{\mu},\ \mu = 6.8582513 - 4,527217t^{-0.5} + 0.009541996(2R)^{0.5}$$

$$K_{F1} = 5k_F\left(\Theta\sqrt{\kappa\rho V_1} + 1.3 \times 0.1520 \times 10^{-3}t^{1.9358} \times 2L_s\right),\ \Theta = 2,\ \kappa = 2$$

$$K_{F2} = k_F\left(\Theta\sqrt{25\rho V_1} + 1.3 \times 0.1520 \times 10^{-3}t^{1.9358} \times 4 \times 2R\pi\right)$$

$$K_{F3} = n_r k_F \left(3\sqrt{3\rho V_R} + 1.3 \times 0.3394 \times 10^{-3} a_{wr}^2 4\pi (R - h_r)\right)$$

$$K_{F4} = k_F \left(3\left(\sqrt{n_r + 1}\right)\rho(5V_1 + n_r V_R) + 1.3 \times 0.3394 \times 10^{-3} a_{wr}^2 n_r 4R\pi\right)$$

$$K_{F5} = k_F \left(3\sqrt{(n_r + n_s + 1)\rho(5V_1 + n_r V_R + n_s A_s L)} + 1.3 \times 0.3394 \times 10^{-3} a_{ws}^2 n_s 2L\right)$$

$$K_P = k_P \left(2R\pi L + 2R\pi(L - n_r h_r) + 2n_r \pi h_r (R - h_r) + 4\pi n_r h_r \left(R - \frac{h_r}{2}\right) + n_s L(h_1 + 2b)\right)$$

$$V_1 = 2R\pi t L_s$$

$$V_R = 2\pi\delta_r h_r^2 (R - h_r) + 4\pi\delta_r h_r^2 \left(R - \frac{h_r}{2}\right)$$

K_{M1} is the cost factor for plates, K_{F0} is the fabrication cost to form plate elements into cylindrical shapes, K_{F1} is the welding cost of shell segments from 2 curved plates, K_{F2} is the welding cost of whole un-stiffened shell from 5 shell segments, K_{F3} is the welding of ring stiffeners from 3 plates, K_{F4} is the welding of ring stiffeners into the shell, K_{F5} is the welding stringers into the shell, V_1 is the volume of shell segment, V_R is the volume of ring stiffener, K_M is the material cost, ρ is the material density of steel.
The above objective function includes the material, manufacturing and painting cost. The design is subjected to the shell buckling, panel stiffener buckling, panel ring buckling and manufacturing limitations. Design is followed according to DNV-RP-C202. They are explained as below.

Shell buckling

For the shell buckling the equivalent stress should satisfy the following condition.

$$g_1(X) = \sigma_e = \sqrt{\sigma_a^2 - \sigma_a\sigma_p + \sigma_p^2} \le \frac{f_{y1}}{\sqrt{1 + \lambda_s^4}} \tag{4.208}$$

where,

$$\sigma_a = \frac{N_F}{2R\pi t_e},\ t_e = t + \frac{A_s}{s},\ s = \frac{2R\pi}{n_s},\ \sigma_p = \frac{p_F R}{t(1 + \alpha)},\ \alpha = \frac{A_R}{L_{e0} t},$$

$L_{e0} = \min\left(L_r, L_{er} = 1.56\sqrt{Rt}\right), L_r = \frac{L}{n_r - 1},\ \lambda_s^2 = \frac{f_{y1}}{\sigma_e}\left(\frac{\sigma_a}{\sigma_{Eas}} + \frac{\sigma_p}{\sigma_{Eps}}\right),\ \sigma_{Eas} = c_{as}\frac{\pi^2 E}{12(1-\nu^2)}$ $\left(\frac{t}{s}\right)^2,\quad c_{as} = \psi_{as}\sqrt{1 + \left(\frac{\rho_{as}\xi_{as}}{\psi_{as}}\right)^2},\qquad \psi_{as} = 4, z_{as} = \frac{s^2}{Rt}\sqrt{1 - \nu^2},\ \xi_{as} = 0.702Z_{as},$

$\rho_{as} = 0.5\left(1 + \frac{R}{150t}\right)^{-0.5},\quad \sigma_{Eps} = C_{ps}\frac{\pi^2 E}{10.92}\left(\frac{t}{s}\right)^2,\qquad C_{ps} = \psi_{ps}\sqrt{1 + \left(\frac{\rho_{ps}\xi_{ps}}{\psi_{ps}}\right)^2},$

$$\rho_{ps} = 0.6, \quad \xi_{ps} = 1.04\tfrac{s}{L_r}\sqrt{z_{ps}},\ z_{ps} = z_{as},\ \psi_{ps} = \left[1 + \left(\tfrac{s}{L_r}\right)^2\right]^2$$

f_y is the yield stress, N_F is the factored compression force, R is the shell radius, t is the shell thickness, A_s is the cross-sectional area of stringer, n_s is the number of longitudinal stiffeners, p_F is the factored external pressure intensity, A_r is the cross-sectional area of ring stiffener, L is the shell length, n_r is the number of ring stiffeners, L_r is the distance between rings, σ_{Eas} is the elastic buckling strength for axial force, σ_{Eps} is the elastic buckling strength for lateral pressure, σ_a is the design axial stress due to axial forces, σ_p is the design cicumferential stress in the shell due to external pressure, C_{as} is the reduced buckling coefficient, λ_s is the reduced shell slenderness, γ is the poissons ratio, s is the distance between longitudinal stiffeners.

Panel stiffener buckling

For panel stiffener buckling the equivalent stress should satisfy the following condition.

$$g_2(X) = \sigma_e \le \frac{f_{y1}}{\sqrt{1 + \lambda_p^4}} \tag{4.209}$$

where

$$\lambda_p^2 = \frac{f_{y1}}{\sigma_e}\left(\frac{\sigma_a}{\sigma_{Eap}} + \frac{\sigma_p}{\sigma_{Epp}}\right),\ \sigma_{Eap} = c_{ap}\frac{\pi^2 E}{10.92}\left(\frac{t}{L_r}\right)^2,$$

$$c_{ap} = \psi_{ap}\sqrt{1 + \left(\frac{\rho_{ap}\xi_{ap}}{\psi_{ap}}\right)^2},\ \rho_{ap} = 0.5,$$

$$\xi_{ap} = 0.702Z_{ap},\ z_{ap} = \frac{L_r^2}{Rt}0.9539,\ \psi_{ap} = \frac{1+\gamma_s}{1+\frac{A_s}{s_e t}},\ \gamma_s = 10.92\frac{I_{sef}}{st^3},$$

$$s_E = 1.9t\sqrt{\frac{E}{f_y}},\ s_E \le s,\ s_e = s_E$$

and if $\quad s_E \ge s,\ s_e = s,\ Z_G = \dfrac{\frac{h_1}{2}t_w\left(\frac{h_1}{4}+\frac{t}{2}\right) + bt_f}{\left(\frac{h_1+t+t_f}{2}\right)s_e t + bt_f + \frac{h_1 t_w}{2}},$

$$I_{sef} = s_e t z_G^2 + \left(\tfrac{h_1}{2}\right)^3\tfrac{t_w}{12} + \tfrac{h_1 t_w}{2}\left(\tfrac{h_1}{4}+\tfrac{t}{2}-z_G\right)^2 + bt_f\left(\tfrac{h_1+t+t_f}{2}-z_G\right)^2\ A_s = bt_f + \tfrac{h_1 t_w}{2}$$

$$\sigma_{Epp} = c_{pp}\frac{\pi^2 E}{10.92}\left(\frac{t}{L_r}\right)^2,\quad c_{pp} = \psi_{pp}\sqrt{1+\left(\frac{\rho_{pp}\xi_{pp}}{\psi_{pp}}\right)^2},\ \xi_{pp} = 1.04\sqrt{z_{pp}},\ z_{pp} = z_{ap},$$

$\rho_{pp} = 0.6$ and $\psi_{pp} = 2(1+\sqrt{1+\gamma_s})$

I_{sef} is the moment of inertia of stiffener including effective shell plating se, b is the flange width, t_f is the flange thickness, $h_1/2$ is the web height, t_w is the web thickness.

Panel ring buckling

The ring stiffeners should satisfy following constraints

$$g_3(X) = A_{Rreq} \leq A_R \tag{4.210}$$

$$g_4(X) = I_{Rreq} \leq I_R \tag{4.211}$$

where,

$$t_r \geq \delta_r h_r, \ \frac{1}{\delta_r} = 42\varepsilon, \ \varepsilon = \sqrt{\frac{235}{f_y}}, \ f_y = 355, \ \delta_r = \frac{1}{34}, \ A_R = 3h_r t_r = 3\delta_r h_r^2$$

$$A_{Rreq} = \left(\frac{2}{Z^2} + 0.06\right) L_r t, \text{ where } Z = \frac{L_r^2}{Rt} 0.9539, \ L_e = \min\left(L_r, 2 \times 1.56\sqrt{Rt}\right)$$

$$y_E = \frac{L_e t\left(h_r + \frac{t}{2}\right) + \delta_r h_r^3}{3\delta_r h_r^2 + L_e t}, \ I_R = \frac{\delta_r h_r^4}{6} + 2\delta_r h_r^2 \left(\frac{h_r}{2} - y_E\right)^2 + \delta_r h_r^2 y_E^2 + L_e t\left(h_r + \frac{t}{2} - y_E\right)^2$$

$$I_{Rreq} = I_a + I_p, \ I_a = \frac{\sigma_a t\left(1 + \frac{A_s}{st}\right) R_0^4}{500 E L_R},$$

$$R_0 = R - (h_r - y_E), \ I_p = \frac{p_F R R_0^2 L_r}{3E}\left[2 + \frac{3E y_E \delta_0}{R_0^2\left(\frac{f_y}{2} - \sigma_p\right)}\right],$$

$$\delta_0 = 0.005R,$$

I_a is the required moment of inertia, A_{Rreq} is the required cross-sectional area of a ring stiffener, y_E is the distance of centroid,

$$a_{ws} = 0.4t_w, \ a_{ws.\min} = 3mm \text{ and } a_{wr} = 0.4t_r, \ a_{wr.\min} = 3mm$$

Manufacturing limitation

Manufacturing constraint is imposed to ensure the welding of the webs of the halved rolled I-section stringers. This is possible if following constraint is satisfied.

$$g_5(X) = n_s \leq \frac{2\left(R - \frac{h_r}{2}\right)\pi}{b + 300} \tag{4.212}$$

For this problem there are five design variables, shell thickness (t), number of longitudinal stiffeners (stringers) (n_s), number of ring stiffeners (n_r), box height (h_r) and stringer stiffness height (h). Design vector can be defined as $X = (x_1, x_2, x_3, x_4, x_5) = (t, n_s, n_r, h_r, h)$. The above problem is solved by considering the design variables as continuous and second by considering desing variables as discrete. The design with continuous design variables will be referred to as Example 12A in this book. For the discrete design variables value of t, n_s and n_r can take integer value, h_r varies in the step of 10 and h can take any value from 152, 203, 254, 305, 356, 406, 457, 533, 610, 686, 762, 838 and 914. The design with discrete variables will be referred to as Example 12B in this book. The best value reported by Jarmai et al. [35] is $f(X) = 54444.62$ with $X = (13.82, 26.85, 8.31, 260.96, 225.79)$ by considering continuous design variables and $f(X) = 55326.3$ with $X = (14, 27, 10, 250, 203)$ by considering discrete design variables.

4.3.6 Example 13: Design of Step Cone Pulley

The objective is to design a 4-step cone pulley for minimum weight with 5 design variables consisting of four design variables for the diameters (d_1, d_2 d_3 and d_4) of each step and fifth one is the width of the pulley (w). Design vector can be defined as $X = (x_1, x_2, x_3, x_4, x_5) = (d_1, d_2, d_3, d_4, w)$. Figure 4.7 shows a step cone pulley. It is assumed in this example that width of cone pulley and belt is same. There are 11 constraints out of which 3 are equality constraints and the rest are inequality constraints. Constraints are for the assurance of same belt length for all the steps, tension ratios and power transmitted by the belt. Step pulley is designed to transmit at least 0.75hp (0.75 × 745.6998 W), with the input speed of 350 rpm and output speeds as 750, 450, 250 and 150 rpm. The problem is taken from Rao [36].

$$Minimize \quad f(x) = \rho w \left[d_1^2 \left\{ 1 + \left(\frac{N_1}{N} \right)^2 \right\} + d_2^2 \left\{ 1 + \left(\frac{N_2}{N} \right)^2 \right\} + d_3^2 \left\{ 1 + \left(\frac{N_3}{N} \right)^2 \right\} + d_4^2 \left\{ 1 + \left(\frac{N_4}{N} \right)^2 \right\} \right] \tag{4.213}$$

Subject to:

$$h_1(x) = C_1 - C_2 = 0 \tag{4.214}$$

$$h_2(x) = C_1 - C_3 = 0 \tag{4.215}$$

$$h_3(x) = C_1 - C_4 = 0 \tag{4.216}$$

$$g_{1,2,3,4}(x) = R_i \geq 2 \tag{4.217}$$

$$g_{5,6,7,8}(x) = P_i \geq (0.75 * 745.6998) \tag{4.218}$$

where, C_i indicates the length of the belt to obtain speed output N_i and it is given by:

$$C_i = \frac{\pi d_i}{2} \left(1 + \frac{N_i}{N} \right) + \frac{\left(\frac{N_i}{N} - 1 \right)^2}{4a} + 2a \quad i = (1, 2, 3, 4)$$

Fig. 4.7 Step cone pulley

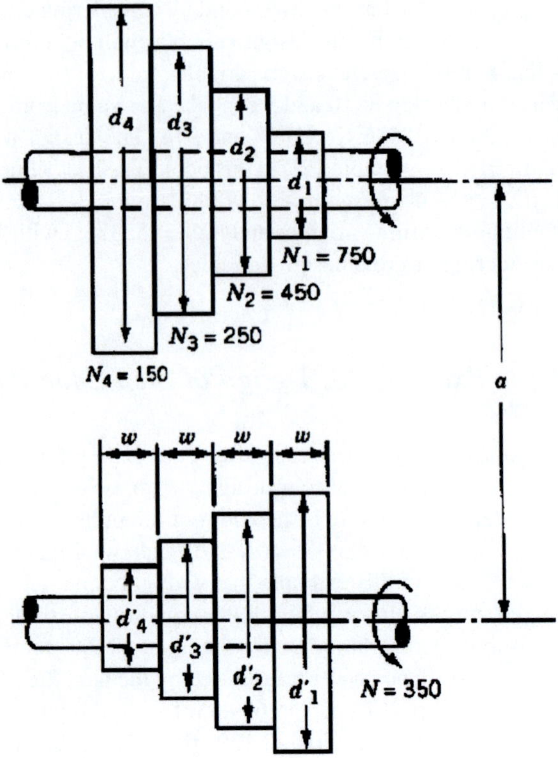

R_i is the tension ratios and it is given by:

$$R_i = \exp\left[\mu\left\{\pi - 2\sin^{-1}\left\{\left(\frac{N_i}{N} - 1\right)\frac{d_i}{2a}\right\}\right\}\right] \quad i = (1, 2, 3, 4)$$

And P_i is the power transmitted at each step which is given by:

$$P_i = stw\left[1 - \exp\left[-\mu\left\{\pi - 2\sin^{-1}\left\{\left(\frac{N_i}{N} - 1\right)\frac{d_i}{2a}\right\}\right\}\right]\right]\frac{\pi d_i N_i}{60} \quad i = (1, 2, 3, 4)$$

$\rho = 7{,}200$ kg/m^3, $a = 3$ m, $\mu = 0.35$, $s = 1.75$ MPa, $t = 8$ mm, $40 \le x_1$, x_2, x_3, $x_4 \le 500$, $16 \le x_5 \le 100$, μ is the coefficient of friction between belt and pulley, ρ is the density of material, N is the input speed, a is the center distance, s is the maximum allowable stress in the belt and t is the thickness of belt.

4.3.7 Example 14: Design of Screw Jack

The screw jack as shown in Fig. 4.8 is to be designed such that the weight of the screw is minimum [36]. The constraints imposed on the design are: design of screw jack should be self locking; torsional shear stress induced in the screw

Fig. 4.8 Screw jack

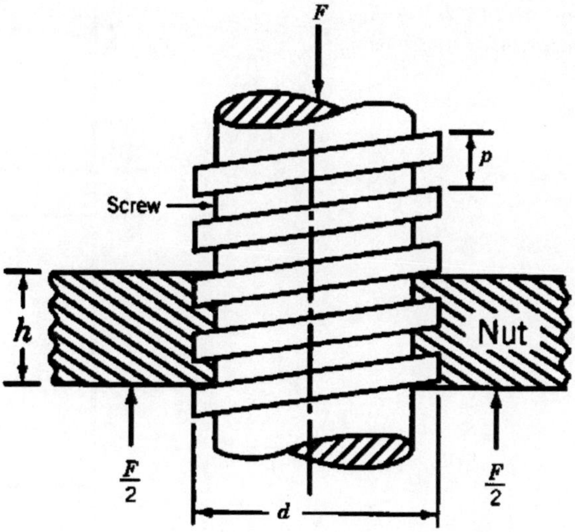

should not increase the torsional shear strength of screw material; shear stress in the threads of screw and nut should not increase the torsional shear strength of screw and nut material, respectively; bearing pressure should not exceed the bearing strength and the buckling strength of the screw should be more than the load lifted by the screw jack. Design variables are the outer diameter of the screw (d), height of nut (h) and length of screw (l). Design vector can be defined as $X = (x_1, x_2, x_3) = (d, h, l)$.

The problem can be stated as

$$f(X) = \pi/4(d - p/2)^2(l + h) \tag{4.219}$$

Subject to:

$$g_1(X) = \phi \geq \alpha \tag{4.220}$$

$$g_2(X) = \frac{16T}{\pi d_c^3} \leq 0.577\sigma_y \tag{4.221}$$

$$g_3(X) = \frac{F}{\pi n d_c t} \leq 0.577\sigma_y \tag{4.222}$$

$$g_4(X) = \frac{F}{\pi n d t} \leq 0.5\sigma_{yn} \tag{4.223}$$

$$g_5(X) = \frac{F}{\pi/4\left(d^2 - d_c^2\right)n} \leq \sigma_b \tag{4.224}$$

$$g_6(X) = A_c\sigma_y\left[1 - \frac{\sigma_y}{4C\pi^2 E}(l/k)^2\right] \geq F \tag{4.225}$$

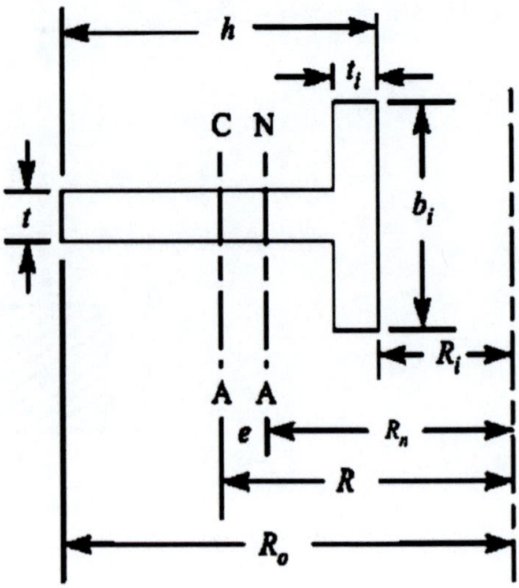

Fig. 4.9 T-cross section and its geometry for C-clamp

where,

$p = 5$ if $22 \leq d \leq 28; 6$ if $30 \leq d \leq 36; 7$ if $38 \leq d \leq 44; 8$ if $46 \leq d \leq 52; 9$ if $55 \leq d \leq 60$,

$F = 80$ kN, $\sigma_y = 220$ MPa, $\sigma_{yn} = 150$ MPa, $E = 206 \times 10^5$ MPa, $d = [22, 24, 26,\ldots 60]$, $10 \leq h \leq 100$, $350 \leq l \leq 500$, $\mu = 0.18$, $\mu_1 = \mu/\cos\beta$, $\beta = 0.252944$, $\Phi = \tan \quad \mu_1$, $\alpha = \tan - 1(p/\pi d_m)$, $d_m = d - p/2$, $T = F\tan(\alpha + \Phi)d_m/2$, $d_c = d - p/2$, $n = h/p$, $t = p/2$, $Ac = \pi/4d_c^2$, $C = 0.25$, $k = 0.25d_c$, p is the pitch of thread, Φ *is the* friction angle, α *is the* helix angle of thread, T is the torque applied on screw, d_c is the core diameter, σ_y is the allowable tensile stress for screw, σ_{yn} is the allowable tensile stress for nut, n is the numbers of threads in contact, F is the load acting on screw, t is the thickness of thread, σ_b is the allowable bearing pressure, A_c is the cross-sectional area at core diameter, C is the end fixity condition, l/k is the slenderness ratio, E is the modulus of elasticity, k is the radius of gyration, μ is the coefficient of friction, μ_1 is the equivalent coefficient of friction, β is the thread angle and dm is the mean diameter of screw.

4.3.8 Example 15: Design of C-Clamp

This problem is taken from Rao [36]. In this problem the objective is to minimize the weight of a C-clamp subjected to the static load. C-clamp is having ‘T’ cross section Fig. 4.9. There are five design variables and two inequality constraints. Design variables are for the geometric dimension of the ‘T’ cross section (b, h, t_i and t) and the distance of inner edge from the center of curvature (R_i). Design vector can be defined as $X = (x_1, x_2, x_3, x_4, x_4) = (b, h, t_i, t, R_i)$.

The problem can be stated as:

$$f(x) = \min weight = \min volume$$

$$f(x) = area \times length$$

$$f(x) = (150 + \pi R_i)\,(x_1 x_3 + (x_2 - x_3)x_4) \tag{4.226}$$

where, x_1 is the length of flange, x_2 is the thickness of flange, x_3 is the height of T is the section, x_4 is the thickness of web.

Subject to:

$$g_1(x) = \sigma_{\text{total},i} \le \sigma_y \tag{4.227}$$

$$g_2(x) = \sigma_{\text{total},0} \le \sigma_y \tag{4.228}$$

where,

$$R_0 = R_i + x_2,\; R_N = \frac{x_3(x_1 - x_4) + x_1 x_2}{(x_1 - x_4)\ln\left(\frac{R_i + x_3}{R_i}\right) + x_4 \ln\left(\frac{R_0}{R_i}\right)},$$

$$R = R_i + \left[\frac{\frac{1}{2}x_2^2 x_4 + \frac{1}{2}x_3^2(x_1 - x_4)}{x_2 x_4 + x_3(x_1 - x_4)}\right],$$

$$e = R - R_N,\; x = 50 + R,\; M = W \times x,\; y_i = R_N - R_i,\; y_0 = R_0 - R_N,$$

$$\sigma_{bi} = \frac{My_i}{AeR_i},\; \sigma_{b0} = \frac{My_0}{AeR_0}$$

$\sigma_t = \frac{W}{A}$, $\sigma_{\text{total},i} = \sigma_{bi} + \sigma_t$, $\sigma_{\text{total},0} = abs(\sigma_t - \sigma_{b0})$, $W = 1{,}200$ N, $\sigma_y = 200$ N/mm^2, $20 \le R_i \le 50$, $1 \le x_1 \le 50$, $1 \le x_2 \le 50$, $0.1 \le x_3 \le 10$, $0.1 \le x_4 \le 10$, R_n is the radius of curvature of neutral axis, R is the radius of curvature of centroidal axis, M is the bending moment acting about centroidal axis, e is the distance of centroidal axis and neutral axis, R_o is the radius of curvature of outside fiber of T-section, y_o is the distance of neutral axis to outer fiber, y_i is the distance of neutral axis to inner fiber, σ_t is the axial stress acting on the cross section, W is the load, σ_y is the allowable stress.

4.3.9 Example 16: Design of Hydrodynamic Bearing

This problem is also taken from Rao [36]. A hydrodynamic bearing is to be designed to minimize a linear combination of frictional moment and angle of twist of the shaft while carrying a load of 1000 lb by considering radius of the bearing (*R*) and half length of bearing (*L*) as the design variables. Design vector can be defined as $X = (x_1, x_2) = (R, L)$. The angular velocity of the shaft is to be greater than 100 rad/sec.

The frictional moment of the bearing (M) and the angle of twist of the shaft (ϕ) are given by:

$$M = \frac{8\pi}{\sqrt{1-n^2}} \frac{\mu\Omega}{c} R^2 L \tag{4.229}$$

$$\phi = \frac{s_e l}{GR} \tag{4.230}$$

where μ is the viscosity of lubricant, n the eccentricity ratio (= e/c), e the eccentricity of the journal relative to bearing, c the radial distance, Ω the angular velocity of the shaft, R the radius of the journal, L the half length of the bearing, S_e the shear stress, l length between the driving point and the rotating mass and G the shear modulus. The load on each bearing (W) is given by

$$W = \frac{2\mu\Omega R L^2 n}{c^2(1-n^2)^2}\left[\pi^2(1-n^2) + 16n^2\right]^{1/2} \tag{4.231}$$

For the data: $W = 1{,}000$ lb, $c/R = 0.0015$, $n = 0.9$, $l = 10$ in., $S_e = 30{,}000$ psi, $\mu = 10^{-6}$ lb-s/in^2 and $G = 12 \times 10^6$ psi.

By considering the objective function as a linear combination of the frictional moment (M), the angle of twist of the shaft (ϕ), and the temperature rise of the oil (T),

$$f = aM + b\phi + cT \tag{4.232}$$

where a, b and c are constants. The temperature rise of the oil in the bearing is given by

$$T = 0.045\frac{\mu\Omega R^2}{c^2 n\sqrt{(1-n^2)}} \tag{4.233}$$

where $\Omega = 11.6RL^{-3}$

By assuming that 1in-lb of frictional moment in bearing is equal to 0.0025 rad of angle of twist, which, in turn, is equivalent to 1° F rise in temperature, the constants a, b and c can be determined. The optimization problem can be stated as

$$\min f(R, L) = 0.44R^3L^{-2} + 10R^{-1} + 0.592RL^{-3} \tag{4.234}$$

Subject to

$$8.62R^{-1}L^3 \le 1 \tag{4.235}$$

where, $0.1 \le x_1, x_2 \le 2$

4.3.10 Example 17: Design of Cone Clutch

This problem is taken from Rao [36]. The objective is to find the minimum volume of the cone clutch shown in Fig. 4.9 such that it can transmit a specified minimum torque. By selecting outer and inner radius of the cone, $R_1(x_1)$ and $R_2(x_2)$, as design variables, the objective function can be expressed as

$$f\left(R_{1,}R_2\right) = \frac{1}{3}\pi h\left(R_1^2 + R_1R_2 + R_2^2\right) \tag{4.236}$$

where the axial thickness, h is given by

$$h = \frac{R_1 - R_2}{\tan\alpha} \tag{4.237}$$

where, α is the half cone angle.

From Eqs. (4.236) and (4.237) objective function reduces to:

$$f\left(R_{1,}R_2\right) = k_1\left(R_1^3 - R_2^3\right) \tag{4.238}$$

where, $k_1 = \dfrac{\pi}{3\tan\alpha}$

The axial force applied (F) and the torque (T) are given by:

$$F = \int p\, dA\, \sin\alpha = \int_{R_2}^{R_1} p\frac{2\pi r\, dr}{\sin\alpha}\sin\alpha = \pi p\left(R_1^2 - R_2^2\right) \tag{4.239}$$

$$T = \int rfp\, dA = \int_{R_2}^{R_1} rfp\frac{2\pi r}{\sin\alpha}dr = \frac{2\pi fp}{3\sin\alpha}\left(R_1^3 - R_2^3\right) \tag{4.240}$$

where p is pressure, f the coefficient of friction and A the area of contact. Substitution of p from Eqs. 4.239 and 4.240 force and torque leads to:

$$T = \frac{k_2\left(R_1^2 + R_1R_2 + R_2^2\right)}{R_1 + R_2} \tag{4.241}$$

where $k_2 = \dfrac{2Ff}{3\sin\alpha}$

Since k_1 is a constant, the objective function can be taken as $f = R_1^3 - R_2^3$.

The minimum torque to be transmitted is assumed to be 5 k_2. In addition, the outer radius R_1 is assumed to be equal to at least twice the inner radius R_2. Thus the optimization problem becomes: (Fig. 4.10)

$$\min f(X) = R_1^3 - R_2^3 \tag{4.242}$$

$$g_1(X) = \frac{R_1}{R_2} \geq 2 \tag{4.243}$$

Fig. 4.10 Cone clutch

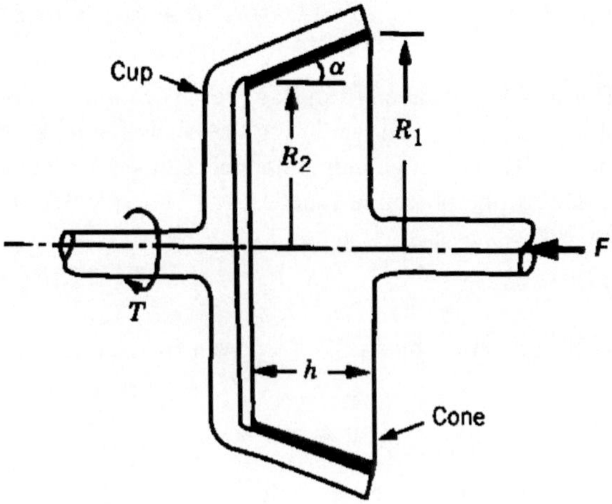

where, $1 \leq x_1, x_2 \leq 10$

4.3.11 Example 18: Design of Cantilever Support

The objective is to determine the cross-sectional dimensions of the cantilever support for minimum weight. The support is having rectangular cross section with dimensions x_1 and x_2. The maximum permissible bending stress is $\sigma_{y.}$ The width and depth of the beam are considered as design variables. The objective function is given by,

$$f(X) = \rho l x_1 x_2 \tag{4.244}$$

where ρ is the weight density and l is the length of the beam. The maximum stress induced at the fixed end is given by

$$\sigma = \frac{Mc}{I} = Pl\frac{x_2}{2}\frac{1}{\frac{1}{12}x_1x_2^2} = \frac{6Pl}{x_1x_2^2} \tag{4.245}$$

Subject to:

$$g_1(X) = \frac{6Pl}{\sigma_y}x_1^{-1}x_2^{-2} \leq 1 \tag{4.246}$$

where, $0.01 \leq x_1, x_2 \leq 0.5$

Table 4.5 Comparison of results for the unconstrained benchmark functions by using different variants of PSO

BM-UC	BEST			MEAN		
	PSO	PSO_M_1	PSO_M_2	PSO	PSO_M_1	PSO_M_2
Sphere	21.484652	1.005329	**0.000015**	29.5030532	19.7039558	**0.0024274**
Schwefel 2.22	21.284104	2.90262	**2.10107**	24.298752	6.2833394	**4.3553452**
Schwefel 1.2	93.156383	129.161132	**33.012155**	143.2574008	148.8475534	**46.8966956**
Schwefel 2.21	2.404852	2.456076	**1.117249**	2.6110444	**2.597043**	5.360889
Rosenbrock	10,346.89086	2,502.49725	**29.272808**	13,901.52245	9,468.709034	**157.5713984**
Step	18	22	**0**	37.6	30.2	**7.8**
Quartic	19.981448	0.318036	**0.000001**	83.6416058	61.570859	**0.0001084**
Schwefel 2.26	**−5,393.636691**	−4,545.949844	−4,373.896729	**−4,067.875837**	−3,768.432847	−3,356.256704
Rastrigin	282.822554	295.274568	**33.926799**	305.370863	299.808742	**71.8920874**
Ackley	4.078291	3.021253	**0.020804**	5.5037588	4.772954	**1.0316518**
Griewank	0.412012	**0.045739**	5.786935	0.677997	**0.402989**	10.3230726
Penalty-1	0.065802	**0.005583**	1.139355	0.595423	**0.2127746**	2.5957132
Penalty-2	4.383156	0.226628	**0.001**	5.3438038	**2.170404**	3.3272798

Table 4.6 Comparison of results for the constrained benchmark functions by using different variants of PSO

BM-C	BEST			MEAN		
	PSO	PSO_M_1	PSO_M_2	PSO	PSO_M_1	PSO_M_2
G01	−13	−11	**−14.999996**	−6.5	−5	**−9.6729557**
G02	−0.310396	−0.323015	**−0.771263**	−0.2223897	−0.2530784	**−0.6187956**
G03	0	0	**−0.10203**	0	0	**−0.010203**
G04	**−30,665.53867**	**−30,665.53867**	**−30,665.53867**	**−30,665.53867**	**−30,665.53867**	**−30,665.53867**
G05	2,015,610.666	6,112.223965	**5,146.042753**	2,632,602.481	1,879,094.211	**491,465.7649**
G06	−6,877.436372	−6,940.480548	**−6,960.623006**	238,472.3322	−6,712.2915	**−6,953.036875**
G07	37.88342	28.611548	**25.403776**	51.3171341	52.7532428	**29.159275**
G08	**−0.095825**	**−0.095825**	**−0.095825**	**−0.095825**	**−0.095825**	−0.0891569
G09	686.116556	681.237319	**680.66421**	693.7287022	684.4327281	**680.7920656**
G10	**9,998.514828**	10,823.76	16,916.82492	**12,112.13699**	13,287.98258	4,583,679.75
G11	**1**	**1**	**1**	**1**	**1**	**1**
G12	**−1**	**−1**	**−1**	**−1**	**−1**	**−1**
G13	201.05012	301.25812	**0.455257**	288.4824832	304.8582036	**44.701568**
G14	296.368692	274.065247	**56.31768**	3,449.24048	349.1313162	**79.2477014**
G15	972.317012	967.520587	**963.833392**	4,918.24595	1,768.419819	**966.4543516**
G16	**−1.888024**	**−1.888024**	**−1.888024**	−1.8880116	**−1.888024**	**−1.888024**
G17	3,034,175.978	3,736,774.269	**9,155.954574**	**3,481,334.056**	3,986,356.782	62,944,325,851
G18	−0.573722	−0.719853	**−0.865871**	−0.494204	19.8782376	**−0.8035284**
G19	40.15962	**38.638187**	48.234623	**48.2093652**	48.5822916	56.4683014
G20	274,346,094.6	**9,330,093.777**	15,000,126,481	21,291,326,210	**17,048,160,044**	48,892,294,911
G21	162,044.7972	70,235.17243	**40,518.18561**	471,564,167	140,731,539.5	**84,995.99152**
G22	**2.54912E + 22**	5.2955E + 22	2.35927E + 23	**2.25719E + 23**	5.23406E + 23	1.45216E + 24
G23	43,610.26831	29,647.56123	**10,311.8776**	**249,430.0779**	345,353.8813	4,485,256.94
G24	**−5.508013**	**−5.508013**	**−5.508013**	**−5.508013**	**−5.508013**	**−5.508013**

Table 4.7 Comparison of results (BEST solutions) for the mechanical design problems by using different variants of PSO

MD		BEST		
		PSO	PSO_M_1	PSO_M_2
Gear	Example 1A	3,135.535546	3,135.516885	**3,135.515853**
	Example 1B	**3,142.712756**	**3,142.712756**	**3,142.712756**
	Example 1C	2,993.674547	2,993.66892	**2,993.665606**
	Example 1D	**2,994.844118**	**2,994.844118**	**2,994.844118**
Bearing	Example 2A	−6,032.249216	−6,032.312611	**−6,032.315095**
	Example 2B	−3,792.409694	−3,792.418263	**−3,792.420036**
	Example 2C	−0.223958	−0.223968	**−0.223972**
Bellivelle spring	Example 3	2.382568	2.183703	**2.020371**
Multiplate clutch	Example 4	**0.313657**	**0.313657**	**0.313657**
Robot gripper	Example 5	5.382161	5.491666	**5.222902**
Hydrostatic bearing	Example 6	3,554.114041	3,168.090122	**1,924.945668**
4-stage Gear train	Example 7	NFS	NFS	NFS
Pressure vessel	Example 8	6,059.714992	6,059.714398	**6,059.714372**
Welded beam	Example 9	1.728235	**1.725362**	1.725573
Spring	Example 10	0.012683	0.012684	**0.012665**
Speed reducer	Example 11	2,996.375892	2,996.348256	**2,996.348165**
Stiffened shell	Example 12A	54,810.54271	54,493.64116	**54,474.79333**
	Example 12B	55,662.45291	**55,326.29341**	**55,326.29341**
Step cone pulley	Example 13	64.576689	52.738806	**43.992012**
Screw jack	Example 14	**419,187.0574**	**419,187.0574**	**419,187.0574**
C-clamp	Example 15	5,637.390548	5,624.007065	**5,617.476052**
Hydrodynamic bearing	Example 16	16.207067	16.206447	**16.205838**
Cone clutch	Example 17	68.877553	**68.877551**	**68.877551**
Cantilever support	Example 18	49.062267	49.062257	**49.062256**
Hydraulic cylinder	Example 19	0.024306	0.024306	**0.024265**
Planetary gear box	Example 20	0.53	**0.525769**	0.528033

Table 4.8 Comparison of results (MEAN solutions) for the mechanical design problems by using different variants of PSO

MD		MEAN		
		PSO	PSO_M_1	PSO_M_2
Gear	Example 1A	3,138.812971	3,138.768313	**3,135.516**
	Example 1B	3,145.727315	3,154.770991	**3,142.713**
	Example 1C	2,996.568691	2,996.562879	**2,993.666**
	Example 1D	**2,994.844118**	2,997.481697	**2,994.844**
Bearing	Example 2A	−6,011.23938	−5,990.851069	**−6,027.12**
	Example 2B	−3,788.895771	−3,775.165792	**−3,792.42**
	Example 2C	−0.2238488	−0.2239565	**−0.22396**
Bellivelle spring	Example 3	2.7837953	2.5972945	**2.085875**
Multiplate clutch	Example 4	**0.313657**	**0.313657**	**0.313657**
Robot gripper	Example 5	**6.0179145**	6.0583745	6.839436
Hydrostatic bearing	Example 6	1,634,070,589	30,512,494.2	**2,768.391**
4-stage Gear train	Example 7	NFS	NFS	NFS
Pressure vessel	Example 8	6,059.729819	6,059.716392	**6,059.715**
Welded beam	Example 9	1.7417705	**1.7291357**	1.879378
Spring	Example 10	0.01271	0.012802	**0.012679**
Speed reducer	Example 11	3,020.273747	3,014.277058	**3,005.312**
Stiffened shell	Example 12A	55,651.25519	**55,533.41976**	56,385.6
	Example 12B	56,142.59128	**55,605.03029**	56,220.83
Step cone pulley	Example 13	4,138.975827	1,080.558404	**157.5067**
Screw jack	Example 14	420,224.4689	**419,631.6623**	482,621.2
C-clamp	Example 15	5,659.174389	5,655.78883	**5,635.794**
Hydrodynamic bearing	Example 16	16.2213032	16.2098181	**16.20601**
Cone clutch	Example 17	68.8776593	68.8775517	**68.87755**
Cantilever support	Example 18	49.0625043	49.0622602	**49.06226**
Hydraulic cylinder	Example 19	0.024306	0.024306	**0.024293**
Planetary gear box	Example 20	0.5361934	**0.532883**	0.53718

Table 4.9 Comparison of results for the unconstrained benchmark functions by using different variants of ABC

BM-UC	BEST		MEAN	
	ABC	ABC_M	ABC	ABC_M
Sphere	0.000104	**0**	0.0002284	**0.0000002**
Schwefel 2.22	0.010707	**0.000154**	0.0121562	**0.0007274**
Schwefel 1.2	1,498.139116	**140.068412**	2,240.262653	**339.1271168**
Schwefel 2.21	14.051419	**13.374404**	17.4708172	**15.5585162**
Rosenbrock	24.629982	**15.608202**	**78.5741148**	90.0214244
Step	**0**	**0**	**0**	1.6
Quartic	0.000002	**0**	0.0000054	**0**
Schwefel 2.26	−12,239.35435	**−12,282.32144**	−11,651.23524	**−12,035.7655**
Rastrigin	25.777577	**8.234402**	30.8445478	**15.864832**
Ackley	0.067207	**0.004777**	**0.0753638**	0.3416602
Griewank	0.249771	**0.008374**	0.3749822	**0.0205818**
Penalty-1	0.097597	**0.000164**	0.68805	**0.0642904**
Penalty-2	0.156342	**0.016479**	**0.411938**	1.036791

4.3.12 Example 19: Design of Hydraulic Cylinder

The objective is to minimize the volume of a hydraulic cylinder subjected to internal pressure, by taking the piston diameter (d), hydraulic pressure (p) and the cylinder wall thickness (t) as design variables. Design vector can be defined as $X = (x_1, x_2, x_3) = (d, p, t)$.

Minimum force required is F, that is,

$$f = p\frac{\pi d^2}{4} \geq F \tag{4.247}$$

Hoop stress induced should be less than allowable stress S,

$$s = \frac{pd}{2t} \geq S \tag{4.248}$$

Constraints on diameter, pressure and thickness imposed on the design of hydraulic cylinder are: $d + 2t \leq D$, $p \leq P$, $t \leq T$, where D is the maximum outside diameter permissible, P the maximum pressure of the hydraulic system and T the minimum cylinder wall thickness required. The normalized form of all the constraints can be stated as:

$$g_1(X) = \frac{4}{\pi}Fp^{-1}d^{-2} \leq 1 \tag{4.249}$$

$$g_2(X) = \frac{1}{2}S^{-1}pdt^{-1} \leq 1 \tag{4.250}$$

$$g_3(X) = D^{-1}d + 2D^{-1}t \leq 1 \tag{4.251}$$

$$g_4(X) = P^{-1}p \leq 1 \tag{4.252}$$

$$g_5(X) = Tt^{-1} \leq 1 \tag{4.253}$$

The volume of the cylinder per unit length to be minimized is given by

$$f(X) = \pi t(d + t) \tag{4.254}$$

where, $200 \leq x_1 \leq 500$, $1 \leq x_2 \leq 5$, $5 \leq x_3 \leq 30$

4.3.13 Example 20: Design of Planetary Gear Train

The gear teeth number of an automatic planetary transmission used in automobiles is formulated as a constrained optimization problem by Simionescu et al. [37]. A planetary transmission of the Ravigneaux type with three forward and one reverse gears used in automobiles is considered.

The objective is to minimize the gear ratio errors which can be stated as:

$$f_{1(\ldots)} = \max|i_k - i_{0k}| \ (k = \{1, 2, R\}) \tag{4.255}$$

where $i_1 = \dfrac{N_6}{N_4}$ and $i_{01} = 3.11$

$$i_2 = \frac{N_6(N_1N_3 + N_2N_4)}{N_1N_3(N_6 - N_4)} \text{ and } i_{02} = 1.84$$

$$i_R = \left(\frac{N_2N_6}{N_1N_3}\right) \text{ and } i_{0R} = -3.11$$

The problem contains the number of teeth as the design variables which are six in number (N_1, N_2, N_3, N_4, N_5 and N_6) and can only take integer value. Moreover, there are three more discrete design variables, number of planet (P) and module of gears (m_1 and m_2) which can only take specified discrete values. Design vector can be defined as $X = (x_1, x_2, x_3, x_4, x_5, x_6, x_7, x_8, x_9, = (N_1, N_2, N_3, N_4, N_5, N_6, P, m_1, m_2)$. The design is subjected to the following constraints:

For the avoidance of under cut:

$$N_{1,4} \geq N_{S\min} = 17 \tag{4.256}$$

$$N_{2,3,5} \geq N_{P\min} = 14 \tag{4.257}$$

Limiting the maximum outer diameter for ring gear, planet-2 and idler-5:

$$m_3(N_6 + 2.5) \leq D_{\max} \tag{4.258}$$

$$m_1(N_1+N_2)+m_1(N_2+2)\leq D_{\max} \tag{4.259}$$

$$m_3(N_4+N_5)+m_3(N_5+2)\leq D_{\max} \tag{4.260}$$

To avoid the contact between neighboring gears:

$$|m_1(N_1+N_2)-m_3(N_6-N_3)|\leq m_1+m_3 \tag{4.261}$$

$$(N_1+N_2)\cdot\sin(\pi/p)-N_2-2-\delta_{22}\geq 0 \tag{4.262}$$

$$(N_6-N_3)\cdot\sin(\pi/p)-N_3-2-\delta_{33}\geq 0 \tag{4.263}$$

$$(N_4+N_5)\cdot\sin(\pi/p)-N_5-2-\delta_{55}\geq 0 \tag{4.264}$$

$$(N_6-N_3)^2+(N_4+N_5)^2-2(N_6-N_3)(N_4+N_5)\cos\left(\frac{2\pi}{\rho}-\beta\right)\geq(N_3+N_5+2+\delta_{35})^2 \tag{4.265}$$

where,

$$\beta=\cos^{-1}\frac{(N_6-N_3)^2+(N_4+N_5)^2-(N_3+N_5)^2}{2(N_6-N_3)(N_4+N_5)}$$

$$N_6-2N_3-N_4-4-2\delta_{34}\geq 0 \tag{4.266}$$

$$N_6-N_4-2N_5-4-2\delta_{56}\geq 0 \tag{4.267}$$

For assembly of ideal gears and to ensure equally spaced planet gears:

$$(N_6-N_4)/p=\text{integer} \tag{4.268}$$

$$Fraction\left(\frac{1}{p}\left|\frac{N_1}{N_2}+\frac{N_6}{N_3}\right|\right)=\left|\frac{A}{N_2}\pm\frac{B}{N_3}\right| \tag{4.269}$$

where,

$$0\leq A<N_2/p,$$

$$0\leq B<N_3/p,$$

$$N_6\leq D_{\max}/m_{3\,\min}-2.5$$

The limitations on design variables are:

$$N_1\leq D_{\max}/m_{1\,\min}-2N_{P\min}-2 \tag{4.270}$$

$$N_2\leq(D_{\max}/m_{1\,\min}-N_{S\min}-2)/2 \tag{4.271}$$

$$N_3\leq(D_{\max}/m_{3\,\min}-N_{S\min}-6.5-2\delta_{34})/2 \tag{4.272}$$

$$N_4\leq(D_{\max}/m_{3\,\min}-2N_{P\min}-6.5-2\delta_{56}) \tag{4.273}$$

Table 4.10 Comparison of results for the constrained benchmark functions by using different variants of ABC

BM-C	BEST		MEAN	
	ABC	ABC_M	ABC	ABC_M
G01	**−15**	−14.9999952	**−15**	−14.9999952
G02	**−0.78972**	−0.6774415	**−0.7443135**	−0.6774415
G03	**−0.906291**	−0.5366163	**−0.5896032**	−0.5366163
G04	**−30,665.53867**	−30,665.53244	**−30,665.53863**	−30,665.5324
G05	**1,005,184.9**	1,597,346.203	**1,225,029.85**	1,597,346.203
G06	**−6,961.788268**	−6,960.476311	**−6,961.634454**	−6,960.47631
G07	**24.446895**	29.4085934	**24.8121042**	29.4085934
G08	**−0.095825**	**−0.095825**	**−0.095825**	**−0.095825**
G09	**680.679911**	680.7168249	**680.7082881**	680.7168249
G10	**7,076.165884**	7,850.8018	**7,449.202949**	7,850.8018
G11	**0.749908**	0.7509324	**0.750004**	0.7509324
G12	**−1**	**−1**	**−1**	**−1**
G13	**0.968644**	161.070554	**40.9768238**	161.070554
G14	**−43.226255**	55.9678366	56.3309436	**55.9678366**
G15	**962.343214**	1,767.259148	**966.5421996**	1,767.259148
G16	**−1.888024**	−1.8581444	**−1.888024**	−1.8581444
G17	**2,010,833.506**	2,856,096.227	**2,615,854.331**	2,856,096.227
G18	**−0.86373**	−0.862204	−0.8594644	**−0.862204**
G19	**35.038677**	41.767342	**37.6985894**	41.767342
G20	**13,161,567.49**	15,483,761.91	16,206,675.49	**15,483,761.91**
G21	**10,409.03455**	149,707.5714	**143,532.0626**	149,707.5714
G22	**4,330,091,720**	9.0292E + 16	**9.90095E + 15**	9.0292E + 16
G23	3,896.779936	**3,856.182249**	39,329.43275	**3,856.182249**
G24	**−5.508013**	**−5.508013**	**−5.508013**	**−5.508013**

$$N_5 \leq (D_{\max}/m_{3\,\min} - N_{S\min} - 6.5 - 2\delta_{56})/2 \quad (4.274)$$

$D_{\max} = 220$, p can be 3, 4 or 5, and m_1 and m_3 can have following discrete values : 1.75, 2.0, 2.25, 2.5, 2.75 or 3.0. The relative clearance between the adjacent gears δ_{22}, δ_{33}, δ_{55}, δ_{35} andδ_{56} are considered as 0.5. The best value reported by Simionescu et al. [37] is $f(X) = 0.525$ with $X = (40, 21, 14, 19, 16, 69, 2.25, 2.5, 5)$.

4.4 Applications of Modified PSO

Basic version of PSO and modified PSO are compared based on a performance evaluations. For the comparison following performance criteria evaluations are considered.

- Population size = 50
- Number of generations = 500 (for unconstrained and constrained benchmark functions), 200 (mechanical design problems)
- For PSO : $c_1 = c_2 = 2$, $V_{\max} = 4$, for basic PSO w varies linearly from 0.9 to 0.4 and for modified PSO w varies as per the modifications suggested in Chap. 2.
- Number of runs = 25

Comparison is based on the best and the mean solution achieved in 25 runs. The comparison of results for the unconstrained benchmark functions, constrained benchmark functions and mechanical element design optimization problems is given in Tables 4.5, 4.6, 4.7, 4.8, respectively. In all these Tables, the values indicated in 'bold' indicate the superior result and 'BEST' and 'MEAN' indicate the best solution and mean solution obtained in 25 runs. It is observed from the results that the modification in PSO has shown better results than the basic PSO. PSO_M_1 has shown better results for the best solution for two unconstrained benchmark functions and PSO_M_2 has shown better results for the best solution for ten benchmark functions. For mean solution, PSO_M_1 has shown better results for four unconstrained benchmark functions and PSO_M_2 has shown better results for eight unconstrained benchmark functions. Moreover, PSO_M_2 has given near optimal solution up to the accuracy of four decimal for Sphere and Quartic functions and global solution for the Step function.

It is observed from the results that the modification is also advantageous for the constrained benchmark functions. Out of two variants of PSO, PSO_M_2 is better when compared to PSO_M_1. PSO_M_2 has shown better results for twenty and seventeen constrained benchmark functions for the best and the mean solutions, respectively. PSO_M_2 has given near optimal solution for the five constrained benchmark functions (G01, G04, G08, G12 and G24), while PSO and PSO_M_1 has shown near optimal solutions for four constrained benchmark functions. There is no major difference in the results for the PSO and PSO_M_1 for the constrained benchmark functions.

Table 4.11 Comparison of results for the mechanical design problems by using different variants of ABC

MD		BEST		MEAN	
		ABC	ABC_M	ABC	ABC_M
Gear	Example 1A	**3,135.515852**	**3,135.515852**	**3,135.515852**	**3,135.515852**
	Example 1B	**3,142.712756**	**3,142.712756**	**3,142.712756**	**3,142.712756**
	Example 1C	**2,993.665605**	**2,993.665605**	**2,993.665605**	**2,993.665605**
	Example 1D	**2,994.844118**	**2,994.844118**	**2,994.844118**	**2,994.844118**
Bearing	Example 2A	**−6,032.315099**	**−6,032.315099**	**−6,032.315099**	−6,032.315098
	Example 2B	**−3,792.420036**	**−3,792.420036**	**−3,792.420036**	**−3,792.420036**
	Example 2C	**−0.223974**	**−0.223974**	**−0.223974**	**−0.223974**
Bellivelle spring	Example 3	**2.01321**	2.023958	**2.0360104**	2.0615033
Multiplate clutch	Example 4	**0.313657**	**0.313657**	**0.313657**	**0.313657**
Robot gripper	Example 5	**4.247644**	**4.247644**	7.7149531	**4.7584009**
Hydrostatic bearing	Example 6	2,335.942363	**2,143.391233**	**2,551.202371**	2,630.783273
4-stage Gear train	Example 7	3,723,758.868	**49,836.16542**	2,790,742,547	**138,706,089**
Pressure vessel	Example 8	6,059.714387	**6,059.714352**	6,059.714853	**6,059.714586**
Welded beam	Example 9	**1.725411**	1.725642	**1.730008**	1.7304741
Spring	Example 10	0.012678	**0.012671**	0.0126821	**0.0126804**
Speed reducer	Example 11	**2,996.348165**	2,996.348186	**2,996.348165**	2,996.348317
Stiffened shell	Example 12A	54,554.82261	**54,504.59192**	55,318.81618	**55,033.43415**
	Example 12B	55,501.30976	**55,326.29341**	**55,883.37249**	56,379.32598
Step cone pulley	Example 13	43.442871	**16.720818**	72.3768391	**67.2064085**
Screw jack	Example 14	**419,187.0574**	**419,187.0574**	**419,187.0574**	**419,187.0574**
C-clamp	Example 15	**5,617.708161**	5,617.838734	**5,620.498802**	5,622.477553
Hydrodynamic bearing	Example 16	16.205844	**16.205841**	**16.2058718**	16.205939
Cone clutch	Example 17	**68.877551**	68.877552	**68.877551**	68.8775575
Cantilever support	Example 18	**49.062256**	49.062263	**49.0622579**	49.0623523
Hydraulic cylinder	Example 19	0.024175	**0.02414**	0.0242424	**0.024174**
Planetary gear box	Example 20	**0.525769**	0.53	**0.5272922**	0.532793

For the mechanical design problems, both the variants of PSO are better than basic PSO. PSO_M_2 has shown better results for twenty-three and nineteen mechanical design problems for the best and the mean solutions, respectively. Moreover, PSO_M_1 has also shown better results than basic PSO. No feasible solutions were obtained for the Example 7 by any variants of PSO. It is observed that PSO_M_2 has shown better performance in finding the best solutions for 10 unconstrained benchmark functions, 20 constrained benchmark functions and 23 mechanical element design problems. PSO_M_1 has shown better performance for 2, 8 and 8 unconstrained benchmark functions, constrained benchmark functions and mechanical element design problems respectively and PSO has shown better performance for 1, 8 and 4 unconstrained benchmark functions, constrained benchmark functions and mechanical element design problems, respectively. So, the performance of PSO_M_2 is approximately 4 times and 3 times better than PSO and PSO_M_1 to find the best solutions. Similarly, it is observed that PSO_M_2 is approximately 3 times and 2.5 times better than PSO and PSO_M_1 to find the mean solutions.

4.5 Applications of Modified ABC

Basic version of ABC and modified ABC are also compared based on a common performance evaluation. For the comparison following performance criteria evaluations are considered.

- Population size = 50
- Number of generations = 500 (for unconstrained and constrained benchmark functions), 200 (mechanical design problems)
- For ABC: number of employed bees = number of onlooker bees, limit = number of generations. ABC_M updates the solutions as per the modifications suggested in Chap. 2.
- Number of runs = 25

Comparison is based on the best and the mean solutions achieved in 25 runs. The comparison of results for the unconstrained benchmark functions, constrained benchmark functions and mechanical element design optimization problems are given in Tables 4.9, 4.10, and 4.11, respectively.

It is observed from the results that modification in ABC has shown better results than the basic ABC for the unconstrained benchmark functions. For all the thirteen unconstrained benchmark problems ABC_M has outperformed basic ABC in finding the best solution and also it has given nine times better results for the best solutions. ABC_M has given global solution for the Sphere, Step and Quartic functions. It has shown near optimal solution for Schwefel 2.22, Ackley, Griewank, Penalty-1 accurate up to two decimal. Compared with PSO_M_2, ABC_M has shown better results for nine benchmark problems. Moreover, ABC has also shown better results than PSO_M_2 for the unconstrained benchmark functions.

Table 4.12 Comparison of results for the unconstrained benchmark functions by using different variants of HEA

BM-UC	BEST		MEAN	
	HEA	HEA_M	HEA	HEA_M
Sphere	**30.297161**	36.397414	**35.10702**	37.60604
Schwefel 2.22	**22.629791**	23.110424	**24.6817**	26.19577
Schwefel 1.2	294.619348	**225.755718**	350.7362	**259.1098**
Schwefel 2.21	**2.628382**	2.738248	2.829465	**2.78994**
Rosenbrock	13,642.06328	**7,851.377546**	17,442.74	**14,298.37**
Step	**34**	37	**38.66667**	42
Quartic	60.814456	**50.60093**	**100.005**	170.177
Schwefel 2.26	−3,304.72498	**−3,972.831203**	−3,107.25	**−3,504.37**
Rastrigin	**297.924276**	306.126284	330.6954	**327.367**
Ackley	**5.337689**	5.377059	**5.625639**	5.68304
Griewank	75.79477	**3.051871**	77.92893	**4.311237**
Penalty-1	1.202707	**0.567072**	1.368613	**0.914487**
Penalty-2	**3.399486**	4.372659	**4.326432**	4.95895

For the constrained benchmark functions ABC has outperformed ABC_M for the 23 functions. For the mean solutions also ABC has shown better results than ABC_M. The performance of ABC_M is nearly same as PSO_M_2 for the constrained benchmark function for finding the best solutions and the mean solutions. So it can be said that modification in ABC is not effective for constrained benchmark functions than that for unconstrained benchmark functions. It can further be observed that for mechanical design problems ABC_M has slightly outperformed ABC in finding the best solutions and ABC has slightly outperformed ABC_M in finding the mean solutions for the considered mechanical design problems. It is observed that ABC has shown better performance in finding the best solutions for 1 unconstrained benchmark functions, 23 constrained benchmark functions and 17 mechanical element design problems. ABC_M has shown better performance for 13, 4 and 19 unconstrained benchmark functions, constrained benchmark functions and mechanical element design problems, respectively. So, the performance of ABC is approximately 1.13 times better than ABC_M to find the best solutions. Similarly, it is observed that ABC is approximately 1.38 times better than ABC_M to find the mean solutions.

4.6 Applications of Modified HEA

Basic version of HEA and modified HEA are also compared based on a common experimental platform. For the comparison following performance criteria evaluations are considered.

- Population size = 50

Table 4.13 Comparison of results for the constrained benchmark functions by using different variants of HEA

BM-C	BEST		MEAN	
	HEA	HEA_M	HEA	HEA_M
G01	−5.999997	**−7.996523**	−4	**−6.33209**
G02	−0.33966	**−0.412949**	−0.30054	**−0.36311**
G03	**0**	**0**	**0**	**0**
G04	−30,665.0375	**−30,665.29292**	−30,664.2	**−30,665.2**
G05	9,285,375,501	**371,684,672**	8.87E + 10	**2.84E + 09**
G06	**−5,622.08828**	−5,330.095601	786,124.8	**452,644**
G07	100.522406	**47.95103**	123.7154	**53.02583**
G08	**−0.095825**	**−0.095825**	**−0.09583**	**−0.09583**
G09	702.026349	**689.1031**	722.21	**691.4605**
G10	**9,479,214.098**	15,358,378.85	1.36E + 08	**33,916,227**
G11	**1**	**1**	**1**	**1**
G12	**−0.991641**	−0.971995	32.48861	**−0.90406**
G13	**301.579417**	305.36216	355.3025	**337.4035**
G14	67,031.75549	**4,266.870089**	91,069.31	**32,859.67**
G15	**3,444.881182**	17,996.00486	2,106,182	**81,095.97**
G16	−1.700703	**−1.847098**	−1.64062	**−1.80343**
G17	1.28523E + 11	**122,093,420.3**	1.71E + 12	**1.13E + 11**
G18	**−0.380726**	−0.25802	**140.5401**	1,306.674
G19	82.776476	**48.774431**	162.7461	**53.90053**
G20	41,173,962,188	**8,386,866,029**	5.66E + 10	**2.11E + 10**
G21	**304,630.7611**	1,361,438,325	2.83E + 09	**1.74E + 09**
G22	5.46746E + 23	**1.27864E + 23**	1.31E + 24	**2.48E + 23**
G23	**212,003.2403**	263,317.6344	693,535.9	**388,170.9**
G24	−5.500894	**−5.507728**	−5.49904	**−5.50761**

Table 4.14 Comparison of results for the mechanical design problems by using different variants of HEA

MD		BEST		MEAN	
		HEA	HEA_M	HEA	HEA_M
Gear	Example 1A	3,135.854	**3,135.624**	3,143.548	**3,139.231**
	Example 1B	3,145.601	**3,145.519**	3,148.813	**3,147.932**
	Example 1C	2,994.953	**2,994.433**	2,998.894	**2,996.544**
	Example 1D	2,995.928	**2,995.668**	2,998.81	**2,997.673**
Bearing	Example 2A	**−6,026.2**	−5,980.48	**−5,995.39**	−5,980.48
	Example 2B	−3,632.41	**−3,732.42**	−3,658.9	**−3,697.17**
	Example 2C	−0.22376	**−0.2239**	**−0.22245**	−0.22216
Bellivelle spring	Example 3	**2.231749**	2.78098	3.455094	**3.052552**
Multiplate clutch	Example 4	**0.313657**	**0.313657**	**0.313657**	**0.313657**
Robot gripper	Example 5	5.814479	**4.874892**	6.787027	**6.435739**
Hydrostatic bearing	Example 6	**3,384.818**	3,899.731	**27,949,033**	5.26E + 08
4-stage Gear train	Example 7	NFS	NFS	NFS	NFS
Pressure vessel	Example 8	6,064.13	**6,059.991**	6,068.828	**6,060.103**
Welded beam	Example 9	1.828208	**1.794189**	1.862864	**1.802354**
Spring	Example 10	0.012863	**0.012723**	0.013095	**0.012756**
Speed reducer	Example 11	3,035.896	**3,008.879**	3,036.877	**3,026.716**
Stiffened shell	Example 12A	56,160.62	**54,928.95**	56,904.28	**55,261.06**
	Example 12B	56,432.45	**55,532.45**	57,349.57	**56,983.87**
Step cone pulley	Example 13	10,135.74	**66.01338**	17478.76	**165.7784**
Screw jack	Example 14	419,187.7	**419,187.1**	419,188.3	**419,187.1**
C-clamp	Example 15	5,666.589	**5,663.389**	5,672.865	**5,663.536**
Hydrodynamic bearing	Example 16	16.22135	**16.21244**	16.24194	**16.21867**
Cone clutch	Example 17	69.22121	**68.88543**	69.69834	**68.90321**
Cantilever support	Example 18	49.28362	**49.08784**	49.58969	**49.12982**
Hydraulic cylinder	Example 19	0.024309	**0.024306**	0.024313	**0.024307**
Planetary gear box	Example 20	NFS	NFS	NFS	NFS

- Number of generations = 500 (for unconstrained and constrained benchmark functions), 200 (mechanical design problems)
- For HEA: codal length = 10, HEA_M updates the solutions as per the modifications suggested in Chap. 2 for which the mutation rate was kept as 0.2.
- Number of runs = 25

Comparison is based on the best and the mean solution achieved in 25 runs. The comparison of results for the unconstrained benchmark functions, constrained benchmark functions and mechanical design problems are given in Tables 4.12, 4.13 and 4.14, respectively.

It is observed from the results that modification is not so effective for the unconstrained benchmark functions for finding the best and the mean solutions. Modification is effective in finding the mean solutions for the constrained benchmark functions. Moreover, modification is very effective for the mechanical design problems for finding the best and the mean solutions. It is observed that HEA_M has shown better performance in finding the best solutions for 6 unconstrained benchmark functions, 16 constrained benchmark functions and 21 mechanical element design problems. HEA has shown better performance for 7, 11 and 7 unconstrained benchmark functions, constrained benchmark functions and mechanical element design problems, respectively. So, the performance of HEA_M is approximately 1.9 times better than HEA to find the best solutions. Similarly, it is observed that HEA_M is approximately 3.6 times better than HEA to find the mean solutions. The main drawback of HEA is that it describes the design variables in the form of a string which requires the complicated coding and decoding process resulting in the increase of computational efforts. Moreover, it is observed from the results that variants of PSO and ABC are better than the variants of HEA.

It is observed that ABC is approximately 1.1 times and 2.9 times better than PSO_M_2 in finding the best and the mean solutions and 32 times and 31 times better than HEA and HEA_M in finding the best solutions and the mean solutions. So, out of all the algorithms considered for the modifications, modification in PSO and HEA is effective than their basic version. Modifications in ABC are not so effective for constrained benchmark functions, but it is effective for unconstrained benchmark functions and mechanical design problems. Overall performance indicates that basic ABC is better than all the variants of PSO and HEA.

The next chapter presents the applications of four different hybrid algorithms to unconstrained and constrained benchmark functions and also on mechanical element design optimization problems.

References

1. Li R, Chang X (2006) A modified genetic algorithm with multiple subpopulations and dynamic parameters applied in CVAR model. Comput Intell for Model, Control and Autom, Sydney, NSW 151

2. Liu J, Tang LA (1999) Modified genetic algorithm for single machine scheduling. Comput Ind Eng 37:43–46
3. Preechakul C, Kheawhom S (2009) Modified genetic algorithm with sampling techniques for chemical engineering optimization. J Ind Eng Chem 15:101–107
4. Montalvo I, Izquierdo J, Perez-Garcia R, Herrera M (2010) Improved performance of PSO with self-adaptive parameters for computing the optimal design of water supply systems. Eng Appl Artif Intell 23:727–735
5. Cai X, Cui Y, Tan Y (2009) Predicted modified PSO with time-varying accelerator coefficients. Int J Bio Inspired Comput 1:50–60
6. Cui H, Turan O (2010) Application of a new multi-agent hybrid co-evolution based particle swarm optimisation methodology in ship design. Comput Aided Des 2:1013–1027
7. Yildiz AR (2009) A novel particle swarm optimization approach for product design and manufacturing. Int J Adv Manuf Technol 40:617–628
8. Shen Q, Jiang J, Tao J, Shen G, Yu R (2005) Modified ant colony optimization algorithm for variable selection in QSAR modeling: QSAR studies of cyclooxygenase inhibitors. J chem inf model 45:1024–1029
9. Karaboga D, Akay B (2010) A modified artificial bee colony (ABC) algorithm for constrained optimization problems. Appl Soft Comput doi: 10.1016/j.asoc.2010.12.001
10. Mouti FSA, Hawary MEE (2009) Modified artificial bee colony algorithm for optimal distributed generation sizing and allocation in distribution systems. IEEE Electrical Power and Energy Conference (EPEC), Montreal, QC, pp 1–9
11. Yue H, Gu G, Liu H, Shen J, Zhao J (2009) A modified ant colony optimization algorithm for tumor marker gene selection. Genomics, Proteomics Bioinf 7:200–208
12. Hui L, Zixing C, Yong W (2010) Hybridizing particle swarm optimization with differential evolution for constrained numerical and engineering optimization. Appl Soft Comput 10:629–640
13. Wen YL (2010) A GA–DE hybrid evolutionary algorithm for path synbook of four-bar linkage. Mech Mach Theor 45:1096–1107
14. Yannis M, Magdalene M (2010) Hybrid multi-swarm particle swarm optimization algorithm for the probabilistic travelling salesman problem. Comput Oper Res 37:432–442
15. Ying PC (2010) An ant direction hybrid differential evolution algorithm in determining the tilt angle for photovoltaic modules. Expert Sys Appl 37:5415–5422
16. Shahla N, Mohammad EB, Nasser G, Mehdi HA (2009) A novel ACO–GA hybrid algorithm for feature selection in protein function prediction. Expert Sys Appl 36:12086–12094
17. Tung Y, Erwie Z (2008) A hybrid genetic algorithm and particle swarm optimization for multimodal functions. Appl Soft Comput 8:849–857
18. Dong HK, Ajith A, Jae HC (2007) A hybrid genetic algorithm and bacterial foraging approach for global optimization. Inf Sci 177:3918–3937
19. Simon D (2008) Biogeography-based optimization. IEEE Trans on Evol Comput 12:702–713
20. Liang JJ, Runarsson TP, Montes EM, Clerc M, Suganthan PN, Coello CAC, and Deb K (2006) Problem definitions and evolution criteria for the CEC 2006 special session on constrained real-parameter optimization. Tech Rep, Nanyang Technol Univ, Singapore. http://www.ntu.edu.sg/home/EPNSugan
21. Sandgren E (1988) Nonlinear integer and discrete programming in mechanical design. In: Proceedings of the ASME design technology conference, Kissimine, FL, pp 95–105
22. Kannan BK, Kramer SN (1994) An augmented Lagrange multiplier based method for mixed integer discrete continuous optimization and its applications to mechanical design. ASME J Mech Des 116:318–320
23. Deb K (1997) GeneAS: a robust optimal design technique for mechanical component design. Evol Algorithms in Eng Appl. Springer, Berlin, pp 497–514
24. Coello CAC (2000) Use of a self-adaptive penalty approach for engineering optimization problems. Comput Ind 41:113–127
25. Ray T, Liew K (2003) Society and civilization: an optimization algorithm based on the simulation of social behavior. IEEE Trans Evol Comput 7:386–396

26. Montes ME, Coello CAC (2005) A simple multimembered evolution strategy to solve constrained optimization problems. IEEE Trans Evol Comput 9:1–17
27. Parsopoulos K, Vrahatis M (2005) Unified particle swarm optimization for solving constrained engineering optimization problems. In: Proceedings of advanced in natural computation, LNCS 3612. Springer-Verlag, Berlin, pp 582–591
28. He Q, Wang L (2007) An effective co-evolutionary particle swarm optimization for constrained engineering design problems. Eng Appl Artif Intell 20:89–99
29. Huang FA, Wang L, He Q (2007) An effective co-evolutionary differential evolution for constrained optimization. Appl Math Comput 186(1):340–356
30. Liu H, Cai Z, Wang Y (2010) Hybridizing particle swarm optimization with differential evolution for constrained numerical and engineering optimization. Appl Soft Comput 10:629–640
31. Akay B, Karaboga D (2010) Artificial bee colonyArtificial bee colony algorithm for large-scale problems and engineering design optimization. J Intell Manuf. Doi: 10.1007/s10845-010-0393-4
32. Ragsdell KM, Phillips DT (1976) Optimal design of a class of welded structures using geometric programming. ASME J Eng Ind 98(3):1021–1025
33. Belegundu AD (1982) A study of mathematical programming methods for structural optimization. Doctoral Dissertation, Department of Civil and Environmental Engineering, University of Iowa, USA
34. Leandro SC, Viviana CM (2008) Use of chaotic sequences in a biologically inspired algorithm. Expert Syst Appl 34(3):1905–1913
35. Jarmai K, Snyman JA, Farkas J (2006) Minimum cost design of a welded orthogonally stiffened cylindrical shell. Comput Struct 84:787–797
36. Rao SS (2002) Engineering optimization: theory and practice. New Age International, New Delhi
37. Simionescu PA, Beale D, Dozier GV (2006) Teeth-number synbook of a multispeed planetary transmission using an estimation of distribution algorithm. J Mech Des 128:108–115

Chapter 5
Applications of Hybrid Optimization Algorithms to the Unconstrained and Constrained Problems

Hybridization is one of the effective ways to improve the effectiveness of the algorithms. Hybridization combines the searching capabilities of different algorithms. There is continuous research going on to hybridize different algorithms and to improve its effectiveness for the particular application. In this book experiment is conducted to hybridize ABC with other optimization algorithms. ABC is chosen because, as observed from the Chaps. 2 and 3, it has better searching tendency than other optimization algorithm. The other reason is that ABC is the recent optimization algorithm and no literature is available for its hybridization. Five different optimization algorithms, PSO, BBO, AIA, DE and GA, are chosen to hybridize it with ABC. All these five algorithms are checked for the unconstrained benchmark functions. Following experimental setup was used for the performance evaluation of the algorithms:

- Population size = 50
- Number of generations = 500
- For PSO: w varies linearly from 0.9 to 0.4, $c_1 = 2$, $c_2 = 2$, $V_{\max} = 4$
- For ABC: Number of employed bees = number of onlooker bees = Population size/2, limit = number of generation.
- For DE: $F = 0.5$, $C = 0.5$
- For BBO: Immigration rate = emigration rate = 1, mutation factor = 0.01,
- For AIA: Clone size = population size, $\beta = 1$, recipoteir rate = 0.25
- For GA: Crossover probability = 0.9, mutation probability = 0.01, crossover = single point crossover, selection = roulette wheel selection.

Results for the best and the mean solutions are given in Tables 5.1 and 5.2. It is observed from the results that DE has outperformed all the algorithms by showing better results for nine benchmark functions in finding best and the mean solutions. Moreover, PSO, BBO and GA have shown better results for two, one and one benchmark functions, respectively. AIA fails to overcome any algorithm for the best and the mean solutions. From the above observation, it is decided to hybridize ABC with PSO, BBO, DE and GA. So four different hybrid algorithms developed in

R. V. Rao and V. J. Savsani, *Mechanical Design Optimization Using Advanced Optimization Techniques*, Springer Series in Advanced Manufacturing,
DOI: 10.1007/978-1-4471-2748-2_5, © Springer-Verlag London 2012

Table 5.1 Comparison of results (best solution) for the unconstrained benchmark functions by using different advanced optimization techniques

BEST	PSO	BBO	AIA	DE	GA
Sphere	21.484652	0.045006	17.200328	**0.000003**	1.619544
Schwefel 2.22	21.284104	1.55874	25.069013	**0.010054**	12.487964
Schwefel 1.2	**93.156383**	2,140.598506	12,289.79509	13,592.41811	5,841.431504
Schwefel 2.21	**2.404852**	10.548412	47.523489	10.46417	38.571335
Rosenbrock	10,346.89086	183.920626	49,748.02868	**28.477708**	649.100591
Step	18	25	2,597	**0**	420
Quartic	19.981448	0.000078	1.38547	**0**	0.000854
Schwefel 2.26	−5,393.636691	−12,520.61382	−9,127.631341	−6,943.756589	**−12,566.83067**
Rastrigin	282.822554	**7.960961**	127.764394	138.753269	73.522086
Ackley	4.078291	2.449824	11.222243	**0.009833**	12.987812
Griewank	0.412012	1.365593	51.118327	**0.006983**	7.185487
Penalty-1	0.065802	0.248276	2,407,957.522	**0.002088**	2.285853
Penalty-2	4.383156	2.333334	9,812,890.749	**0.002769**	13.911089

Table 5.2 Comparison of results (mean solution) for the unconstrained benchmark functions by using different advanced optimization techniques

MEAN	PSO	BBO	AIA	DE	GA
Sphere	29.5030532	0.132851	21.6168734	**0.0000074**	2.3644714
Schwefel 2.22	24.298752	1.9037872	32.3413056	**0.013291**	15.5768742
Schwefel 1.2	**143.2574008**	3069.06681	14648.73684	15772.55927	9611.765148
Schwefel 2.21	**2.6110444**	13.434259	52.3192566	11.5690966	47.1027266
Rosenbrock	13901.52245	294.6904326	82858.74906	**30.589422**	890.7224288
Step	37.6	43.8	3604.2	**0**	718.4
Quartic	83.6416058	0.0002128	5.6798458	**0**	0.007514
Schwefel 2.26	−4067.875837	−12496.96246	−8750.007459	−6539.734366	**−12565.55067**
Rastrigin	305.370863	**8.5160554**	148.473993	160.866342	91.0348762
Ackley	5.5037588	2.978083	13.367403	**0.0143356**	13.9423348
Griewank	0.677997	1.4751216	84.8755844	**0.046504**	8.2381338
Penalty-1	0.595423	0.379269	10893269.48	**0.003993**	4.3328856
Penalty-2	5.3438038	3.4917536	49503835.25	**0.0123458**	20.3437036

Table 5.3 Comparison of results (BEST solution) for the unconstrained benchmark functions by using different hybrid optimization techniques

BM-UC	BEST			
	HPABC	HBABC	HDABC	HGABC
Sphere	0.000088	0.000004	**0**	0.000044
Schwefel 2.22	1.184095	0.000217	**0.000002**	0.038606
Schwefel 1.2	503.525798	69.475646	**32.331711**	41.768063
Schwefel 2.21	10.042743	4.966752	6.927807	**0.228172**
Rosenbrock	32.546706	83.514598	**17.054065**	27.753121
Step	152	**0**	2	**0**
Quartic	**0**	**0**	**0**	**0**
Schwefel 2.26	−7787.748693	**−12564.88713**	−12332.58572	−12557.72112
Rastrigin	30.860414	**0.013784**	0.99801	1.109163
Ackley	3.519417	**0.016447**	0.018263	0.018495
Griewank	0.316696	0.044606	**0.000013**	0.393688
Penalty-1	1.350518	0.004374	**0**	0.000023

Table 5.4 Comparison of results (MEAN solution) for the unconstrained benchmark functions by using different hybrid optimization techniques

BM-UC	MEAN			
	HPABC	HBABC	HDABC	HGABC
Sphere	0.003524	0.0151116	**0.0000004**	0.0000666
Schwefel 2.22	4.0940878	0.0014506	**0.0000476**	0.0735882
Schwefel 1.2	1100.577885	184.5669032	**49.1297914**	106.969643
Schwefel 2.21	21.9817796	9.5648362	8.5768576	**0.4599282**
Rosenbrock	104.0428604	94.0994142	69.4546108	**28.590297**
Step	244.4	10.4	6.2	**0**
Quartic	0.0000006	0.000149	**0**	**0**
Schwefel 2.26	−7643.581618	−12491.38836	−12128.93538	**−12554.96307**
Rastrigin	55.7848378	**2.2034244**	6.3645678	12.9073396
Ackley	6.812338	0.2793278	0.8349822	**0.1317778**
Griewank	0.6562612	0.0968346	**0.034429**	0.7140134
Penalty-1	5.271248	0.1031668	**0.0000006**	0.0000834
Penalty-2	38.094094	2.9497502	1.737886	**0.0026552**

this book are Hybrid Particle swarm-based Artificial Bee Colony (HPABC), Hybrid Biogeography-based Artificial Bee Colony (HBABC), Hybrid Differential evolution-based Artificial Bee Colony (HDABC) and Hybrid Genetic algorithm-based Artificial bee colony (HGABC). The details of the algorithms are given in Chap. 2.

5.1 Applications of Hybrid Optimization Algorithms

All hybrid optimization algorithms are compared based on a common experimental platform. For the comparison following performance criteria evaluations are considered.

Table 5.5 Comparison of results (BEST solution) for the constrained benchmark functions by using different hybrid optimization techniques

BM-C	BEST			
	HPABC	HBABC	HDABC	HGABC
G01	−11	**−15**	**−15**	−14.962117
G02	−0.38303	−0.748981	−0.743072	**−0.769406**
G03	−0.999933	−1.000116	**−1.00041**	−0.908716
G04	**−30665.53867**	**−30665.53867**	**−30665.53867**	−30661.97624
G05	5127.051873	5127.896622	**5126.496716**	15831988.76
G06	−6961.813874	**−6961.813876**	**−6961.813876**	−6671.485071
G07	24.83686	24.725735	**24.323636**	27.669474
G08	**−0.095825**	**−0.095825**	**−0.095825**	**−0.095825**
G09	680.638643	680.630701	**680.630332**	683.085962
G10	7304.510202	**7106.484459**	7143.890531	7651.225802
G11	0.749909	**0.7499**	**0.7499**	0.751031
G12	**−1**	**−1**	**−1**	**−1**
G13	**0.383755**	0.55963	0.901652	0.991843
G14	61.987375	−46.188182	**−46.968524**	156.039946
G15	961.783308	961.987688	961.764842	**961.716877**
G16	**−1.888024**	**−1.888024**	**−1.888024**	−1.861636
G17	8938.327929	**8928.478227**	8940.407178	3699802.099
G18	**−0.866017**	−0.863618	−0.865976	−0.811193
G19	48.350446	**33.187255**	34.62052	38.507394
G20	41433091.79	13312926.08	14466371.07	**1000349.33**
G21	91442.99424	193.754766	**193.754224**	163816.2771
G22	5.5176E + 18	2.07822E + 14	**139385275.4**	2.31661E + 13
G23	1166.228548	−19.503574	**−321.485457**	−0.001495
G24	**−5.508013**	**−5.508013**	**−5.508013**	−5.497502

- Population size = 50
- Number of generations = 500 (for unconstrained and constrained benchmark functions), 200 (mechanical design problems)
- For HPABC: $w = 0.7$, $c_1 = c_2 = 0.8$, $V_{\max} = 4$, number of employed bees = population size.
- For HBABC: Immigration rate = emigration rate = 1, mutation factor = 0.01, number of employed bees = population size.
- For HDABC: $F = 0.5$, $C = 0.5$, number of employed bees = population size.
- For HGABC: crossover probability = 0.9, mutation probability = 0.01, crossover = single point crossover, selection = roulette wheel selection, number of employed bees = population size.
- Number of runs = 25

Comparison is based on the best and the mean solution achieved in 25 runs. Tables 5.3 and 5.4 show the comparison of results for the unconstrained benchmark functions, Tables 5.5 and 5.6 show the comparison of results for the constrained benchmark functions and Tables 5.7 and 5.8 show the comparison of results for the mechanical element design optimization problems.

Table 5.6 Comparison of results (MEAN solution) for the constrained benchmark functions by using different hybrid optimization techniques

BM-C	MEAN			
	HPABC	HBABC	HDABC	HGABC
G01	−7.6	**−15**	**−15**	−14.8719566
G02	−0.3094417	−0.6780855	−0.6438653	**−0.7501423**
G03	49.2811631	−0.9950221	**−0.996886**	−0.3893368
G04	−30665.53861	−30665.53861	**−30665.53867**	−30602.3601
G05	419969.4498	**5250.854236**	5281.512188	29404976.11
G06	193038.186	−6961.781506	**−6961.813876**	−5385.982478
G07	48.4202269	26.0059673	24.6143301	31.6084807
G08	**−0.095825**	−0.0891569	**−0.095825**	−0.0958187
G09	680.6706487	680.6867908	**680.6344396**	685.9619782
G10	8262.829776	7943.977794	**7237.877898**	8042.291921
G11	0.8834262	**0.7499**	**0.7499**	0.8642834
G12	**−1**	**−1**	**−1**	**−1**
G13	81.3958922	**0.9736198**	0.9954718	104.7640348
G14	141.1351492	−24.3580828	**−45.529184**	157.2434362
G15	1763.425164	964.1383936	**963.604144**	966.3025414
G16	**−1.888024**	−1.8619364	**−1.888024**	−1.6956546
G17	1567982404	9040.261009	**8994.320904**	32060333.84
G18	−0.7473602	1818.180163	**−0.788882**	−0.6725528
G19	149.1784138	41.006861	**36.0241206**	43.3501844
G20	33808003464	15626925.09	15604945.9	**5793202.773**
G21	765071.683	9074.032606	**4252.105379**	208155.2447
G22	3.40929E + 19	3.53243E + 16	9.18624E + 16	**1.38082E + 16**
G23	19633.68237	4847.183274	**841.4603966**	2481.62147
G24	14.4919698	**−5.508013**	**−5.508013**	−5.4785796

It is observed from the results that HDABC has outperformed all the hybrid algorithms for the unconstrained benchmark functions by finding better solutions for the eight functions. HBABC and HGABC have also shown better performance for the best solution but inferior to HDABC. HPABC has performed poorly for the unconstrained benchmark functions. For the mean solution, HDABC and HGABC have shown better performance. HPABC and HBABC have performed poorly to find the mean solutions. All the hybrid algorithms have found global solution for the Quartic function. HBABC and HGABC have found global solution for the step function and moreover, HBABC have shown near-optimal solution for Schwefel 2.26 function. HDABC has found global solution for Step and Penalty-1 functions and near-optimal solution for Schwefel 2.22, Griewank and Penalty-2 functions. Moreover, HDABC and HGABC are consistent in finding the global solution which is reflected by their mean solution and HPABC and HBABC are consistent in finding near-optimal solution for the Step function. HDABC is also consistent in finding near-optimal solution for the Step, Schwefel 2.22 and Griewank functions.

For the constrained benchmark functions also HDABC has outperformed other algorithms by showing better performance for the sixteen functions and eighteen

Table 5.7 Comparison of results (BEST solution) for the mechanical design problems by using different hybrid optimization techniques

MD		BEST			
		HPABC	HBABC	HDABC	HGABC
Gear	Example 1A	3135.515965	3135.5311	**3135.5159**	3136.2004
	Example 1B	**3142.712756**	**3142.712756**	**3142.712756**	**3142.712756**
	Example 1C	2993.665618	2993.6658	**2993.6656**	2993.6659
	Example 1D	**2994.844118**	**2994.844118**	**2994.844118**	**2994.844118**
Bearing	Example 2A	**−6032.315099**	**−6032.315099**	**−6032.315099**	−6017.834159
	Example 2B	**−3792.420036**	**−3792.42**	**−3792.42**	−3780.0856
	Example 2C	**−0.223974**	**−0.223974**	**−0.223974**	**−0.223974**
Belleville spring	Example 3	1.990962	1.979731	**1.979675**	1.981879
Multiplate clutch	Example 4	**0.313657**	**0.313657**	**0.313657**	**0.313657**
Robot gripper	Example 5	4.203501	4.416086	6.285471	**4.20271**
Hydrostatic bearing	Example 6	1848.392827	1640.82959	**1626.553483**	2045.909142
4-stage Gear train	Example 7	37875156133	**46623.2053**	59763.56295	55494.49422
Pressure vessel	Example 8	**6059.714335**	**6059.714335**	**6059.714335**	**6059.714335**
Welded beam	Example 9	**1.724852**	**1.724852**	**1.724852**	1.770658
Spring	Example 10	**0.012665**	**0.012665**	**0.012665**	0.012671
Speed reducer	Example 11	**2996.348165**	**2996.348165**	**2996.348165**	2997.665018
Stiffened shell	Example 12A	54444.6307	54446.19844	**54444.47963**	54868.84219
	Example 12B	55342.91995	**55326.2934**	**55326.29341**	55724.83112
Step cone pulley	Example 13	31.953962	16.642827	**16.634505**	19.411676
Screw jack	Example 14	**419187.0574**	**419187.0574**	**419187.0574**	**419187.0574**
C-clamp	Example 15	5623.630726	5616.151813	**5616.151812**	5673.954031
Hydrodynamic bearing	Example 16	16.205865	**16.205833**	**16.205833**	16.205956
Cone clutch	Example 17	**68.877551**	**68.877551**	**68.877551**	68.880307
Cantilever support	Example 18	**49.062256**	**49.062256**	**49.062256**	49.068421
Hydraulic cylinder	Example 19	0.024245	**0.024124**	**0.024124**	0.024213
Planetary gear box	Example 20	**0.526667**	0.52735	0.527814	0.53

Table 5.8 Comparison of results (MEAN solution) for the mechanical design problems by using different hybrid optimization techniques

MD		MEAN			
		HPABC	HBABC	HDABC	HGABC
Gear	Example 1A	3135.682894	3135.807775	**3135.5159**	3136.851
	Example 1B	3148.741874	**3142.7128**	**3142.7128**	**3142.713**
	Example 1C	2996.538989	2993.737825	**2993.6656**	2994.202
	Example 1D	**2994.844118**	**2994.8441**	**2994.8441**	**2994.844**
Bearing	Example 2A	−5968.929762	−6032.315091	**−6032.315099**	−5956.52
	Example 2B	−3788.970347	−3787.47894	**−3792.42**	−3768.12
	Example 2C	**−0.223974**	**−0.22397**	**−0.22397**	−0.22367
Belleville spring	Example 3	2.0478585	1.9819249	**1.9796762**	2.005818
Multiplate clutch	Example 4	**0.313657**	0.3144411	**0.313657**	0.316291
Robot gripper	Example 5	30839.45805	**4.8081054**	8.774491714	5.261168
Hydrostatic bearing	Example 6	2624.378662	1761.923195	**1724.559019**	2625.568
4-stage Gear train	Example 7	2.58148E + 11	47023887602	443507531.4	**1.53E + 08**
Pressure vessel	Example 8	**6059.714335**	**6059.714335**	**6059.714335**	**6059.714**
Welded beam	Example 9	1.7248593	1.7304695	**1.724852**	1.800531
Spring	Example 10	0.0126817	0.0126712	**0.0126651**	0.012681
Speed reducer	Example 11	3012.992535	2996.348532	**2996.348165**	3000.367
Stiffened shell	Example 12A	55634.7773	55072.73776	**54639.84478**	56135.99
	Example 12B	56245.27169	55921.85604	**55852.80969**	56920.4
Step cone pulley	Example 13	58.8119921	56.3661545	**24.2746931**	1027.797
Screw jack	Example 14	419335.259	**419187.0574**	**419187.0574**	**419187.1**
C-clamp	Example 15	5706.239857	5645.067208	**5616.151816**	5793.687
Hydrodynamic bearing	Example 16	16.2059317	**16.205833**	**16.205833**	16.20674
Cone clutch	Example 17	**68.877551**	**68.877551**	**68.877551**	68.90157
Cantilever support	Example 18	**49.062256**	**49.062256**	**49.062256**	49.11617
Hydraulic cylinder	Example 19	0.024293	0.0241579	**0.024124**	0.024282
Planetary gear box	Example 20	0.534544	**0.52908**	0.544793	0.536693

functions in finding the best and the mean solutions. HBABC has shown better performance than HPABC and HGABC in finding best solutions. HDABC has found global solutions for G01, G04, G05, G06, G08, G11, G12 and G24, which is better in comparison of other hybrid algorithms. Also HDABC has shown consistent performance in finding the global solution for G01, G04, G06, G08, G11, G12 and G24.

For the mechanical design problems also HDABC has shown better performance than other hybrid algorithms by showing better performance for 23 examples in finding best and the mean solutions. HBABC has also outperformed HPABC and HGABC in finding the best solutions and the mean solutions. Most of the solutions found by HDABC are either the global solutions or the near-global solutions.

It is observed that HDABC has shown better performance for 8, 16 and 23 unconstrained benchmark functions, constrained benchmark functions and mechanical element design problems, respectively for the best solution. While, HPABC, HBABC and HGABC has shown better performance for 1, 7 and 14, 5, 11 and 17 and 3, 5 and 7 unconstrained benchmark functions, constrained benchmark functions and mechanical element design problems, respectively for the best solution. So, overall performance of HDABC is approximately 2.1, 1.4 and 3.1 times better than HPABC, HBABC and HGABC, respectively in finding the best solutions. Similarly, performance of HDABC is approximately 5.8, 2.7 and 2.9 times better than HPABC, HBABC and HGABC, respectively for finding the mean solution.

Further comparison of the overall performance of hybrid algorithms with the basic and modified algorithms is made and it is observed that hybridization of ABC and PSO is effective than the basic PSO and its variants. For searching the best solutions, hybridization of ABC with PSO and GA is not so effective than the basic ABC. Hybridization of ABC with BBO and DE is effective than basic ABC in finding the best solutions. Basic ABC is better than HBABC. Moreover, Hybridization of ABC with DE is very effective than the basic ABC and other algorithms.

The next chapter presents the development of a new optimization algorithm, Teaching–Learning-Based optimization (TLBO) and it applications to unconstrained benchmark functions, constrained benchmark functions and mechanical element design optimization problems.

Chapter 6
Development and Applications of a New Optimization Algorithm

The main limitation of all the algorithms mentioned in previous chapters is that different parameters are required for proper working of these algorithms. Proper selection of the parameters is essential for the searching of the optimum solution by these algorithms. A change in the algorithm parameters changes the effectiveness of the algorithm. Most commonly used evolutionary optimization technique is genetic algorithm (GA). However, GA provides a near optimal solution for a complex problem having large number of variables and constraints. This is mainly due to the difficulty in determining the optimum controlling parameters such as crossover rate and mutation rate. The same is the case with PSO, which uses inertia weight, social and cognitive parameters. Similarly, ABC requires optimum controlling parameters of number of bees (employed, scout and onlookers), limit, etc. HS requires harmony memory consideration rate, pitch adjusting rate and the number of improvisations. Sometimes, the difficulty for the selection of parameters increases with modifications and hybridization. Therefore, the efforts must be continued to develop an optimization technique which is free from the algorithm parameters, i.e. no algorithm parameters are required for the working of the algorithm. This aspect is considered in the present work. An optimization method, Teaching–Learning-Based Optimization (TLBO), is proposed in this book to obtain global solutions for continuous nonlinear functions with less computational effort and high consistency. The TLBO method works on the philosophy of teaching and learning. The TLBO method is based on the effect of the influence of a teacher on the output of learners in a class. Here, output is considered in terms of results or grades. The teacher is generally considered as a highly learned person who shares his or her knowledge with the learners. The quality of a teacher affects the outcome of learners. It is obvious that a good teacher trains learners such that they can have better results in terms of their marks or grades. Moreover, learners also learn from interaction between themselves, which also helps in their results.

R. V. Rao and V. J. Savsani, *Mechanical Design Optimization Using Advanced Optimization Techniques*, Springer Series in Advanced Manufacturing,
DOI: 10.1007/978-1-4471-2748-2_6, © Springer-Verlag London 2012

6.1 Teaching–Learning-Based Optimization

Teaching–learning process is the heart of education. The fulfillment of the aims and objectives of the education depends on Teaching–learning process. Based on the above fact of teaching–learning process, mathematical model is prepared and it is implemented for the optimization process. Assume two different teachers, T_1 and T_2, teaching a subject with same content to the same merit level learners in two different classes. The distribution of marks obtained by the learners of two different classes evaluated by the teachers follows some distribution depending on the group of learners. A Normal distribution is assumed for the obtained marks. Normal distribution is defined as,

$$f(X) = \frac{1}{\sigma\sqrt{2\pi}} e^{\frac{-(x-\mu)^2}{2\sigma^2}} \tag{6.1}$$

where σ^2 is the variance, μ is the mean and x is any value of which normal distribution function is required.

Like other nature-inspired algorithms, TLBO is also a population based method which uses a population of solutions to proceed to the global solution. For TLBO population is considered as a group of learners or a class of learners. In optimization algorithms population consists of different design variables. In TLBO different design variables will be analogous to different subjects offered to learners and the learners' result is analogous to the 'fitness' as in other population-based optimization techniques. The teacher is considered as the best solution obtained so far.

The process of working of TLBO is divided into two parts. The first part consists of 'Teacher Phase' and the second part consists of 'Learner Phase'. The 'Teacher Phase' means learning from the teacher and the 'Learner Phase' means learning due through the interaction between learners.

6.1.1 Teacher Phase

It is the first part of the algorithm where learners learn through the teacher. During this phase a teacher tries to increase the mean result of the class in the subject taught by him or her depending on his or her capability. At any iteration i, assume that there are m number of subjects (i.e. design variables), n number of learners (i.e. population size, $k = 1, 2, \ldots, n$) and $M_{j,i}$ be the mean result of the learners in a particular subject j ($j = 1, 2, \ldots, m$). The best overall result $X_{\text{total-kbest},i}$ considering all the subjects together obtained in the entire population of learners can be considered as the result of best learner kbest. However, as the teacher is usually considered as a highly learned person who trains learners so that they can have better results, the best learner identified is considered by the algorithm as the teacher. The difference between the existing mean result of each subject and the corresponding result of the teacher for each subject is given by Eq. 6.2 as,

$$\text{Difference\_Mean}_{j,k,i} = r_i\left(X_{j,\text{kbest},i} - T_F M_{j,i}\right) \tag{6.2}$$

where, $X_{j,\text{kbest},i}$ is the result of the best learner (i.e. teacher) in subject j. T_F is the teaching factor which decides the value of mean to be changed and r_i is the random number in the range [0, 1]. Value of T_F can be either 1 or 2. The value of T_F is decided randomly with equal probability as,

$$T_F = \text{round}[1 + \text{rand}(0, 1)\{2 - 1\}] \tag{6.3}$$

T_F is not a parameter of the TLBO algorithm. The value of T_F is not given as an input to the algorithm and its value is randomly decided by the algorithm using Eq. 6.2. After conducting a number of experiments on many benchmark functions it is concluded that the algorithm performs better if the value of T_F is between 1 and 2. However, the algorithm is found to perform much better if the value of TF is either 1 or 2 and hence to simplify the algorithm, the teaching factor is suggested to take either 1 or 2 depending on the rounding up criteria given by Eq. 6.3.

Based on the Difference_Mean$_{j,k,i}$, the existing solution is updated in the teacher phase according to the following expression.

$$X'_{j,k,i} = X_{j,k,i} + \text{Difference\_Mean}_{j,k,i} \tag{6.4}$$

where $X'_{j,k,i}$ is the updated value of $X_{j,k,i}$. Accept $X'_{j,k,i}$ if it gives better function value. All the accepted function values at the end of the teacher phase are maintained and these values become the input to the learner phase.

6.1.2 Learner Phase

It is the second part of the algorithm where learners increase their knowledge by interaction among themselves. A learner interacts randomly with other learners for enhancing his or her knowledge. A learner learns new things if the other learner has more knowledge than him or her. Considering a population size of n, the learning phenomenon of this phase is expressed below.

Randomly select two learners P and Q such that $X'_{\text{total-}P,i} \neq X'_{\text{total-}Q,i}$ (where, $X'_{\text{total-}P,i}$ and $X'_{\text{total-}Q,i}$ are the updated values of $X'X_{\text{total-}P,i}$ and $X_{\text{total-}Q,i}$ respectively at the end of teacher phase)

$$X''_{j,P,i} = X'_{j,P,i} + r_i\left(X'_{j,P,i} - X'_{j,Q,i}\right), \text{ If } X'_{\text{total}-P,i} < X'_{\text{total}-Q,i} \tag{6.5a}$$

$$X''_{j,P,i} = X'_{j,P,i} + r_i\left(X'_{j,Q,i} - X'_{j,P,i}\right), \text{ If } X'_{\text{total}-Q,I} < X'_{\text{total}-P,i} \tag{6.5b}$$

Accept $X''_{j,P,i}$ if it gives a better function value. All the accepted function values at the end of the learner phase are maintained and these values become the input to the teacher phase of the next iteration. The values of r_i used in Eqs. 6.2 and 6.5a, 6.5b can be different. The flowchart of TLBO algorithm is given in Fig. 6.1.

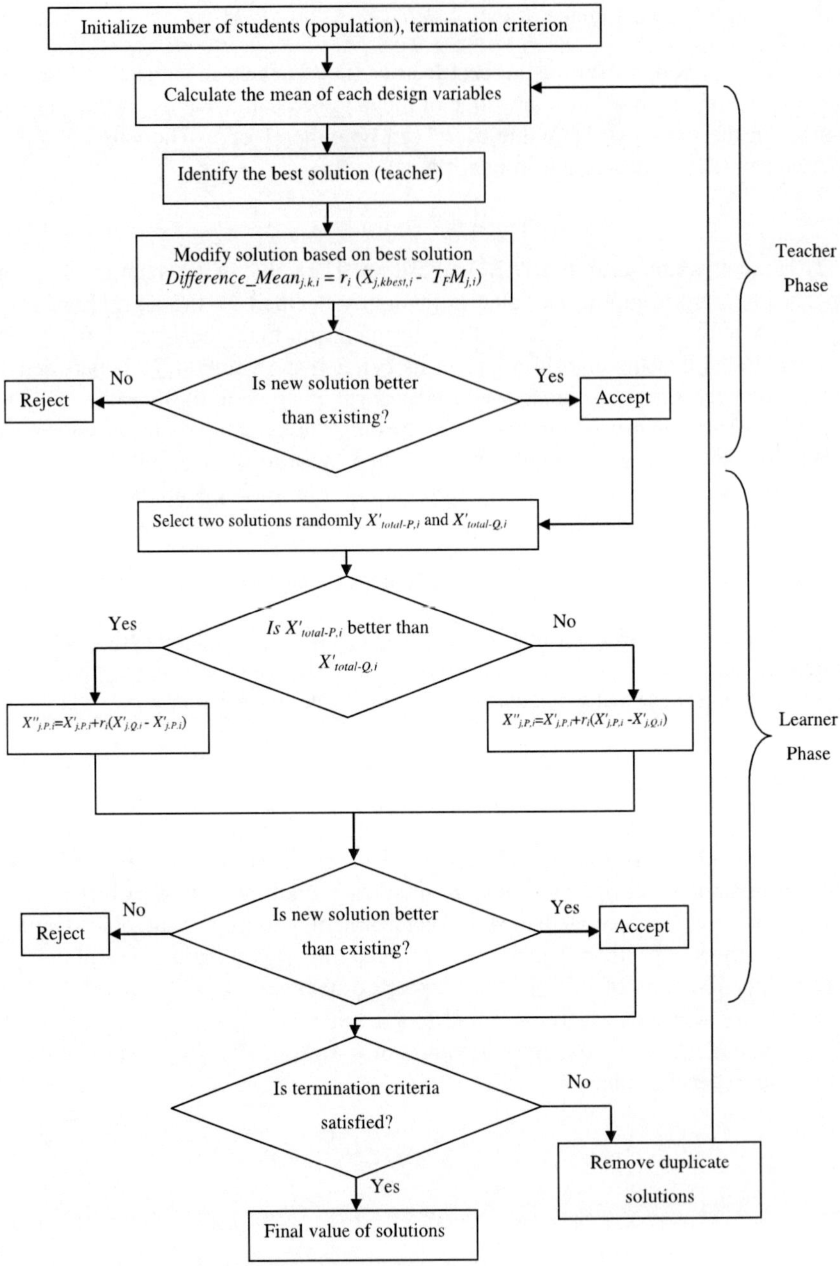

Fig. 6.1 Flowchart of TLBO algorithm

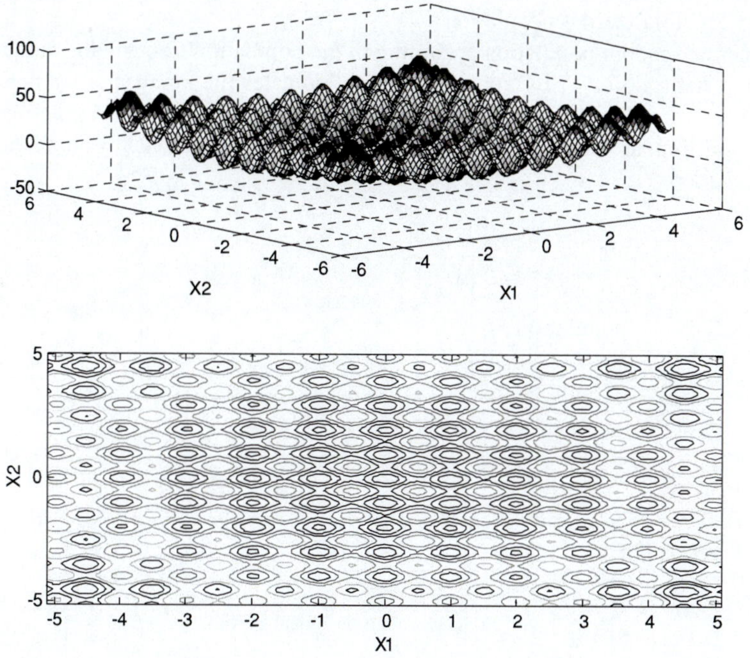

Fig. 6.2 Three-dimensional and contour plot for the Rastrigin function

6.2 Demonstration of TLBO for Optimization

Step-wise procedure for the demonstration of TLBO is given in this section. For demonstration Rastrigin function $\left[f(x) = \sum_{i=1}^{n} \left[x_i^2 - 10\cos(2\pi x_i) + 10\right]\right]$ is considered. Rastrigin is the multimodal, separable and regular function. Three-dimensional plot and the contour plot for the Rastrigin function is shown in Fig. 6.2.

The procedure is demonstrated as follows:

Step 1: Define the optimization problem and initialize the optimization parameters

Initialize population size = 10
Number of generations = 20
Number of design variables $(D) = 2$
Limits of design variables $(L_{L,i} \leq x_{,i} \leq U_{L,I}) = -5.12 \leq x_{,1,2} \leq 5.12$
Define optimization problem as:

$$= \text{Minimize } f(x) = \sum_{i=1}^{n} \left[x_i^2 - 10\cos(2\pi x_i) + 10\right]$$

Step 2: Initialize the population

Generate random population according to the population size and the number of design variables. For TLBO, population size indicates the number of learners and the design variables indicate the subjects (i.e. courses) offered. Short the population according to their respective objective function value. This population is expressed as

$$\text{population} = \begin{bmatrix} x_{1,1} & x_{1,2} & \cdots\cdots & x_{1,D} \\ x_{2,1} & x_{2,2} & \cdots\cdots & x_{2,D} \\ \vdots & \vdots & & \vdots \\ x_{P_n,1} & x_{P_n,2} & \cdots\cdots & x_{P_n,D} \end{bmatrix}$$

$$= \begin{bmatrix} 0.7654 & -1.1465 \\ -1.1075 & 2.7974 \\ -1.0483 & 0.6283 \\ -0.9548 & 2.4612 \\ -1.9326 & 3.7264 \\ -4.8123 & -1.0392 \\ -0.3295 & 3.2264 \\ -4.9845 & -1.5195 \\ 4.7380 & -0.4940 \\ -4.6556 & -2.3371 \end{bmatrix} \text{ and the corresponding objective function value } = \begin{bmatrix} 14.8839 \\ 18.3136 \\ 18.8732 \\ 27.0735 \\ 29.9786 \\ 30.7256 \\ 33.8296 \\ 47.1267 \\ 53.4364 \\ 57.9284 \end{bmatrix}$$

Step 3: Teacher phase

Calculate the mean of the population column wise, which will give the mean for the particular subject as,

$$\begin{aligned} M_{,D} &= [m_1, m_2, \ldots m_D] \\ &= [-1.4322,\ 0.6303] \end{aligned} \tag{6.6}$$

The best solution will act as a teacher for that iteration

$$\begin{aligned} X_{\text{teacher}} &= X_{f(X)=\min} \\ &= [0.7654 - 1.1465] \end{aligned} \tag{6.7}$$

The teacher will try to shift the mean and the difference is expressed by Eq. 6.2. The value of T_F is randomly selected as 1 or 2. The obtained difference is added to the current solution to update its values using Eq. 6.4

$$= \begin{bmatrix} 3.4434 & -2.5854 \\ 1.5705 & 1.3585 \\ 1.6297 & -0.8106 \\ 1.7232 & 1.0223 \\ 0.7454 & 2.2876 \\ -2.1343 & -2.4781 \\ 2.3485 & 1.7875 \\ -2.3065 & -2.9583 \\ 5.1200 & -1.9329 \\ -1.9776 & -3.7760 \end{bmatrix} \text{ and the corresponding objective function value } = \begin{bmatrix} 56.5089 \\ 39.6501 \\ 26.4588 \\ 15.7869 \\ 28.4185 \\ 33.9543 \\ 32.1771 \\ 27.8865 \\ 33.5362 \\ 26.6438 \end{bmatrix}$$

Accept $X'_{j,k,i}$ if it gives better function value.

$$X_{\text{new}} = \begin{bmatrix} 0.7654 & -1.1465 \\ -1.1075 & 2.7974 \\ -1.0483 & 0.6283 \\ 1.7232 & 1.0223 \\ 0.7454 & 2.2876 \\ -4.8123 & -1.0392 \\ 2.3485 & 1.7875 \\ -2.3065 & -2.9583 \\ 5.1200 & -1.9329 \\ -1.9776 & -3.7760 \end{bmatrix} \text{ and the corresponding objective function value } = \begin{bmatrix} 14.8839 \\ 18.3136 \\ 18.8732 \\ 15.7869 \\ 28.4185 \\ 30.7256 \\ 32.1771 \\ 27.8865 \\ 33.5362 \\ 26.6438 \end{bmatrix}$$

Step 4: Learner phase

As explained above, learners increase their knowledge with the help of their mutual interaction. The mathematical expression is explained under Sect. 6.1.2. Obtain X_{new} after the student phase using Eq. 6.5a or 6.5b.

$$X_{\text{new}} = \begin{bmatrix} 0.7654 & -1.1465 \\ -1.1479 & -1.1465 \\ 1.7232 & 1.0223 \\ -1.1075 & 2.7974 \\ 0.6497 & 2.7974 \\ -1.0483 & 0.6283 \\ 0.4677 & 1.0451 \\ -1.9397 & -2.7420 \\ -1.6623 & 0.8464 \\ -1.9776 & -3.7760 \end{bmatrix} \text{ and the corresponding objective function value } = \begin{bmatrix} 14.8839 \\ 14.8839 \\ 15.7869 \\ 18.3136 \\ 18.3136 \\ 18.8732 \\ 21.5048 \\ 22.4935 \\ 23.0245 \\ 26.6438 \end{bmatrix}$$

Step 5: Termination criterion

Stop if the maximum generation number is achieved; otherwise repeat from step 3.

Detailed variation of the design variables and the objective function are given in Table 6.1 for two generations. It is observed from the Table 6.1 that the average ($F(X)$average) and the best value (given in bold) of the objective function decreases as the algorithm precedes from teacher phase to the student phase in the same generation and it also decreases with the generations. So it can be concluded that the algorithm guarantees convergence. Moreover, the visualization for the convergence is presented in Fig. 6.3 for Generation-1, 2, 3, 5, 10 and 20.

6.3 Comparison of TLBO with Other Optimization Techniques

Like GA, PSO, ABC, DE, BBO, etc., TLBO is also a population-based technique which implements a group of solutions to proceed for the optimum solution. Many optimization methods require algorithm parameters that affect the performance of the algorithm. GA requires crossover probability, mutation rate and selection method; PSO requires learning factors, variation of weight and maximum value of velocity; ABC requires number of employed bees, onlooker bees and value of limit; HS requires harmony memory consideration rate, pitch adjusting rate and number of improvisations: SFLA requires number of memeplexes, iteration per memeplexes: ACO requires exponent parameters, pheromone evaporation rate and reward factor. Unlike other optimization techniques TLBO does not require any algorithm parameters to be tuned, thus making the implementation of TLBO simpler and easy. As in PSO, TLBO uses the best solution of the iteration to change the existing solution in the population thereby increasing the convergence rate. TLBO does not divide the population like ABC and SFLA. Like GA which uses selection, crossover and mutation phase and ABC which uses employed, onlooker and scout bees phase, TLBO uses two different phases, 'teacher phase' and 'learner phase'. TLBO uses the mean value of the population to update the solution. TLBO implements greediness to accept the good solution like ABC.

The strength of TLBO is that it does not require any algorithm-specific parameter setting for the working of the algorithm. Future research will consist of checking TLBO for real life problems and on more challenging benchmark problems.

6.4 Implementation of TLBO for the Optimization of Unconstrained Problems

In the field of optimization it is a common practice to compare different algorithms by using different benchmark problems. In this book also different benchmark problems are considered having different characteristics such as separability, multimodality and regularity. A function is multimodal if it has two or more local optima. A function is separable if it can be written as a sum of functions of variable separately. A function is regular if it is differentiable at each point of their domain. Non-separable functions are more difficult to optimize and difficulty increases if the function is multi-model. Complexity increases when the local optima are randomly distributed. Moreover, complexity increases with the increase in

Table 6.1 Variation of design variables and objective function (Rastrigin) for the first two generations

				Teacher phase			Student phase		
	X		$f(X)$	X		$f(X)$	X		$f(X)$
Iteration-1 $TF = 2$	0.7654	−1.1465	14.8839	0.7654	−1.1465	14.8839	0.7654	−1.1465	14.8839
	−1.1075	2.7974	18.3136	−1.1075	2.7974	18.3136	−1.1479	−1.1465	14.8839
	−1.0483	0.6283	18.8732	−1.0483	0.6283	18.8732	1.7232	1.0223	15.7869
	−0.9548	2.4612	27.0735	1.7232	1.0223	15.7869	−1.1075	2.7974	18.3136
	−1.9326	3.7264	29.9786	0.7454	2.2876	28.4185	0.6497	2.7974	18.3136
	−4.8123	−1.0392	30.7256	−4.8123	−1.0392	30.7256	−1.0483	0.6283	18.8732
	−0.3295	3.2264	33.8296	2.3485	1.7875	32.1771	0.4677	1.0451	21.5048
	−4.9845	−1.5195	47.1267	−2.3065	−2.9583	27.8865	−1.9397	−2.7420	22.4935
	4.7380	−0.4940	53.4364	5.1200	−1.9329	33.5362	−1.6623	0.8464	23.0245
	−4.6556	−2.3371	57.9284	−1.9776	−3.7760	26.6438	−1.9776	−3.7760	26.6438
Mean	−1.4322	0.6303							
$F_{average}$			33.21695			24.72453			19.47217
Iteration-2 $TF = 1$	0.7654	−1.1465	14.8839	0.7654	−1.1465	14.8839	0.0794	0.0270	1.3684
	−1.1479	−1.1465	14.8839	−0.0202	−1.7478	13.2735	1.9560	1.0181	5.3067
	1.7232	1.0223	15.7869	1.7232	1.0223	15.7869	0.0579	2.0991	6.9402
	−1.1075	2.7974	18.3136	0.0202	2.1961	11.5833	0.0395	2.1092	7.0185
	0.6497	2.7974	18.3136	0.6497	2.7974	18.3136	0.0202	2.1961	11.5833
	−1.0483	0.6283	18.8732	0.0794	0.0270	1.3684	−0.0202	−1.7478	13.2735
	0.4677	1.0451	21.5048	0.4677	1.0451	21.5048	0.6497	2.7974	18.3136
	−1.9397	−2.7420	22.4935	−1.9397	−2.7420	22.4935	−1.9397	−2.7420	22.4935
	−1.6623	0.8464	23.0245	−1.6623	0.8464	23.0245	−1.6623	0.8464	23.0245
	−1.9776	−3.7760	26.6438	−1.9776	−3.7760	26.6438	−1.9776	−3.7760	26.6438
Mean	−0.5277	0.0326							
$F_{average}$			19.47217			16.88762			13.5966

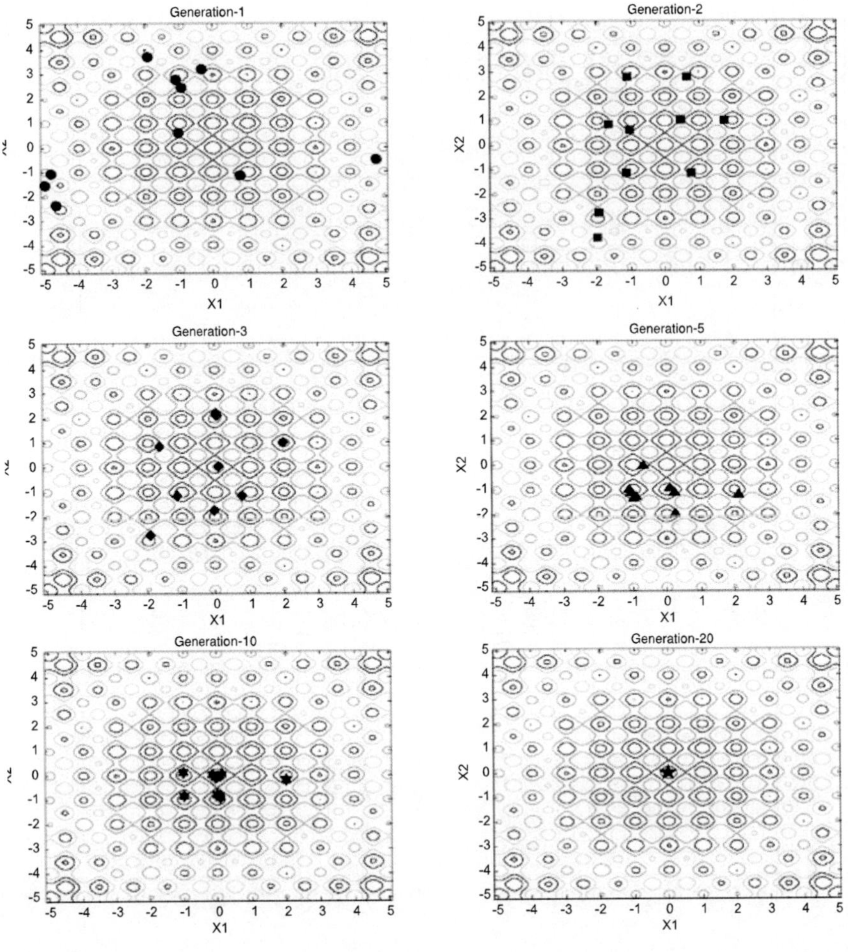

Fig. 6.3 Visualization of convergence of solutions for Rastigin function for different generations

dimensionality. To validate the proposed algorithm its results are compared with the results of different algorithms for different benchmark problems available in the literatures. Details of benchmark functions which are not discussed in previous chapters are given below.

1. **De Jong function**

$$\max f(X) = 3905.93 - 100(x_1^2 - x_2)^2 - (1 - x_1)^2 \tag{6.8}$$

Table 6.2 Details of benchmark functions considered for experiment 1

Sr. No.	Function	Interval	Global optimum
1	De Jong	[−2.048, 2.048]	$f(X) = 3{,}905.93$, $X = (1,1)$
2	Goldstein and Price	[−2, 2]	$f(X) = 3$, $X = (1,1)$
3	Martin and Gaddy	[0, 10]	$f(X) = 0$, $X = (5,5)$
4	Rosenbrock ($D = 1$)	(a)[−1.2, 1.2] (b) [−10, 10]	$f(X) = 0$, $X = (1,1)$
5	Rosenbrock ($D = 3$)	[−1.2, 1.2]	$f(X) = 0$, $X = (1,1,1,1)$
6	Hyper Sphere ($D = 6$)	[−5.12, 5.12]	$f(X) = 0$, $X = (0,0,0,0,0,0)$

2. **Goldstein and Price**

$$\min f(X) = (1 + (x_1 + x_2 + 1)^2(19 - 14x_1 + 3x_1^2 - 14x_2 + 6x_1x_2 + 3x_2^2)) \\ (30 + (2x_1 - 3x_2)^2(18 - 32x_1 + 12x_1^2 + 48x_2 - 36x_1x_2 + 27x_2^2)) \tag{6.9}$$

3. **Martin and Gaddy**

$$\min f(X) = (x_1 - x_2)^2 + [(x_1 + x_2 - 10)/3]^2 \tag{6.10}$$

4. **Powell badly scaled function**

$$\max f(X) = (10x_1x_2 - 1)^2 + [\exp(-x_1) + \exp(-x_2) - 1.0001]^2 \tag{6.11}$$

5. **B2 function**

$$\min f(X) = x_1^2 + 2x_2^2 - 0.3\cos(3\pi x_1) - 0.4\cos(4\pi x_2) + 0.7 \tag{6.12}$$

6. **Booth function**

$$\min f(X) = (x_1 + 2x_2 - 7)^2 + (2x_1 + x_2 - 5)^2 \tag{6.13}$$

6.4.1 Experiment 1

In this experiment five different benchmark problems presented by Ahrari and Atai [1] are considered and optimized by using the proposed method, TLBO. The results are compared with the results of well-known optimization techniques like GA, Ant colony system (ANTS), Bee Algorithm (BA) and Grenade Explosion Method (GEM). Description of the benchmark problems is given in Table 6.2.

Two different criteria are taken for the comparison of algorithms, viz. success percentage and mean function evaluations required. Success percentage indicates the consistency of the algorithm to find the results in different runs and mean function evaluations required indicate the computational effort of the algorithm. In this experiment, an algorithm is considered as successful if the difference between

Table 6.3 Comparison of the results for the success percentage and mean number of function evaluation for GA, ANTS, Bee colony, GEM and TLBO

	GA		ANTS		Bee colony		GEM		TLBO	
	A*	B*	A*	B*	A*	B*	A*	B*	A*	B*
1	100	10,160	100	6,000	100	868	100	746	100	676
2	100	5,662	100	5,330	100	999	100	701	100	649
3	100	2,488	100	1,688	100	526	100	258	100	243
4a	100	10,212	100	6,842	100	631	100	572	100	541
4b	–	–	100	7,505	100	2,306	100	2,289	100	1,082
5	–	–	100	8,471	100	28,529	100	82,188	100	2,563
6	100	15,468	100	22,050	100	7,113	100	423	100	308

*A** Success percentage, *B** Mean number of function evaluations

Table 6.4 Details of benchmark functions considered for experiment 2

Sr. No.	Function	Interval	Global optimum
1	Powell badly scaled function	[−50,50]	$f(X) = 0$, $X = (1.098e-5, 9.106)$
2	B2 function	[−50,50]	$f(X) = 0$, $X = (0,0)$
3	Booth function	[−50,50]	$f(X) = 0$, $X = (1,3)$
4	Griewank ($D = 10$)	[−50,50]	$f(X) = 0$, $X = (0,0\ldots.)$
5	Rastrigin ($D = 10$)	[−50,50]	$f(X) = 0$, $X = (0,0\ldots.)$
6	Sphere ($D = 30$)	[−50,50]	$f(X) = 0$, $X = (0,0,\ldots)$
7	Griewank ($D = 50$)	[−50,50]	$f(X) = 0$, $X = (0,0\ldots.)$

the best value obtained by the algorithm and global optimum value is less then 0.1% of global optimum or less than 0.001, whichever is smaller [1].

An algorithm is tested for 100 independent runs with the population size of 20 (except for Rosenbrock function with $D = 3$, for which population size is taken as 50) and maximum number of generations as 50. Average of the function evaluation required and the success percentage is presented in Table 6.3. Results of the other algorithms except TLBO are directly taken from [1].

It can further be observed that for all the considered benchmark functions TLBO requires less number of mean function evaluations with very high consistency of 100% success. The results for the functions 1–4a and 6 is better but nearer to the results of GEM given by Ahrari and Atai [1]. For functions 4b and 5 TLBO has shown much better results (twice better than those for function 4b and 40 times better than those for function 5) than the results given by Ahrari and Atai [1]. This experiment shows that the TLBO method is effective in terms of the computational effort and consistency.

6.4.2 Experiment 2

In this experiment six different benchmark functions are taken and its results are compared with those given by particle swarm optimization (PSO) and a hybrid

Table 6.5 Comparision of the results for the success percentage, mean number of function evaluation and error for PSO, NM-PSO and TLBO

	PSO			NM-PSO			TLBO		
	A*	B*	C*	A*	B*	C*	A*	B*	C*
1	94	20,242	9.89E − 06	100	2,971	3.78E − 06	100	2,867	4.0036E − 06
2	100	4,188	1.4E − 08	100	1,124	3.23E − 10	100	1,048	0
3	100	3,848	2.6E − 08	100	1,065	1.26E − 09	100	654	3.8293E − 12
4	0	(504,657)		82	14,076	1.04E − 11	100	1,059	0
5	30	510,050	1.08E − 04	60	12,353	1.91E − 11	100	1,134	0
6	0	(4,530,150)		100	87,004	2.76E − 11	100	1,543	0
7	0	(2,550,250)		82	378,354	9.96E − 12	100	1,857	0

*A** Success percentage, *B** Mean number of function evaluations, *C** Error

Nelder-Mead simplex search with PSO called NM-PSO [2]. Details of the problems taken are shown in Table 6.4.

Three different criteria are considered in this experiment viz, success percentage, mean function evaluations required and error. Error is the average difference between the obtained best solution and the global solution, which indicates the ability of the algorithm to reach the global optimum solution. The algorithm is considered successful if the difference between the obtained best solution and global optimum is less than 0.001 [2]. The mean function evaluation is obtained only for the successful runs.

For this experiment results were obtained for 100 independent runs with population size of 20 and maximum number of generation as 200. Results other than TLBO are taken from [2] and are presented in Table 6.5.

It is observed from the results that for Powell badly scaled function TLBO is better for mean number of function evaluations but error value is better for NM-PSO. TLBO requires approximately 1/10th function evaluation for Powell badly scaled function and B2 function, 1/50th function evaluation for the Sphere function and 1/200th function evaluations for the Griewank function with $D = 50$. Also success rate of TLBO is better than NM-PSO for the Griewank (with $D = 10$ and 30) function and the Rastrigin function. This experiment shows that TLBO is effective in terms of the computational effort, consistency and obtaining the optimum solution.

6.4.3 Experiment 3

In this experiment five different benchmark problems [3] are considered and comparison is made with the Harmony Search Algorithm (HS), Improved Bee Algorithm (IBA) and Artificial Bee Colony (ABC) optimization.

Comparison criteria are the mean solution and the standard solution for different independent runs. The mean solution describes the average ability of the algorithm to find the global solution and the standard deviation describes the variation in solution from the mean solution. In this experiment the algorithm runs

Table 6.6 Details of benchmark functions considered for experiment 3

Sr. No.	Function	Interval	Global optimum
1	Sphere	[−100,100]	$f(X) = 0, X = (0,0,\ldots)$
2	Rosenbrock	[−30,30]	$f(X) = 0, X = (1,1,\ldots)$
3	Rastrigin	[−5.12,5.12]	$f(X) = 0, X = (0,0\ldots.)$
4	Griewank	[−600,600]	$f(X) = 0, X = (0,0\ldots.)$
5	Ackley	[−32,32]	$f(X) = 0, X = (0,0\ldots.)$

for a specified maximum number of function evaluations. Moreover, the results are obtained for different independent runs and the values of the mean and the standard deviation are calculated for the results obtained in different runs. Description of the functions is given in Table 6.6.

In this experiment, dimensions (D) of the benchmark functions are taken as 5, 10, 30, 50 and 100 for all the considered problems. So experiment is performed for small-scale to large-scale problems. Maximum number of function evaluation was set as 50,000 by Karaboga and Akay [3] for HS, IBA and ABC. For TLBO maximum function evaluation is set as 2,000 (with population size of 10) for all the functions except Rosenbrock function for which its value is set as 50,000 (with population size of 50). Except Rosenbrock function, maximum number of function evaluations is 1/25th of the maximum number of function evaluation set for the HS, IBA and ABC algorithms. Table 6.7 shows the results of TLBO and the other considered algorithms. It is observed from the results that TLBO has outperformed all the algorithms except for Rosenbrock. For Rosenbrock TLBO is still better than HS and IBA. It is also further observed that as Dimension increases to 100 for Rosenbrock function TLBO gives better results than those given by the ABC algorithm. This experiment shows that TLBO is effective in finding the optimum solution with increase in dimensions.

6.4.4 Experiment 4

In this experiment eight different benchmark functions [4] are optimized and the results are compared with those given by PSO, DE and ABC algorithms. Detail for the algorithm is given in Table 6.8.

A comparison criterion for this experiment is the mean of results obtained for different runs. This experiment is conducted for very high dimension of 500 for all the considered functions. In this experiment results are obtained for 30 independent runs. Maximum function evaluation considered by Akay and Karaboga [4] equals to 100,000. For TLBO maximum function evaluation was set as 2,000 (1/50th of that given in [4], with population size of 10) except for Rosenbrock, Schwefel and Penalised for which maximum function evaluation is set as 100,000 (with population size of 10 for Schwefel and Penalised and 50 for Rosenbrock). Results are given in Table 6.9.

Table 6.7 Comparison of the results for mean and standard deviation for HS, IBA, ABC and TLBO

	D	HS Mean	SD	IBA Mean	SD	ABC Mean	SD	TLBO Mean	SD
Sphere	5	3.20E − 10	2.89E − 10	3.91E − 17	1.24E − 17	4.30E − 17	1.07E − 17	5.03E − 33	1.27E − 32
	10	6.45E − 08	3.07E − 08	4.95E − 17	2.30E − 17	7.36E − 17	4.43E − 17	2.26E − 29	3.61E − 29
	30	7.21E + 00	3.62E + 00	2.92E − 16	6.77E − 17	4.69E − 16	1.07E − 16	6.90E − 26	3.18E − 25
	50	5.46E + 02	9.27E + 01	5.39E − 16	1.07E − 16	1.19E − 15	4.68E − 16	8.71E − 26	1.86E − 25
	100	1.90E + 04	1.78E + 03	1.45E − 15	1.63E − 16	1.99E − 06	2.26E − 06	9.42E − 26	3.70E − 25
Rosenbrock	5	5.94E + 00	6.71E + 00	4.55E − 01	1.54E + 00	2.33E − 01	2.24E − 01	1.80E − 01	8.04E − 02
	10	6.52E + 00	8.16E + 00	1.10E + 01	2.55E + 01	4.62E − 01	5.44E − 01	5.58E + 00	6.18E − 01
	30	3.82E + 02	5.29E + 02	7.57E + 01	1.16E + 02	9.98E − 01	1.52E + 00	2.71E + 01	1.14E + 00
	50	2.47E + 04	1.02E + 04	6.30E + 02	1.20E + 03	4.33E + 00	5.48E + 00	4.78E + 01	1.01E + 00
	100	1.45E + 07	2.16E + 06	6.42E + 02	8.20E + 02	1.12E + 02	6.92E + 01	9.81E + 01	3.61E − 01
Ackley	5	2.68E − 05	1.24E − 05	6.35E − 10	9.77E − 11	9.64E − 17	5.24E − 17	0.00E + 00	0.00E + 00
	10	2.76E − 04	7.58E − 05	6.71E − 02	3.61E − 01	3.51E − 16	6.13E − 17	0.00E + 00	0.00E + 00
	30	9.43E − 01	5.63E − 01	1.75E + 00	9.32E − 01	3.86E − 15	3.16E − 15	7.11E − 16	1.82E − 15
	50	5.28E + 00	4.03E − 01	8.43E + 00	7.70E + 00	4.38E − 08	4.65E − 08	1.24E − 15	1.95E − 15
	100	1.32E + 01	4.90E − 01	1.89E + 01	8.50E − 01	1.32E − 02	1.30E − 02	2.13E − 15	1.19E − 15
Griewank	5	2.60E − 02	1.38E − 02	3.14E + 00	1.41E + 00	4.04E − 17	1.12E − 17	0.00E + 00	0.00E + 00
	10	0.00E + 00	3.02E − 02	1.04E + 00	1.13E + 00	6.96E − 17	4.06E − 17	0.00E + 00	0.00E + 00
	30	1.09E + 00	3.92E − 02	6.68E + 00	6.43E + 00	5.82E − 06	3.13E − 05	0.00E + 00	0.00E + 00
	50	5.81E + 00	9.16E − 01	1.34E + 02	2.41E + 01	5.72E − 01	9.22E − 01	0.00E + 00	0.00E + 00
	100	1.78E + 02	1.98E + 01	7.93E + 02	7.96E + 01	1.31E + 01	6.30E + 00	0.00E + 00	0.00E + 00
Rastrigin	5	6.07E − 08	5.52E − 08	4.58E + 00	2.31E + 00	4.34E − 17	1.10E − 17	0.00E + 00	0.00E + 00
	10	1.05E − 05	5.23E − 06	2.20E + 01	7.46E + 00	5.77E − 17	2.98E − 17	0.00E + 00	0.00E + 00
	30	7.40E − 01	7.00E − 01	1.28E + 02	2.49E + 01	4.80E − 05	2.43E − 04	0.00E + 00	0.00E + 00
	50	3.76E + 01	4.87E + 00	2.72E + 02	3.27E + 01	4.72E − 01	4.92E − 01	0.00E + 00	0.00E + 00
	100	3.15E + 02	2.33E + 01	6.49E + 02	4.52E + 01	1.46E + 01	4.18E + 00	0.00E + 00	0.00E + 00

Table 6.8 Details for benchmark functions considered for experiment 4

Name	Function
Sphere	$-100 \le x_i \le 100$
Rosenbrock	$-100 \le x_i \le 100$
Step	$-100 \le x_i \le 100$
Schwefel	$-500 \le x_i \le 500$
Rastrigin	$-5.12 \le x_i \le 5.12$
Ackley	$-32 \le x_i \le 32$
Griewank	$-600 \le x_i \le 600$
Penalty 1	$-50 \le x_i \le 50$

Table 6.9 Comparison of the results for mean solution for PSO, DE, ABC and TLBO

	PSO	DE	ABC	TLBO
Sphere	181.16	20.33	8.71E − 07	2.02073E − 52
Step	1,621	1,998.03	0	0
Schwefel	−98,168.1	−138,152.03	−190,906.66	−184,297.381
Rosenbrock	1.09E + 06	8.72E + 10	1,007.87	497.91
Rastrigin	1,033.04	594.69	87.96	0
Griewank	2.2	0.645	0	0
Ackley	3.69	13	0.058	0
Penalty-1	5.29	1.48E + 10	3.46E − 08	0.06292

It is observed from the results that with only 2,000 maximum function evaluations TLBO has outperformed all the algorithms for Sphere, Rastrigin and Ackley functions. For Step and Griewank functions, TLBO and ABC have shown same result but TLBO requires only 1/50th of the function evaluations than those required by the ABC algorithm. For same function evaluation of 100,000 TLBO has shown better result for Rosenbrock function than all the algorithms. However, TLBO has shown inferior result for Penalty 1 function compared to that given by ABC, but still the result of TLBO is better than those given by PSO and DE algorithms. For Schwefel results of TLBO and ABC are nearly same. This experiment shows that TLBO is effective at very high dimensions for functions having different characteristics like seperability, multimodality and regularity.

6.4.5 Experiment 5

This experiment is conducted to check the convergence rate of the TLBO and ABC algorithms. Comparison is done for TLBO and ABC. Six different benchmark functions are taken for the experiment viz. Sphere, Rosenbrock, Schwefel 2.26, Rastrigin, Ackley and Griewank with the dimension of 30. Maximum number of function evaluations is taken as 2,000 with population size of 10 and maximum number of generations of 100. The results of ABC are obtained by using the code given in the website dedicated to ABC (http://mf.erciyes.edu.tr/abc/). A

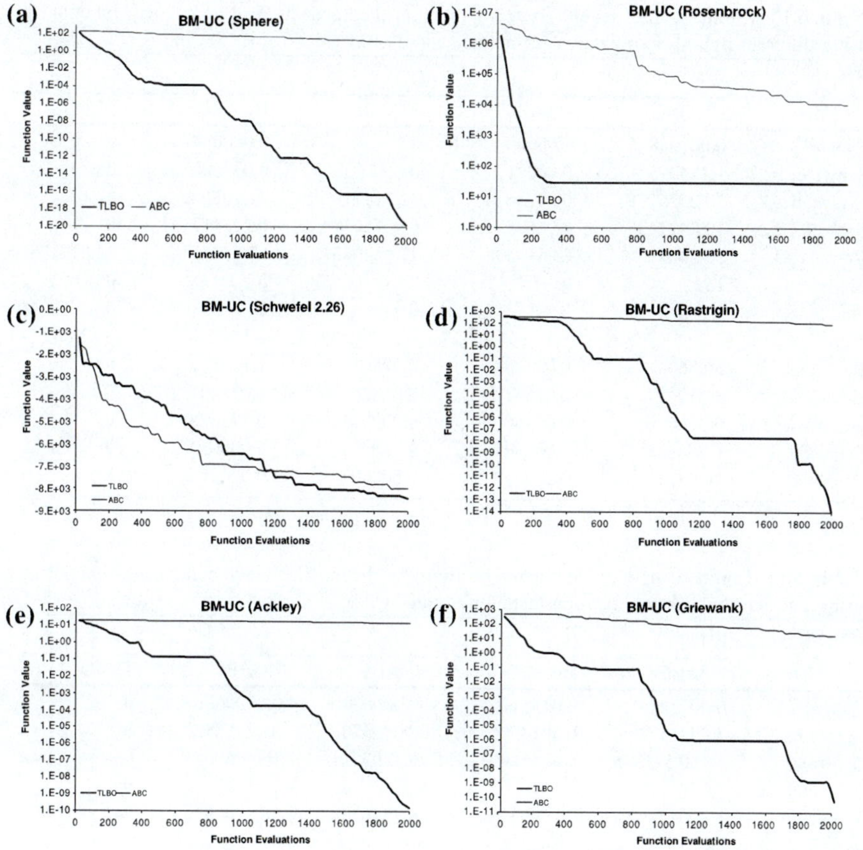

Fig. 6.4 Convergence curve for different benchmark problems for the comparison of TLBO and ABC algorithms. **a** Sphere. **b** Rosenbrock. **c** Schwefel 2.26. **d** Rastrigin. **e** Ackley and **f** Griewank

graph is plotted between the function value and the function evaluations. The function value considered is the average of function value for 10 different independent runs. Figure 6.4 shows the convergence graphs for different benchmark problems. It is clear from Fig. 6.4 that the convergence rate of TLBO is higher than ABC for the considered problems except for Schwefel 2.26 for which the convergence is nearly the same.

6.4.6 Experiment 6

In this experiment all the unconstrained benchmark functions considered in Chaps. 4 and 5 are tested by using TLBO. TLBO is implemented by using the population size of 50 and number of generations of 500. It is observed from the Chap. 5 that

Table 6.10 Comparison of results (best solution) for the unconstrained benchmark functions by using different hybrid optimization techniques and TLBO

BM-UC	Best				
	HPABC	HBABC	HDABC	HGABC	TLBO
Sphere	0.000088	0.000004	**0**	0.000044	**0**
Schwefel 2.22	1.184095	0.000217	0.000002	0.038606	**0**
Schwefel 1.2	503.525798	69.475646	32.331711	41.768063	**0**
Schwefel 2.21	10.042743	4.966752	6.927807	0.228172	**0**
Rosenbrock	32.546706	83.514598	**17.054065**	27.753121	24.012708
Step	152	**0**	2	**0**	**0**
Quartic	**0**	**0**	**0**	**0**	**0**
Schwefel 2.26	−7,787.748693	**−12,564.88713**	−12,332.58572	−12,557.72112	−11,869.841
Rastrigin	30.860414	0.013784	0.99801	1.109163	**0**
Ackley	3.519417	0.016447	0.018263	0.018495	**0**
Griewank	0.316696	0.044606	0.000013	0.393688	**0**
Penalty-1	1.350518	0.004374	**0**	0.000023	**0**
Penalty-2	20.142417	0.277422	0.000001	0.00028	**0**

Table 6.11 Comparison of results (mean solution) for the unconstrained benchmark functions by using different hybrid optimization techniques and TLBO

BM-UC	Mean				
	HPABC	HBABC	HDABC	HGABC	TLBO
Sphere	0.003524	0.0151116	0.0000004	0.0000666	**0**
Schwefel 2.22	4.0940878	0.0014506	0.0000476	0.0735882	**0**
Schwefel 1.2	1,100.577885	184.5669032	49.1297914	106.969643	**0**
Schwefel 2.21	21.9817796	9.5648362	8.5768576	0.4599282	**0**
Rosenbrock	104.0428604	94.0994142	69.4546108	28.590297	**24.6990298**
Step	244.4	10.4	6.2	**0**	**0**
Quartic	0.0000006	0.000149	**0**	**0**	**0**
Schwefel 2.26	−7,643.581618	−12,491.38836	−12,128.93538	**−12,554.96307**	−11,485.29748
Rastrigin	55.7848378	2.2034244	6.3645678	12.9073396	**0**
Ackley	6.812338	0.2793278	0.8349822	0.1317778	**0**
Griewank	0.6562612	0.0968346	0.034429	0.7140134	**0**
Penalty-1	5.271248	0.1031668	0.0000006	0.0000834	**0**
Penalty-2	38.094094	2.9497502	1.737886	0.0026552	**0**

hybridization of ABC with other optimization algorithms has shown better performance than the other algorithm. So in this book TLBO is compared with the results of hybrid algorithms. Comparison of results for the hybrid algorithms and TLBO for the best and the mean solutions are given in Tables 6.10 and 6.11. It is observed from the results that TLBO has shown better results for seven unconstrained benchmark problems and equivalent results for three unconstrained benchmark functions for finding the best solutions. TLBO has shown better mean results for ten benchmark functions. Moreover, TLBO has obtained global solutions for 11 unconstrained benchmark functions with 100% consistency which can be observed from the results of mean solutions.

6.5 Implementation of TLBO for the Optimization of Constrained Benchmark Functions

6.5.1 Experiment 7

In this experiment the performance of TLBO is compared with different existing optimization algorithms, modified optimization algorithms and hybrid optimization algorithms for the multimodal constrained problem. It is discussed in the earlier chapters that multimodal problems (problems having many local optima) are difficult to solve than unimodal problem. Difficulty further increases if the problem is the constrained optimization problem. Following problem is considered for the evaluation of the performance of TLBO.

Minimize:

$$f(X) = -\frac{\sin^3(2\pi x_1)\sin(2\pi x_2)}{x_1^3(x_1 + x_2)} \tag{6.14}$$

Subjected to:

$$g_1(X) = x_1^2 - x_2 + 1 \leq 0 \tag{6.15}$$

$$g_2(X) = 1 - x_1 + (x_2 - 4)^2 \leq 0 \tag{6.16}$$

where, $0 \leq x_1 \leq 10$ and $0 \leq x_2 \leq 10$

The problem is having two continuous design variables and two inequality constraints. The feasible region is only nearly 0.85% of the total solution space which can be considered as very less in comparison of the total solution space. The contour plot of the above problem is shown in the Fig. 6.5 in which the contours of objective function and constrains with their contour values are given. Constrains are having inequality sign having value less than or equal to zero, so the feasible direction is shown with the arrow pointing toward the feasible region. The optimum point is shown with a circle. The shaded portion shows the feasible region.

The above problem is solved by using TLBO by considering the population size as 10 and maximum generation as 50. The performance of TLBO is compared with GA, PSO, modified PSO (PSO_M_1, PSO_M_2), ABC, modified ABC (ABC_M), HPABC, HBABC, HDABC and HGABC. The population size and the number of generations are considered same as that of TLBO for all the other optimization techniques mentioned above. The performance is checked based on the best solutions, mean solutions and the convergence rate shown by the algorithms. The result for the best and the mean solutions are given in Table 6.12. It is observed from the results that optimum solution is obtained by PSO, PSO_M_1, PSO_M_2, ABC, ABC_M, HPABC, HBABC, HDABC and TLBO. The consistency of the algorithm is shown by the mean solution and for finding the mean solution ABC and TLBO has outperformed all the other considered optimization algorithms. GA, BBO and AGABC have performed poorly for the multimodal constrained

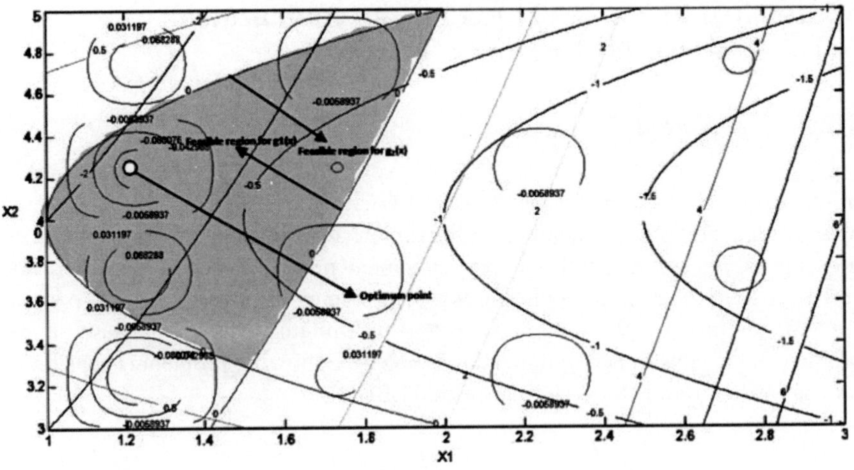

Fig. 6.5 Contour plot for the multimodal constrained optimization problem

Table 6.12 Comparison of the results for the multimodal constrained optimization problem

	Best	Mean
GA	−0.010353	−0.006479
PSO	−0.095825	−0.079155
BBO	−0.089556	−0.044378
DE	−0.095823	−0.087874
ABC	−0.095825	−0.095825
PSO_M_1	−0.095825	−0.085808
PSO_M_2	−0.095825	−0.077529
ABC_M	−0.095825	−0.095824
HPABC	−0.095825	−0.079155
HBABC	−0.095825	−0.079155
HDABC	−0.095825	−0.079155
HGABC	−0.075350	−0.037802
TLBO	−0.095825	−0.095825

optimization problems. The reason for performing poorly is that, multimodal problems have many local optima and due to this algorithm get stuck in those local optima. The convergence plots for all the considered optimization techniques can be obtained by averaging the results for all the runs. It is observed that the convergence of GA and BBO is poor in comparison with other techniques. PSO_M_1 has shown unsteady convergence during the initial generations and it becomes steady with the increase in generations, while all the other algorithms have shown steady convergence. During the initial generations, ABC, PSO, PSO_M_2, HGABC, HPABC and TLBO, have shown better convergence and so, for these algorithms the convergence speed is fast. But out of these algorithms only TLBO and ABC have shown better performance in finding the optimum solution

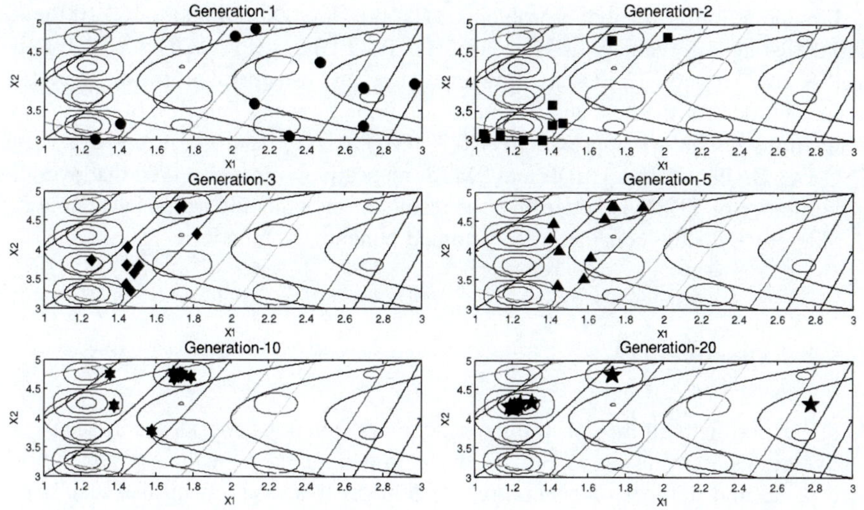

Fig. 6.6 Visualization of the convergence of solutions with different generations for the multimodal constrained optimization problem

consistently. The visualization for the convergence of the solutions for 20 generations is shown in Fig. 6.6. It is observed that, there are no feasible solutions in the first generation. But as TLBO algorithm proceeds the infeasible solutions try to move toward the feasible region and so in generation-2 there are two feasible solutions and other solutions have moved nearer to the feasible region. In generation-3 there are only two infeasible solutions. In generation-10, it is observed that the solutions are different local optima with only one solution in the contour of global optimum. In generation-20 most of the solutions have reached near to the global optimum.

6.5.2 Experiment 8

In this experiment different constrained benchmark functions (G01–G04, G06–G12) are considered and the performance of TLBO is compared with the results available in the literature obtained by using different optimization algorithms. Results of TLBO are compared with different optimization techniques like, hybrid PSO-DE (PSO-DE) [5], changing range genetic algorithm (CRGA) [6], self-adaptive-based penalty function-based optimization (SAPF) [7], co-evolutionary differential evolution (CDE) [8], cultured differential evolution (CULDE) [9], co-evolutionary PSO (CPSO-GD) [10] and simple multi-membrane evolutionary strategies (SMES) [11]. The results are shown in Tables 6.13 and 6.14 for the best and the mean solutions, respectively. TLBO is implemented by considering maximum function evaluations of 100,000 with the population size of 50.

For the solution of above problems 140,100, 500,000, 248,000, 100,100 and 240,000 function evaluations were used for PSO-DE, SAPF, CDE, CULDE and SMES, respectively. TLBO has shown better results or equivalent results than the other methods for the best and the mean solutions by using only 100,000 function evaluations which are approximately 0.7, 0.2, 0.4, 0.99 and 0.41 times less than PSO-DE, SAPF, CDE, CULDE and SMES, respectively. It is observed that overall performance of TLBO is better than the results of the other algorithms available in the literature for the considered constrained benchmark functions.

6.5.3 *Experiment 9*

In this experiment all the constrained benchmark functions considered in Chaps. 4 and 5 are tested by using TLBO. TLBO is implemented by using the population size of 50 and number of generations of 500. In this experiment also results of TLBO is compared with the results of all the hybrid algorithms discussed in Chap. 5. Comparison of results for the hybrid algorithms and TLBO for the best and the mean solutions are given in Tables 6.15 and 6.16. It is observed from the results that TLBO have shown better results for the ten and eight constrained benchmark functions in finding the best and the mean solutions respectively. TLBO is successful in finding the global solution for G01, G04, G05, G08, G11, G12, G15, G16, G18 and G24. TLBO have shown near optimal solution for G03, G06, G07, G09 and G21. So, TLBO has given optimum or near optimum solution solutions for 15 constrained benchmark functions. Moreover, TLBO is consistent in finding the global solutions for G01, G04, G08, G11, G12 and G24.

6.6 Implementation of TLBO for the Design Optimization of Mechanical Elements

6.6.1 *Experiment 10*

In this experiment four different constrained benchmark mechanical design problems, welded beam design, pressure vessel design, tension compression spring and speed reducer, with different characteristics of objective function and constraints (linear and nonlinear) are experimented. Some of the problems are having mixed discrete–continuous design variables. These problems are used by many researchers to test the performance of different algorithms. All these problems are discussed in detail in Chap. 4 (Examples 8, 9, 10 and 11).

The above mentioned mechanical benchmark problems were attempted by many researchers, but in this book the effectiveness of the results of TLBO is compared with the results of researchers published after the year 2004. As PSO,

Table 6.13 Comparison of the results for best solution by using PSO-DE, CRGA, SAPF, CDE, CULDE, CPSO-GD, SMES and TLBO

	PSO-DE	CRGA	SAPF	CDE	CULDE	CPSO-GD	SMES	TLBO
G01	**−15**	−14.9977	**−15**	**−15**	**−15**	**−15**	**−15**	**−15**
G02	−0.803614	−0.802959	−0.803202	−0.794669	**−0.803619**	NA	−0.803601	**−0.803619**
G03	**−1.0005**	−0.9997	−1	NA	−0.995413	NA	−1	**−1.0005**
G04	**−30,665.539**	−30,665.52	−30,665.401	**−30,665.539**	**−30,665.539**	NA	−30,665.539	**−30,665.539**
G06	**−6,961.814**	−6,956.251	−6,961.046	**−6,961.814**	−6,961.8139	NA	−6,961.814	**−6,961.814**
G07	**24.306209**	24.882	24.838	NA	**24.306209**	24.711	24.327	**24.306209**
G08	−0.095826	**−0.095825**	**−0.095825**	**NA**	**−0.095825**	**NA**	**−0.095825**	**−0.095825**
G09	**680.63**	680.726	680.773	680.771	**680.63**	680.678	680.632	**680.63**
G10	**7,049.248**	7,114.743	7,069.981	NA	7,049.2481	7,055.6	7,051.903	**7,049.248**
G11	**0.749999**	0.75	**0.749**	NA	**0.7499**	NA	0.75	**0.7499**
G12	**−1**	**−1**	**−1**	**−1**	**−1**	NA	**−1**	**−1**

Table 6.14 Comparison of the results for mean solution by using PSO-DE, CRGA, SAPF, CDE, CULDE, CPSO-GD, SMES and TLBO

	PSO-DE	CRGA	SAPF	CDE	CULDE	CPSO-GD	SMES	TLBO
G01	**−15**	−14.985	−14.552	**−15**	−14.999996	−14.997	**−15**	**−15**
G02	−0.756678	−0.764494	−0.755798	−0.78548	−0.724886	NA	−0.785238	**−0.79753**
G03	**−1.00501**	−0.9972	−0.964	NA	−0.788635	NA	−1	−0.997445
G04	**−30,665.54**	−30,664.398	−30,665.922	−30,665.536	**−30,665.539**	NA	**−30,665.539**	**−30,665.539**
G06	**−6,961.814**	−6,740.288	−6,953.061	−6,960.603	**−6,961.8139**	NA	−6,761.284	**−6,961.8139**
G07	**24.30621**	25.746	27.328	NA	**24.30621**	25.709	24.475	**24.30621**
G08	−0.095825	−0.095819	−0.095635	NA	−0.095825	NA	−0.095825	**−0.095825**
G09	680.6301	681.347	681.246	681.503	680.63006	680.781	680.643	**680.63**
G10	**7,049.248**	8,785.149	7,238.964	NA	7,049.2483	8,464.2	7,253.047	**7,049.248**
G11	0.74999	0.752	0.751	NA	0.757995	NA	0.75	**0.7499**
G12	**−1**	**−1**	−0.99994	**−1**	**−1**	NA	**−1**	**−1**

Table 6.15 Comparison of results (best solution) for the constrained benchmark functions by using different hybrid optimization techniques and TLBO

BM-C	Best				
	HPABC	HBABC	HDABC	HGABC	TLBO
G01	−11	**−15**	**−15**	−14.962117	**−15**
G02	−0.38303	−0.748981	−0.743072	−0.769406	**−0.7984**
G03	−0.999933	−1.000116	−1.00041	−0.908716	**−1.00049**
G04	**−30,665.53867**	**−30,665.53867**	**−30,665.53867**	−30,661.97624	**−30,665.53867**
G05	5,127.051873	5,127.896622	5,126.496716	15,831,988.76	**5,126.49671**
G06	−6,961.813874	**−6,961.813876**	**−6,961.813876**	−6,671.485071	−6,961.813791
G07	24.83686	24.725735	24.323636	27.669474	**24.317487**
G08	**−0.095825**	**−0.095825**	**−0.095825**	**−0.095825**	**−0.095825**
G09	680.638643	680.630701	**680.630332**	683.085962	680.6305114
G10	7,304.510202	7,106.484459	7,143.890531	7,651.225802	**7,078.400036**
G11	0.749909	**0.7499**	**0.7499**	0.751031	**0.7499**
G12	**−1**	−0.99999	**−1**	**−1**	**−1**
G13	**0.383755**	0.55963	0.901652	0.991843	0.885
G14	61.987375	−46.188182	**−46.968524**	156.039946	−46.008214
G15	961.783308	961.987688	961.764842	961.716877	**961.715022**
G16	−1.888024	−1.888024	−1.888024	−1.861636	**−1.905155**
G17	8,938.327929	8,928.478227	8,940.407178	3,699,802.099	**8,907.513734**
G18	**−0.86601**	−0.863618	−0.865976	−0.811193	**−0.86601**
G19	48.350446	33.187255	34.62052	38.507394	**33.169223**
G20	41,433,091.79	13,312,926.08	14,466,371.07	1,000,349.33	**0.147429**
G21	91,442.99424	193.754766	**193.754224**	163,816.2771	193.76128
G22	5.5176E + 18	2.07822E + 14	**139,385,275.4**	2.31661E + 13	1,205,888,791
G23	1,166.228548	−19.503574	**−321.485457**	−0.001495	−0.065672

DE, ES and ABC are some of the well-known optimization algorithms, many researchers had tried to enhance the performance of the basic algorithms by modifying the basic algorithms between the years 2005 and 2010. Still efforts are going on to modify or hybridize such well-known algorithms to increase their effectiveness and efficiency. The above mentioned mechanical design problems were attempted by ($\mu + \lambda$)-Evolutionary Strategy (ES) [11], Unified Particle Swarm Optimization (UPSO) [12], Co-evolutionary Particle Swarm Optimization (CPSO) [13], Co-evolutionary Differential Evolution (CDE) [8], Hybrid PSO-DE [5] and ABC [4].

TLBO is implemented by considering the population size of 50 and maximum function evaluations of 10,000. TLBO is compared with the above mentioned optimization methods for the best solution, mean solution and maximum function evaluations required to find the optimum solution. For TLBO, 25 independent runs are carried out to check the performance of the algorithm. Results for the comparison of all the methods are shown in Table 6.17.

Table 6.16 Comparison of results (mean solution) for the constrained benchmark functions by using different hybrid optimization techniques and TLBO

BM-C	Mean				
	HPABC	HBABC	HDABC	HGABC	TLBO
G01	−7.6	**−15**	**−15**	−14.8719566	**−15**
G02	−0.3094417	−0.6780855	−0.6438653	−0.7501423	**−0.7753**
G03	49.2811631	−0.9950221	−0.996886	−0.3893368	**−0.997445**
G04	−30,665.53861	−30,665.53861	**−30,665.53867**	−30,602.3601	**−30,665.53867**
G05	419,969.4498	5,250.854236	5,281.512188	29,404,976.11	**5,245.4532**
G06	193,038.186	−6,961.781506	**−6,961.813876**	−5,385.982478	−6,961.812117
G07	48.4202269	26.0059673	24.6143301	31.6084807	**24.5609413**
G08	**−0.095825**	−0.0891569	**−0.095825**	−0.0958187	**−0.095825**
G09	680.6706487	680.6867908	**680.6344396**	685.9619782	680.6436311
G10	8,262.829776	7,943.977794	**7,237.877898**	8,042.291921	7,482.529486
G11	0.8834262	**0.7499**	**0.7499**	0.8642834	**0.7499**
G12	**−1**	**−1**	**−1**	**−1**	**−1**
G13	81.3958922	**0.9736198**	0.9954718	104.7640348	0.984562
G14	141.1351492	−24.3580828	**−45.529184**	157.2434362	−24.2488048
G15	1,763.425164	964.1383936	963.604144	966.3025414	**963.1373768**
G16	**−1.888024**	−1.8619364	**−1.888024**	−1.6956546	−1.8850326
G17	1,567,982,404	9,040.261009	**8,994.320904**	32,060,333.84	209,616.2123
G18	−0.7473602	1,818.180163	−0.788882	−0.6725528	**−0.8651452**
G19	149.1784138	41.006861	**36.0241206**	43.3501844	39.4778044
G20	33,808,003,464	15,626,925.09	15,604,945.9	5,793,202.773	**2,766,258.771**
G21	765,071.683	9,074.032606	**4,252.105379**	208,155.2447	6,597.187215
G22	3.40929E + 19	3.53243E + 16	9.18624E + 16	**1.38082E + 16**	9.00299E + 16
G23	19,633.68237	4,847.183274	841.4603966	2,481.62147	**465.1448548**
G24	14.4919698	**−5.508013**	**−5.508013**	−5.4785796	**−5.508013**

It is observed from the results that TLBO finds the best solution for all the problems except for the pressure vessel problem for which $(\mu + \lambda)$-ES has shown better performance. But for the pressure vessel problem the average performance of TLBO is better than $(\mu + \lambda)$-ES. The average performance of TLBO and PSO-DE is same for all the problems except for welded beam problem for which PSO-DE is better. TLBO requires 66, 70 and 66% less function evaluations than $(\mu + \lambda)$-ES, PSO-DE and ABC, respectively. So it can be concluded that TLBO requires less function evaluations and also it does not require any algorithm parameters.

6.6.2 Experiment 11

In this experiment all the mechanical elements design optimization problems considered in Chaps. 4 and 5 are tested by using TLBO. TLBO is implemented by

Table 6.17 Comparison of results obtained by different optimization methods for mechanical design problems

Problem	Criteria of evolution	$(\mu + \lambda)$-ES (A*)	UPSO (B*)	CPSO (C*)	CDE (D*)	PSO-DE (E*)	ABC (F*)	TLBO
Welded	Best	**1.724852**	1.92199	1.728	1.73346	**1.72485**	**1.724852**	**1.724852**
Beam	Mean	1.777692	2.83721	1.74883	1.76815	**1.72485**	1.741913	1.7248561
	Evaluations	30,000	100,000	200,000	240,000	33,000	30,000	**10,000**
Pressure	Best	**6,059.7016**	6,544.27	6,061.077	6,059.734	6,059.714	6,059.714	6,059.714335
Vessel	Mean	6,379.938	9,032.55	6,147.1332	6,085.23	**6,059.714**	6,245.308	**6,059.71434**
	Evaluations	30,000	100,000	200,000	240,000	42,100	30,000	**10,000**
Tension	Best	0.012689	0.01312	0.012674	0.01267	**0.012665**	**0.012665**	**0.012665**
Compression	Mean	0.013165	0.02294	0.01273	0.012703	**0.012665**	0.012709	**0.01266576**
Spring	Evaluations	30,000	100,000	200,000	240,000	24,950	30,000	**10,000**
Gear	Best	**2,996.348**	NA	**NA**	NA	**2,996.348**	2,997.058	**2,996.34817**
Train	Mean	**2,996.348**	NA	NA	NA	**2,996.348**	2,997.058	**2,996.34817**
	Evaluations	30,000	NA	NA	NA	54,350	30,000	**10,000**

*A** [11], *B** [12], *C** [13], *D** [8], *E** [5], *F** [4]

using the population size of 50 and number of generations of 500. In this experiment also results of TLBO is compared with the results of all the hybrid algorithms discussed in Chap. 5. Comparison of results for the hybrid algorithms and TLBO for the best and the mean solutions are given in Tables 6.18 and 6.19. For the mechanical design problems TLBO has outperformed all the hybrid algorithms in finding the best solutions but it has shown slight inferior results for the mean solutions than HDABC (Figs. 6.7, 6.8).

The overall performance of all the hybrid algorithms and TLBO is shown by using bar chart in Figs. 6.9 and 6.10 for the best and the mean solutions. It is observed from the bar charts that TLBO has shown better performance for 11, 17 and 25 unconstrained benchmark functions, constrained benchmark functions and mechanical element design optimization problems respectively for the best solution. Moreover, TLBO has shown better performance for 12, 14 and 17 unconstrained benchmark functions, constrained benchmark functions and mechanical element design optimization problems respectively for the mean solution. So, the overall performance of TLBO is approximately 1.4 and 1.22 times than HDABC for the best and the mean solutions. Results showing the best values of objective function along with the values of its design variable and constraints for all the mechanical elements design optimization problems are given in Tables 6.20, 6.21 and 6.22.

6.6.3 Experiment 12

In this experiment results are compared based on the convergence for HDABC and TLBO for the problems for which both the algorithms have same mean solutions. Same mean solutions are considered as it is the measure of consistency of the algorithm. If two different algorithms are having same mean it implies that both the algorithms are capable of finding the global solutions with the same consistency. So it is required to check that out of these two algorithms which of the algorithms finds the solution rapidly and based on this the convergence plots were obtained. Convergence graphs are plotted for the average solutions obtained in five different runs in each iteration. Convergence graphs were drawn for the Step and Quartic unconstrained benchmark functions, G01, G04, G08 and G12 constrained benchmark functions and Examples 1, 2, 4, 8, 11, 14 and 16 for the mechanical design problems Fig. 6.7. It is seen from the convergence graphs that the convergence rate of TLBO is very high compared to HDABC for the Step and the Quartic functions. There is no much difference in the convergence of HDABC and TLBO for the constrained benchmark functions, but still HDABC is slightly better than TLBO for G01 and G12 and TLBO is slightly better than HDABC for G04 and G08. For mechanical design problems TLBO has shown better convergence for all the examples except Example 11.

Table 6.18 Comparison of results (best solution) for the mechanical design problems by using different hybrid optimization techniques and TLBO

MD		Best				
		HPABC	HBABC	HDABC	HGABC	TLBO
Gear	Example 1A	3,135.515965	3,135.5311	**3,135.515852**	3,136.2004	**3,135.515852**
	Example 1B	**3,142.712756**	**3,142.7128**	**3,142.7128**	**3,142.7128**	**3,142.71275**
	Example 1C	2,993.665618	2,993.6658	**2,993.6656**	2,993.6659	**2,993.6656**
	Example 1D	**2,994.844118**	**2,994.8441**	**2,994.8441**	**2,994.8441**	**2,994.8441**
Bearing	Example 2A	**−6,032.315099**	**−6,032.315099**	**−6,032.315099**	−6,017.834159	**−6,032.3151**
	Example 2B	**−3,792.420036**	**−3,792.420036**	**−3,792.420036**	−3,780.0856	**−3,792.420036**
	Example 2C	**−0.223974**	**−0.223974**	**−0.223974**	**−0.223974**	**−0.223974**
Bellivelle spring	Example 3	1.990962	1.979731	**1.979675**	1.981879	**1.979675**
Multiplate clutch	Example 4	**0.313657**	**0.313657**	**0.313657**	**0.313657**	**0.313657**
Robot gripper	Example 5	**4.247644**	4.416086	6.285471	**4.247644**	**4.247644**
Hydrostatic bearing	Example 6	1,848.392827	1,640.82959	1,626.553483	2,045.909142	**1,625.45809**
4-stage gear train	Example 7	37,875,156,133	46,623.2053	59,763.56295	55,494.49422	**43,792.4331**
Pressure vessel	Example 8	**6,059.714335**	**6,059.714335**	**6,059.714335**	**6,059.714335**	**6,059.71433**
Welded beam	Example 9	**1.724852**	**1.724852**	**1.724852**	1.770658	**1.724852**
Spring	Example 10	**0.012665**	**0.012665**	**0.012665**	0.012671	**0.012665**
Speed reducer	Example 11	**2,996.348165**	**2,996.348165**	**2,996.348165**	2,997.665018	**2,996.348165**
Stiffened shell	Example 12A	54,444.6307	54,446.19844	54,444.47963	54,868.84219	**54,444.4774**
	Example 12B	55,342.91995	**55,326.2934**	**55,326.2934**	55,724.83112	**55,326.2934**
Step cone pulley	Example 13	31.953962	16.642827	**16.63451**	19.411676	**16.63451**
Screw jack	Example 14	**419,187.0574**	**419,187.0574**	**419,187.0574**	**419,187.0574**	**419,187.0574**
C-clamp	Example 15	5,623.630726	5,616.151813	**5,616.151812**	5,673.954031	**5,616.151812**
Hydrodynamic bearing	Example 16	16.205865	**16.205833**	**16.205833**	16.205,956	**16.205,833**
Cone clutch	Example 17	**68.877551**	**68.877551**	**68.877551**	68.880307	**68.877551**
Cantilever support	Example 18	**49.062256**	**49.062256**	**49.062256**	49.068421	**49.062256**
Hydraulic cylinder	Example 19	0.024245	**0.024124**	**0.024124**	0.024213	**0.024124**
Planetry gear box	Example 20	**0.526667**	0.52735	0.527814	0.53	0.52735

Table 6.19 Comparison of results (mean solution) for the mechanical design problems by using different hybrid optimization techniques and TLBO

MD		Mean				
		HPABC	HBABC	HDABC	HGABC	TLBO
Gear	Example 1A	3,135.682894	3,135.807775	**3,135.515852**	3,136.851425	**3,135.515852**
	Example 1B	3,148.741874	3,142.7128	**3,142.712756**	3,142.7128	**3,142.712756**
	Example 1C	2,996.538989	2,993.737825	**2,993.6656**	2,994.202425	**2,993.6656**
	Example 1D	2,994.844118	**2,994.8441**	**2,994.8441**	**2,994.8441**	**2,994.8441**
Bearing	Example 2A	−5,968.929762	−6,032.315091	**−6,032.315099**	−5,956.518396	**−6,032.315099**
	Example 2B	−3,788.970347	−3,787.47894	**−3,792.420036**	−3,768.12423	**−3,792.420036**
	Example 2C	**−0.223974**	**−0.223974**	**−0.223974**	−0.223665	**−0.223974**
Bellivelle spring	Example 3	2.0478585	1.9819249	**1.9796762**	2.0058175	1.9797062
Multiplate clutch	Example 4	**0.313657**	0.3144411	**0.313657**	0.3162912	**0.313657**
Robot gripper	Example 5	30,839.45805	**4.8081054**	8.774491714	5.261168333	4.9377
Hydrostatic bearing	Example 6	2,624.378662	1,761.923195	**1,724.559019**	2,625.568094	1,870.575353
4-stage gear train	Example 7	2.58148E + 11	47,023,887,602	443,507,531.4	152,941,037.2	**29,816,446.6**
Pressure vessel	Example 8	**6,059.714335**	**6,059.714335**	**6,059.714335**	6,059.71434	**6,059.714335**
Welded beam	Example 9	1.7248593	1.7304695	**1.724852**	1.8005308	1.7248561
Spring	Example 10	0.0126817	0.0126712	**0.0126651**	0.0126806	0.0126673
Speed reducer	Example 11	3,012.992535	2,996.348532	**2,996.348165**	3,000.367059	**2,996.348165**
Stiffened shell	Example 12A	55,634.7773	55,072.73776	**54,639.84478**	56,135.9943	55,412.27135
	Example 12B	56,245.27169	55,921.85604	55,852.80969	56,920.40249	**55,667.4087**
Step cone pulley	Example 13	58.8119921	56.3661545	**24.2746931**	1,027.797361	30.1857977
Screw jack	Example 14	419,335.259	**419,187.0574**	**419,187.0574**	**419,187.0574**	**419,187.0574**
C-clamp	Example 15	5,706.239857	5,645.067208	**5,616.151816**	5,793.686586	5,616.151878
Hydrodynamic bearing	Example 16	16.2059317	**16.205833**	**16.205833**	16.2067377	**16.205833**
Cone clutch	Example 17	68.877551	68.877551	68.877551	68.9015692	**68.877551**
Cantilever support	Example 18	**49.062256**	**49.062256**	**49.062256**	49.1161695	**49.062256**
Hydraulic cylinder	Example 19	0.024293	0.0241579	0.024124	0.0242823	**0.024124**
Planetry gear box	Example 20	0.534544	**0.52908**	0.544793	0.5366932	0.5337054

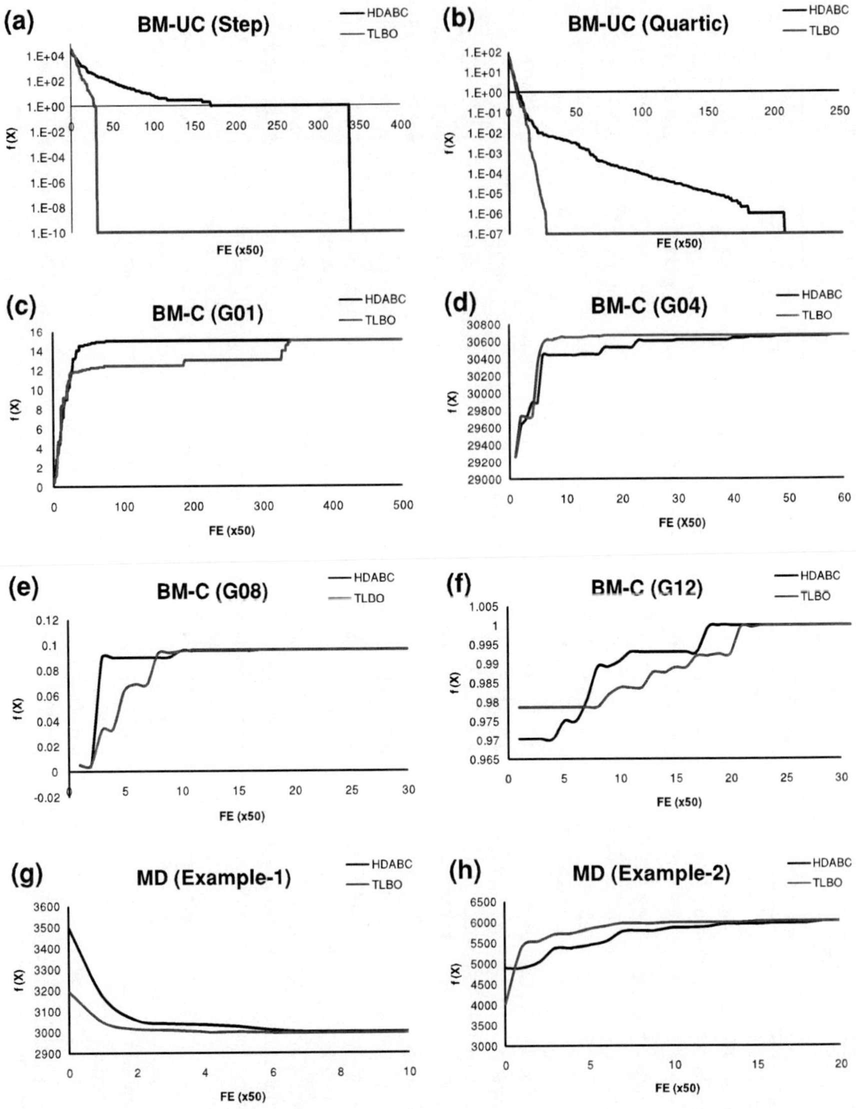

Fig. 6.7 Comparison of HDABC and TLBO based on convergence curve for different benchmark functions and mechanical design problems

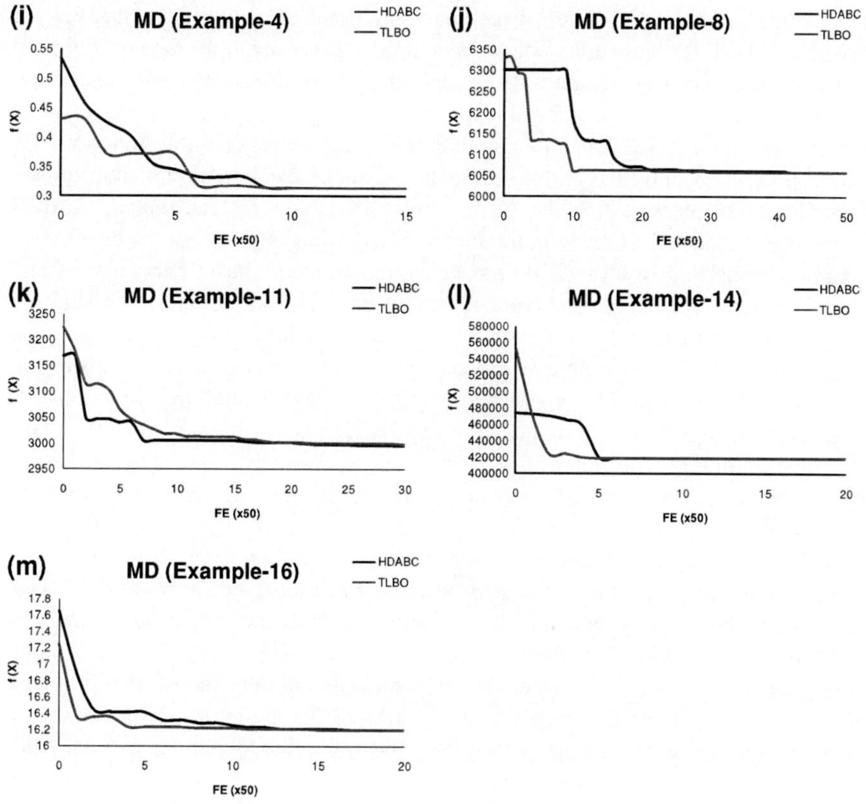

Fig. 6.7 (continued)

6.7 Implementation of TLBO for the Real Parameter Optimization

6.7.1 Experiment 1

It is a common practice in the field of optimization to compare different algorithms by using different benchmark problems. These comparisons are limited to the test problems taken for the study and sometimes the chosen algorithm and the test problems are complimentary to each other and the same algorithm may not show the same performance for the other real parameter optimization problems. So, a common platform is required to compare the performance of different algorithms for different benchmark problems.

Congress on Evolutionary Computation 2005 [14] had provided the common platform for the comparison of the performances of different algorithms by specifying a common termination criterion, size of problem, initialization scheme, etc. In this section, 25 different benchmark problems are experimented using the proposed TLBO algorithm. These benchmark functions are having different characteristics such as separability, scalability and multimodality. A function is multimodal if it has two or more local optima. A function is separable if it can be written as a sum of functions of variable separately. Non-separable functions are more difficult to optimize and difficulty increases if the function is multimodal. Complexity increases when the local optima are randomly distributed. Furthermore, complexity increases with the increase in dimensionality. All the benchmark functions proposed in CEC 2005 are generated from the basic benchmark functions (Sphere, Rastrigin, Rosenbrock, Schwefel, Griewank, etc.) by shifting, rotating or hybridizing different basic benchmark functions. Shifting, rotating and hybridizing add more complexity to the benchmark functions and so testing of algorithms for such problems is a real challenge. These benchmark functions are available on: http://www.ntu.edu.sg/home/EPNSugan.

Some common evolution criteria were presented in CEC 2005 [14] and these criteria are considered to have the completeness. All the benchmark functions are run for 25 times each. Function error value ($f(x) - f(x^*)$) is recorded after 1E3, 1E4, 1E5 number of function evaluations and at termination for each run. Function error is considered as the difference between the global optimum and the best result obtained by the algorithm. If the algorithm finds the global optimum, then the error will be 0E0. Termination is done when the maximum function evaluations equals to 10,000*D, where D indicates the dimension of the problem. Dimensions taken for the experiment are 10 and 30. Function error values are recorded for all the 25 runs and sorted from the best value to the worst value. After sorting, 1st (Best), 7, 13th (Median), 19 and 25th (Worst) values are recorded. Mean and standard deviation for all the runs are also recorded. TLBO is coded in MATLAB 7 and it is run on a laptop machine possessing Intel Pentium processor with 2.2 GHz and 3 GB RAM. TLBO is applied for all the benchmark functions by considering the population size of 20. As TLBO is a parameter-less algorithm no other parameter is required for the working of the algorithm. Function error values for dimension 10 are presented in Tables 6.23, 6.24 and 6.25 and function error values for dimension 30 are given in Tables 6.26, 6.27 and 6.28. The procedure to calculate the complexity of the algorithm is given in [14]. The complexity of TLBO is calculated as ($T2_{(mean)} - T1)/T0$. where, $T0$ is the time to calculate the following

For $i = 1{:}1{,}000{,}000$
$x = $ (double) 5.55
$x = x + x$; $x = x/2$; $x = x * x$; $x = \text{sqrt}(x)$; $x = \ln(x)$; $x = \exp(x)$; $y = x/x$;
end

$T1$ is the time to calculate only function B03 for 200,000 evaluations for certain dimension and $T2$ is the mean time for the optimization algorithm to calculate function B03 for 200,000 function evaluations for the same dimension. $T2_{(mean)}$ is the mean time for $T2$ obtained for 5 times. The complexity of the algorithm is given in Table 6.29.

The performance of TLBO is compared with the other seven optimization algorithms based on the mean value for 1E5 function evaluations with dimension 10. It is seen from Tables 6.30, 6.31 and 6.32 that TLBO has outperformed all the other algorithms or performed equally best for nine benchmark functions, B01, B02, B04, B05, B16, B17, B21, B22 and B24. Moreover, TLBO has outperformed six other algorithms for other six benchmark functions, B07, B08, B13, B14, B19, and B20.

6.7.2 Experiment 2

In this experiment, 22 different constrained benchmark functions from CEC 2006 (Liang et al. [26]) are experimented. The capability of the algorithm to find the global solution for the constrained problem depends on the constraint handling technique. In this experiment four different constraint handling techniques like superiority of feasible solutions (SF) [15], self-adaptive penalty approach (SP) [7], ε-constraint technique (EC) [16] and stochastic ranking technique (SR) [17] are experimented with TLBO. Moreover, a new constraint handling technique, ECHT (ensemble of constraint handling technique) suggested in [18] is also experimented with TLBO. The technique suggested by Montes and Coello [11] ensembles four different constraint handling techniques, namely SF, SP, EC and SR. In this experiment, the algorithm is run for 30 times for each benchmark function with the population size of 50 and maximum number of generations as 2,000. Constrained problems are generally considered more complex than the unconstrained problems and require more population size than that required by the unconstrained problems. Hence the population size of 50 is considered in this experiment. For the performance of the algorithm, best solution, mean solution, median solution, worst solution and standard deviation are recorded for all the functions. The results for all the 22 benchmark functions using TLBO with different constraint handling techniques are given in Table 6.33. The notations B, MD, MN, W and SD in Table 6.33 denote Best, Median, Mean, Worst and Standard deviation, respectively.

All the constraint handling techniques are compared based on the searching capability for the best and the mean solutions. It is observed from Table 6.33 that TLBO with ECHT has produced superior results for searching the best solution than the other constraint handling techniques for four benchmark functions, viz. G14, G19, G21 and G23. TLBO with ECHT has performed equivalent with other constraint handling techniques for the rest 18 benchmark functions. Moreover, TLBO with ECHT has produced superior results for eleven benchmark functions in searching the mean solution, viz. G01, G02, G05, G07, G10, G14, G17, G18, G19, G21 and G23 and for the rest of the benchmark functions TLBO with ECHT has produced equivalent results than the other constraint handling techniques. The graphical representation for the performance of different constraint handling techniques is shown in Fig. 6.8. It is observed from Fig. 6.8 that all the constraint handling techniques except ECHT has shown nearly equivalent performance for

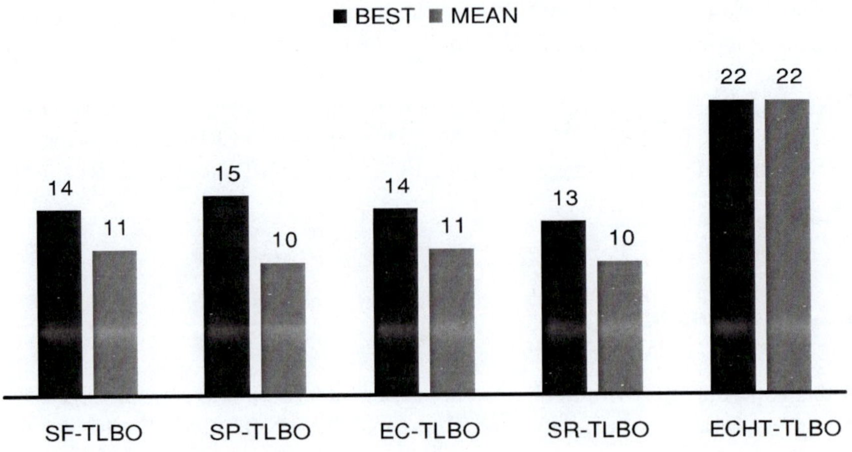

Fig. 6.8 Performance of different constraint handling techniques using TLBO for the 22 constrained benchmark functions in searching the best and the mean solutions

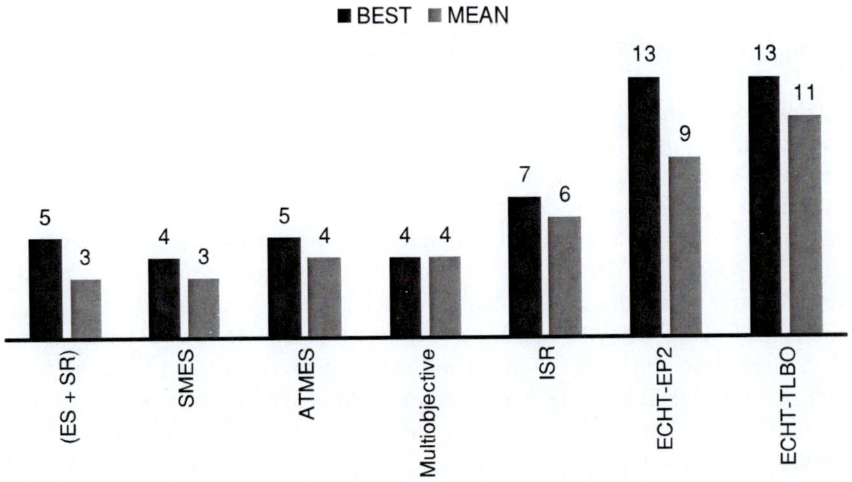

Fig. 6.9 Comparison of TLBO with other optimization techniques for the 22 constrained benchmark functions in searching the best and the mean solutions

searching the best as well as the mean solutions. It is also observed from the results that TLBO with ECHT has performed 1.57 times better than other constraint handling techniques for searching the best solutions and 2.2 times better for searching the mean solutions.

The performance of TLBO is also compared with the other optimization methods with different constraint handling techniques such as evolutionary strategies with stochastic ranking (ES + SR) [17], simple multi-membrane evolutionary strategy (SMES) [11], adaptive tradeoff model evolutionary strategy

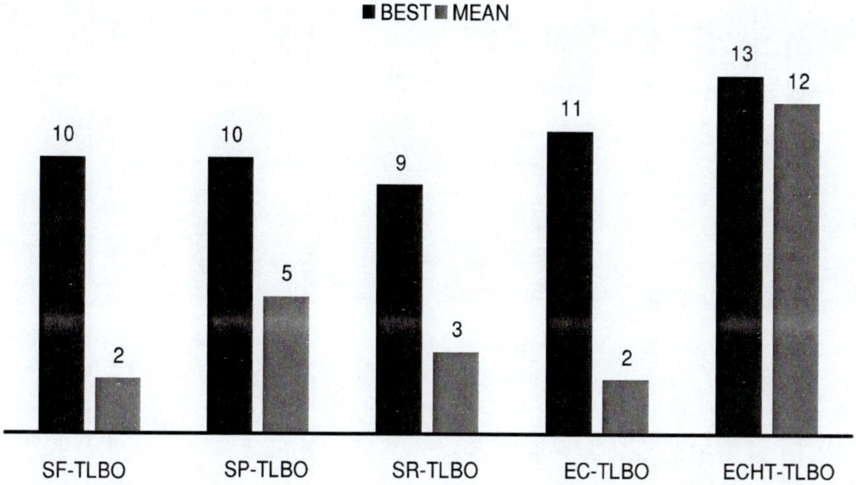

Fig. 6.10 Performance of different constraint handling techniques using TLBO for 13 constrained benchmark functions in searching the best and the mean solutions

(ATMES) [19], multi-objective evolutionary strategy [19], improved stochastic ranking (ISR) [17] and ensemble of constraint handling technique strategy (ECHT-EP) [18]. Comparison of the results for the performance of TLBO with other techniques is given in Table 6.34 and the graphical representation for the comparison in searching the best and the mean solution is shown in Fig. 6.9. It is observed from the results that the ECHT is an effective constraint handling method. ECHT has outperformed all the other techniques in searching the best and the mean solutions. ECHT with ES and TLBO has produced similar results for the best solution and so both the algorithms have performed equivalently with ECHT in searching the best solution. For searching the mean solution, ECHT with TLBO has outperformed by performing 1.22 times better than ES.

6.7.3 Experiment 3

In this experiment, 13 constrained benchmark functions given in [18] are experimented. For the considered 13 problems all the constrained handling techniques mentioned in experiment 2 are used with the TLBO. In this experiment also, the population size is taken as 50 and maximum number of generations as 1,000. The results are compared based on the best solutions, median solutions, mean solutions, worst solutions and the standard deviation. The results of TLBO are compared with the results obtained using DE with all the constrained handling techniques presented by Mallipeddi and Suganthan [18].

For this experiment also the comparison is made for TLBO using all the constraint handling techniques. The results of comparison are given in Table 6.35 and

Table 6.20 Values of objective functions, constraints and design variables for mechanical design problems obtained by using TLBO (Examples 1–3)

	Example 1A	Example 1B	Example 1C	Example 1D	Example 2A	Example 2B	Example 2C	Example 3
Design variables								
x_1	23.91929587	24	22	21.51959325	6.212971019	6.212971019	4.753120254	0.204143354
x_2	30	30	30	30	20.05924637	20.05924637	22.39500759	0.2
x_3	36.7664865	37	36.74744829	37.49778403	7	7	9	10.03047329
x_4	18	18	18	18	0.515	0.515	0.587004569	12.01
x_5			399.9999993	400	0.515	0.515	0.515	
x_6					0.497672764	0.5	0.439582596	
x_7					0.663854869	0.68990733	0.673058506	
x_8					0.3	0.3	0.3	
x_9					0.076575736	0.061620206	0.06965378	
x_{10}					0.700006668	0.7	0.700020204	
Constraints								
$g_1(X)$	0	0.003374018	2.313977942	2.313977945	0	0	0	1.76588E − 06
$g_2(X)$	0.72945272	0.72945272	0.486952997	0.486953003	2.472486758	2.425942038	0.714588587	7.4632E − 08
$g_3(X)$	1.173038229	1.173038229	0.799300351	0.799300351	0.851155333	1.372204554	3.954929606	5.8E − 11
$g_4(X)$	0	0.01917505	0	0	0.087088994	0.087028981	1.54706158	1.595856646
$g_5(X)$	0.616161616	0.616161616	6	6	0.05924637	0.05924637	2.395007593	2.349E − 09
$g_6(X)$	–	–	1.17641742	1.17641742	3.003783064	2.405561868	0.39114362	1.979526711
$g_7(X)$	–	–	0	0.020759922	0	0	0	0.198965748
$g_8(X)$	–	–	0.616161616	0.616161616	0	0	0.072004569	–
$g_9(X)$	–	–	–	–	0	0	0	–
Objective function								
$f(X)$	3,135.515852	3,142.712756	2,993.665605	2,994.844118	6,032.315099	3,792.420036	0.223974155	1.979674758

Table 6.21 Values of objective functions, constraints and design variables for mechanical design problems obtained by using TLBO (Examples 4–11 except 7)

	Example 4	Example 5	Example 6	Example 8	Example 9	Example 10	Example 11
Design variables							
x_1	70	150	5.95578050261541000000000	0.8125	0.20572963	0.051684775	3.5
x_2	90	150	5.38901305194167000000000	0.4375	3.47048834	0.356614639	0.7
x_3	1	200	0.00000535869726706299000	42.0984456	9.036625372	11.29501278	17
x_4	810	0	2.26965597280973000000000	176.6365958	0.205729634	–	7.3
x_5	3	150	–	–	–	–	7.8
x_6	–	100	–	–	–	–	3.350214666
x_7	–	2.339539113	–	–	–	–	5.28668323
Constraints							
$g_1(X)$	0	28.09283911	0.0001374735	0	5.64517E − 05	6E − 10	0.07391528
$g_2(X)$	24	21.90716089	0.0000010103	0.035880829	0.008826575	3E − 10	0.952823443
$g_3(X)$	0.91942781	33.64959994	0.0000000210	0	4.1E − 09	4.053581954	0.499172248
$g_4(X)$	9,830.371094	16.35040006	0.0003243625	63.36340416	3.432983608	0.72780039	0.901471698
$g_5(X)$	7,894.69659	79,999.998	0.5667674507	–	0.08072963	–	0
$g_6(X)$	0.702013203	9.8E − 11	0.0009963614	–	0.235540329	–	0
$g_7(X)$	37,706.25	0.00001	0.0000090762	–	0.000111156	–	0.7025
$g_8(X)$	14.2979868	–	–	–	–	–	0
$g_9(X)$	–	–	–	–	–	–	0.583333333
$g_{10}(X)$	–	–	–	–	–	–	0.051325754
$g_{11}(X)$	–	–	–	–	–	–	0.010852365
Objective function							
$f(X)$	0.313656611	4.247643634	1,625.4427649821	6,059.714335	1.724852455	0.012665233	2,996.348165

Table 6.22 Values of objective functions, constraints and design variables for mechanical design problems obtained by using TLBO (Examples 12–19)

	Example 12A	Example 12B	Example 13	Example 14	Example 15	Example 16	Example 17	Example 18	Example 19
Design variables									
x_1	13.82658378	14.44039743	40	19.10795274	20	1.286677503	4.285714286	0.066038545	468.2785803
x_2	26.8797131	26.64632368	54.76430219	98.4146508	4.000000128	0.53046184	2.142857143	0.033019272	4.064429292
x_3	8.367976691	9.50123373	73.01317731	350	26.34469348	–	–	–	15.86070982
x_4	260.4268329	250	88.42841977	–	2.230498008	–	–	–	–
x_5	224.2938865	203	85.98624273	–	1.000000032	–	–	–	–
Constraints									
$g_1(X)$	2.471E – 07	0.009301856	5.14E – 09	–0.119328121	–1,217.537277	0	0	0	0
$g_2(X)$	6.053E – 07	0.062982226	1.003E – 09	–55.58228047	–0.000000034	–	0	0	0
$g_3(X)$	2.520537709	2.906819152	1E – 10	–98.70645161	–17.42270144	–	–	–	–1E – 10
$g_4(X)$	6.798E – 07	0.055724082	0.986864108	–61.38157895	–8E – 10	–	–	–	–0.187114142
$g_5(X)$	3.1922E – 06	0.00979699	0.997360146	0	–	–	–	–	–0.684755597
$g_6(X)$	–	–	1.010153898	–70,650.66906	–	–	–	–	–
$g_7(X)$	–	–	1.020591661	–	–	–	–	–	–
$g_8(X)$	–	–	698.577268	–	–	–	–	–	–
$g_9(X)$	–	–	475.8271647	–	–	–	–	–	–
$g_{10}(X)$	–	–	209.036936	–	–	–	–	–	–
$g_{11}(X)$	–	–	1.04527E – 06	–	–	–	–	–	–
Objective function									
$f(X)$	54,444.47748	55,326.29341	16.63450513	419,187.0574	5,616.152055	16.20583322	68.87755102	49.06225601	0.024123639

Table 6.23 Error values achieved in 1.00E + 03, 1.00E + 04 and 1.00E + 05 function evaluations for the functions B01–B08 with dimension $D = 10$

$D = 10$		B01	B02	B03	B04	B05	B06	B07	B08
1.00E + 03	Best	4.96E − 01	6.40E + 01	1.89E + 05	1.46E + 02	4.27E + 00	4.83E + 01	4.15E − 02	2.04E + 01
	7th	8.43E + 00	1.92E + 02	5.56E + 05	3.50E + 02	2.32E + 01	1.68E + 04	2.36E − 01	2.06E + 01
	Median	1.48E + 01	4.51E + 02	9.97E + 05	1.49E + 03	6.75E + 01	4.11E + 04	4.29E − 01	2.07E + 01
	19th	2.14E + 01	9.11E + 02	2.10E + 06	2.95E + 03	1.28E + 02	8.83E + 04	7.34E − 01	2.08E + 01
	Worst	1.25E + 02	2.37E + 03	4.83E + 06	6.70E + 03	3.72E + 03	3.68E + 06	2.98E + 00	2.08E + 01
	Mean	2.32E + 01	6.08E + 02	1.55E + 06	2.04E + 03	3.30E + 02	2.47E + 05	5.80E − 01	2.07E + 01
	Std	2.95E + 01	5.71E + 02	1.34E + 06	1.86E + 03	8.33E + 02	7.43E + 05	6.28E − 01	1.08E − 01
1.00E + 04	Best	0.00E + 00	0.00E + 00	4.29E + 03	0.00E + 00	0.00E + 00	2.04E − 01	1.02E − 01	2.03E + 01
	7th	0.00E + 00	0.00E + 00	8.32E + 03	0.00E + 00	0.00E + 00	6.34E − 01	1.23E − 01	2.04E + 01
	Median	0.00E + 00	0.00E + 00	3.87E + 04	0.00E + 00	0.00E + 00	7.92E − 01	1.43E − 01	2.04E + 01
	19th	0.00E + 00	0.00E + 00	8.35E + 04	0.00E + 00	0.00E + 00	1.04E + 00	1.67E − 01	2.05E + 01
	Worst	0.00E + 00	0.00E + 00	9.56E + 04	0.00E + 00	0.00E + 00	1.34E + 00	1.87E − 01	2.06E + 01
	Mean	0.00E + 00	0.00E + 00	3.98E + 04	0.00E + 00	0.00E + 00	6.93E − 01	1.32E − 01	2.04E + 01
	Std	0.00E + 00	0.00E + 00	6.76E + 03	0.00E + 00	0.00E + 00	4.34E − 01	1.16E − 01	1.08E − 01
1.00E + 05	Best	0.00E + 00	0.00E + 00	7.46E + 02	0.00E + 00	0.00E + 00	3.10E − 02	8.30E − 03	2.00E + 01
	7th	0.00E + 00	0.00E + 00	1.25E + 03	0.00E + 00	0.00E + 00	4.00E − 02	9.40E − 03	2.00E + 01
	Median	0.00E + 00	0.00E + 00	2.85E + 03	0.00E + 00	0.00E + 00	4.20E − 02	1.27E − 02	2.00E + 01
	19th	0.00E + 00	0.00E + 00	7.43E + 03	0.00E + 00	0.00E + 00	5.10E − 02	1.32E − 02	2.01E + 01
	Worst	0.00E + 00	0.00E + 00	1.34E + 04	0.00E + 00	0.00E + 00	5.30E − 02	1.43E − 02	2.01E + 01
	Mean	0.00E + 00	0.00E + 00	2.86E + 03	0.00E + 00	0.00E + 00	4.10E − 02	1.26E − 02	2.00E + 01
	Std	0.00E + 00	0.00E + 00	1.54E + 03	0.00E + 00	0.00E + 00	1.00E − 02	3.20E − 02	2.00E − 03

Table 6.24 Error values achieved in 1.00E + 03, 1.00E + 04 and 1.00E + 05 function evaluations for the functions B09–B17 with dimension $D = 10$

$D = 10$		B09	B10	B11	B12	B13	B14	B15	B16	B17
1.00E + 03	Best	7.76E + 00	2.08E + 01	5.03E + 00	1.10E + 04	1.29E + 00	3.50E + 00	1.43E + 02	1.18E + 02	1.64E + 02
	7th	1.75E + 01	3.26E + 01	6.96E + 00	1.68E + 04	1.61E + 00	3.85E + 00	3.05E + 02	1.56E + 02	1.93E + 02
	Median	2.05E + 01	4.51E + 01	7.82E + 00	2.02E + 04	1.83E + 00	4.06E + 00	4.69E + 02	1.97E + 02	2.11E + 02
	19th	2.51E + 01	4.96E + 01	9.71E + 00	3.84E + 04	2.48E + 00	4.13E + 00	5.03E + 02	2.18E + 02	2.36E + 02
	Worst	3.52E + 01	6.92E + 01	1.15E + 01	6.64E + 04	3.96E + 00	4.30E + 00	6.77E + 02	5.10E + 02	3.13E + 02
	Mean	2.10E + 01	4.32E + 01	8.18E + 00	2.75E + 04	2.18E + 00	3.99E + 00	4.18E + 02	1.99E + 02	2.18E + 02
	Std	6.42E + 00	1.35E + 01	1.66E + 00	1.60E + 04	7.82E − 01	2.09E − 01	1.53E + 02	7.67E + 01	3.68E + 01
1.00E + 04	Best	2.14E + 00	1.04E + 01	3.09E + 00	5.48E + 02	4.00E − 01	2.30E + 00	5.04E + 01	9.02E + 01	1.20E + 02
	7th	4.52E + 00	1.13E + 01	4.12E + 00	1.23E + 03	5.12E − 01	2.53E + 00	7.23E + 01	1.27E + 02	1.49E + 02
	Median	1.23E + 01	1.92E + 01	4.97E + 00	3.26E + 03	7.83E − 01	2.67E + 00	1.25E + 02	1.65E + 02	1.84E + 02
	19th	3.87E + 01	2.24E + 01	5.30E + 00	7.32E + 03	9.37E − 01	3.02E + 00	1.67E + 02	1.87E + 02	2.10E + 02
	Worst	5.32E + 01	2.40E + 01	5.97E + 00	3.49E + 04	1.23E + 00	3.40E + 00	1.93E + 02	2.97E + 02	2.43E + 02
	Mean	1.21E + 01	1.86E + 01	4.93E + 00	3.36E + 03	6.93E − 01	2.60E + 00	1.26E + 02	1.69E + 02	1.83E + 02
	Std	1.02E + 01	9.32E + 00	8.00E − 01	9.34E + 03	1.34E − 01	1.65E − 01	5.34E + 02	6.32E + 02	6.39E + 02
1.00E + 05	Best	9.80E − 01	9.23E + 00	2.98E + 00	4.02E + 01	1.15E − 01	2.21E + 00	4.10E + 01	9.02E + 01	9.10E + 01
	7th	9.90E − 01	1.02E + 01	3.23E + 00	1.25E + 03	1.90E − 01	2.38E + 00	5.30E + 01	9.05E + 01	1.02E + 02
	Median	9.90E − 01	1.15E + 01	3.87E + 00	2.34E + 03	2.15E − 01	2.68E + 00	8.70E + 01	9.06E + 01	1.06E + 02
	19th	9.95E − 01	1.32E + 01	4.02E + 00	3.56E + 03	2.45E − 01	2.94E + 00	1.27E + 02	9.13E + 01	1.14E + 02
	Worst	9.96E − 01	1.42E + 01	5.43E + 00	8.74E + 03	2.76E − 01	3.03E + 00	1.46E + 02	9.17E + 01	1.21E + 02
	Mean	9.92E − 01	1.17E + 01	3.88E + 00	2.35E + 03	2.14E − 01	2.67E + 00	8.60E + 01	9.06E + 01	1.07E + 02
	Std	1.00E − 03	3.21E + 00	1.02E + 00	3.35E + 03	5.49E − 02	3.15E − 01	4.07E + 01	5.64E − 01	1.03E + 01

Table 6.25 Error values achieved in 1.00E + 03, 1.00E + 04 and 1.00E + 05 function evaluations for the functions B18–B25 with dimension $D = 10$

$D = 10$		B18	B19	B20	B21	B22	B23	B24	B25
1.00E + 03	Best	8.03E + 02	6.40E + 02	4.88E + 02	5.03E + 02	7.86E + 02	5.59E + 02	2.05E + 02	6.45E + 02
	7th	8.28E + 02	8.34E + 02	8.03E + 02	9.38E + 02	8.23E + 02	9.57E + 02	2.22E + 02	8.46E + 02
	Median	9.78E + 02	9.53E + 02	9.91E + 02	1.16E + 03	8.67E + 02	1.14E + 03	5.61E + 02	8.66E + 02
	19th	1.04E + 03	1.02E + 03	1.01E + 03	1.22E + 03	9.18E + 02	1.20E + 03	8.62E + 02	1.27E + 03
	Worst	1.08E + 03	1.12E + 03	1.07E + 03	1.31E + 03	1.00E + 03	1.28E + 03	1.17E + 03	1.38E + 03
	Mean	9.50E + 02	9.20E + 02	9.01E + 02	1.07E + 03	8.76E + 02	1.03E + 03	5.65E + 02	9.93E + 02
	Std	9.66E + 01	1.21E + 02	1.73E + 02	2.14E + 02	6.26E + 01	2.45E + 02	3.35E + 02	2.32E + 02
1.00E + 04	Best	5.26E + 02	3.45E + 02	3.76E + 02	4.17E + 02	5.22E + 02	5.52E + 02	2.00E + 02	5.83E + 02
	7th	5.87E + 02	4.53E + 02	4.27E + 02	4.67E + 02	5.65E + 02	5.68E + 02	2.00E + 02	6.12E + 02
	Median	6.35E + 02	5.67E + 02	5.33E + 02	4.82E + 02	5.87E + 02	6.12E + 02	2.00E + 02	6.82E + 02
	19th	6.93E + 02	6.32E + 02	5.67E + 02	5.06E + 02	6.12E + 02	6.47E + 02	2.00E + 02	7.23E + 02
	Worst	7.83E + 02	6.94E + 02	5.92E + 02	5.62E + 02	6.66E + 02	7.18E + 02	2.00E + 02	8.76E + 02
	Mean	6.39E + 02	5.66E + 02	5.23E + 02	4.83E + 02	5.89E + 02	6.21E + 02	2.00E + 02	6.93E + 02
	Std	1.23E + 02	8.20E + 01	9.20E + 01	5.12E + 01	7.02E + 01	6.01E + 01	5.23E − 04	1.30E + 02
1.00E + 05	Best	4.40E + 02	3.20E + 02	3.42E + 02	3.70E + 02	5.20E + 02	5.52E + 02	2.00E + 02	5.21E + 02
	7th	4.82E + 02	3.54E + 02	3.89E + 02	3.89E + 02	5.23E + 02	5.63E + 02	2.00E + 02	5.55E + 02
	Median	5.01E + 02	3.78E + 02	4.23E + 02	4.06E + 02	5.28E + 02	5.97E + 02	2.00E + 02	5.78E + 02
	19th	5.20E + 02	3.92E + 02	4.79E + 02	4.23E + 02	5.34E + 02	6.21E + 02	2.00E + 02	5.93E + 02
	Worst	5.80E + 02	4.32E + 02	5.23E + 02	4.44E + 02	5.39E + 02	6.43E + 02	2.00E + 02	6.13E + 02
	Mean	5.02E + 02	3.77E + 02	4.26E + 02	4.06E + 02	5.29E + 02	5.95E + 02	2.00E + 02	5.77E + 02
	Std	6.10E + 01	3.75E + 01	6.41E + 01	2.58E + 01	6.97E + 00	3.42E + 01	2.25E − 04	2.00E + 01

Table 6.26 Error values achieved in 1.00E + 03, 1.00E + 04 and 1.00E + 05 and 3.00E + 05 function evaluations for the functions B01–B08 with dimension $D = 30$ [24]

$D = 30$		B01	B02	B03	B04	B05	B06	B07	B08
1.00E + 03	Best	5.43E + 03	6.39E + 04	3.47E + 06	3.96E + 04	1.24E + 04	5.23E + 08	4.83E + 00	2.09E + 01
	7th	6.54E + 03	7.43E + 04	4.58E + 07	4.25E + 04	1.69E + 04	7.39E + 08	9.83E + 00	2.10E + 01
	Median	1.23E + 04	8.21E + 04	7.93E + 07	4.79E + 04	1.84E + 04	1.46E + 09	1.57E + 01	2.12E + 01
	19th	1.79E + 04	8.94E + 04	1.02E + 08	5.63E + 04	2.11E + 04	2.76E + 09	1.93E + 01	2.13E + 01
	Worst	1.97E + 04	1.10E + 05	2.33E + 08	5.98E + 04	2.65E + 04	3.33E + 09	2.35E + 01	2.13E + 01
	Mean	1.29E + 04	8.43E + 04	8.02E + 07	4.77E + 04	1.85E + 04	1.45E + 09	1.46E + 01	2.11E + 01
	Std	5.75E + 03	1.76E + 04	7.78E + 07	7.79E + 03	4.67E + 03	1.11E + 09	6.65E + 00	1.62E − 01
1.00E + 04	Best	3.67E + 00	1.23E + 02	4.48E + 05	3.24E + 04	3.52E + 03	9.83E + 03	4.45E − 01	2.09E + 01
	7th	4.97E + 00	1.57E + 02	8.23E + 05	7.83E + 04	7.54E + 03	1.34E + 04	6.73E − 01	2.09E + 01
	Median	5.29E + 00	1.87E + 02	9.22E + 05	9.32E + 04	9.22E + 03	1.89E + 04	9.32E − 01	2.10E + 01
	19th	6.72E + 00	2.31E + 02	1.29E + 06	1.14E + 05	1.23E + 04	2.13E + 04	1.29E + 00	2.10E + 01
	Worst	9.33E + 00	2.56E + 02	2.87E + 06	1.32E + 05	1.82E + 04	2.45E + 04	3.43E + 00	2.13E + 01
	Mean	5.28E + 00	1.92E + 02	9.12E + 05	9.00E + 04	9.87E + 04	1.76E + 04	1.35E + 00	2.10E + 01
	Std	2.77E + 00	3.20E + 01	9.43E + 05	3.81E + 04	5.50E + 03	5.94E + 03	1.20E + 00	1.64E − 01
1.00E + 05	Best	0.00E + 00	0.00E + 00	1.76E + 05	3.34E + 02	2.24E + 03	4.46E + 01	3.40E − 03	2.08E + 01
	7th	0.00E + 00	3.00E − 02	1.89E + 05	8.23E + 02	7.22E + 03	9.83E + 01	4.32E − 03	2.09E + 01
	Median	0.00E + 00	7.34E − 02	1.94E + 05	1.65E + 03	8.96E + 03	1.23E + 02	8.72E − 03	2.09E + 01
	19th	0.00E + 00	9.23E − 02	2.26E + 05	5.53E + 03	1.21E + 04	1.97E + 02	1.23E − 02	2.09E + 01
	Worst	0.00E + 00	1.84E − 01	2.87E + 05	9.21E + 03	1.93E + 04	2.13E + 02	3.42E − 02	2.10E + 01
	Mean	0.00E + 00	7.59E − 02	1.96E + 05	1.67E + 03	9.26E + 03	1.24E + 02	8.70E − 03	2.09E + 01
	Std	0.00E + 00	6.30E − 02	4.06E + 04	3.47E + 03	5.66E + 03	6.28E + 01	1.14E − 02	6.32E − 02
3.00E + 05	Best	0.00E + 00	0.00E + 00	1.76E + 05	0.00E + 00	2.17E + 03	1.18E + 01	1.28E − 06	2.08E + 01
	7th	0.00E + 00	0.00E + 00	1.78E + 05	0.00E + 00	3.28E + 03	1.54E + 01	4.58E − 05	2.08E + 01
	Median	0.00E + 00	0.00E + 00	1.81E + 05	0.00E + 00	4.28E + 03	1.84E + 01	3.44E − 03	2.08E + 01
	19th	0.00E + 00	0.00E + 00	1.86E + 05	0.00E + 00	5.33E + 03	5.84E + 01	3.29E − 02	2.09E + 01
	Worst	0.00E + 00	0.00E + 00	1.88E + 05	0.00E + 00	7.95E + 03	6.93E + 01	4.49E − 02	2.09E + 01
	Mean	0.00E + 00	0.00E + 00	1.82E + 05	0.00E + 00	4.60E + 03	3.47E + 01	1.63E − 02	2.08E + 01
	Std	0.00E + 00	0.00E + 00	4.58E + 03	0.00E + 00	1.98E + 03	2.42E + 01	1.89E − 02	4.90E − 02

Table 6.27 Error values achieved in 1.00E + 03, 1.00E + 04 and 1.00E + 05 and 3.00E + 05 function evaluations for the functions B09–B17 with dimension $D = 30$ [24]

$D = 30$		B09	B10	B11	B12	B13	B14	B15	B16	B17
1.00E + 03	Best	1.65E + 02	3.42E + 02	4.36E + 01	6.34E + 05	4.47E + 01	1.39E + 01	6.13E + 02	4.58E + 02	4.38E + 02
	7th	1.86E + 02	3.80E + 02	4.44E + 02	7.13E + 05	4.89E + 01	1.41E + 01	7.54E + 02	4.83E + 02	4.49E + 02
	Median	2.13E + 02	4.17E + 02	4.69E + 02	7.75E + 05	5.18E + 01	1.41E + 01	7.93E + 02	4.96E + 02	5.26E + 02
	19th	2.35E + 02	4.30E + 02	4.92E + 02	8.42E + 05	7.42E + 01	1.42E + 01	8.27E + 02	5.24E + 02	5.37E + 02
	Worst	2.67E + 02	4.44E + 02	5.01E + 02	8.99E + 05	1.34E + 02	1.42E + 01	8.72E + 02	6.24E + 02	5.92E + 02
	Mean	2.13E + 02	4.03E + 02	3.90E + 02	7.73E + 05	7.07E + 01	1.41E + 01	7.72E + 02	5.17E + 02	5.08E + 02
	Std	3.59E + 01	3.70E + 01	1.74E + 02	9.33E + 04	3.32E + 01	1.10E − 01	8.84E + 01	5.76E + 01	5.76E + 01
1.00E + 04	Best	9.80E + 01	1.96E + 02	3.42E + 02	2.34E + 05	1.15E + 01	1.35E + 01	3.93E + 02	3.12E + 02	2.77E + 02
	7th	1.06E + 02	2.18E + 02	3.67E + 02	2.83E + 05	1.26E + 01	1.38E + 01	4.59E + 02	3.89E + 02	3.26E + 02
	Median	1.19E + 02	2.26E + 02	3.82E + 02	3.01E + 05	1.48E + 01	1.39E + 01	5.13E + 02	4.13E + 02	3.34E + 02
	19th	1.26E + 02	2.39E + 02	4.04E + 02	3.26E + 05	1.59E + 01	1.39E + 01	5.88E + 02	4.73E + 02	3.48E + 02
	Worst	1.86E + 02	2.41E + 02	4.12E + 02	3.36E + 05	1.64E + 01	1.40E + 01	6.32E + 02	4.97E + 02	3.52E + 02
	Mean	1.27E + 02	2.24E + 02	3.81E + 02	2.96E + 05	1.42E + 01	1.38E + 01	5.17E + 02	4.17E + 02	3.27E + 02
	Std	3.47E + 01	1.83E + 01	2.83E + 01	4.04E + 04	2.12E + 00	1.92E − 01	9.62E + 01	7.31E + 01	3.01E + 01
1.00E + 05	Best	5.69E + 01	1.48E + 02	3.42E + 01	7.24E + 04	8.21E + 00	1.31E + 01	2.00E + 02	2.63E + 02	2.64E + 02
	7th	6.72E + 01	1.68E + 02	3.46E + 01	1.05E + 05	9.32E + 00	1.34E + 01	2.00E + 02	2.89E + 02	2.64E + 02
	Median	8.63E + 01	1.97E + 02	3.53E + 01	1.32E + 05	9.87E + 00	1.34E + 01	3.00E + 02	3.18E + 02	2.84E + 02
	19th	8.92E + 01	2.17E + 02	3.76E + 01	1.43E + 05	1.12E + 01	1.34E + 01	4.00E + 02	3.27E + 02	3.02E + 02
	Worst	9.01E + 01	2.36E + 02	3.99E + 01	1.58E + 05	1.14E + 01	1.36E + 01	4.00E + 02	3.41E + 02	3.33E + 02
	Mean	7.79E + 01	1.93E + 02	3.63E + 01	1.22E + 05	1.00E + 01	1.34E + 01	3.00E + 02	3.08E + 02	2.89E + 02
	Std	1.34E + 01	3.19E + 01	2.14E + 00	3.03E + 04	1.20E + 00	1.60E − 01	8.94E + 01	2.81E + 01	2.60E + 01
3.00E + 05	Best	2.12E + 01	6.16E + 01	1.38E + 01	1.43E + 03	1.93E + 00	1.22E + 01	2.00E + 02	2.52E + 02	2.64E + 02
	7th	2.25E + 01	8.23E + 01	1.58E + 01	2.66E + 03	2.30E + 00	1.31E + 01	2.00E + 02	2.78E + 00	2.64E + 02
	Median	2.31E + 01	9.96E + 01	1.82E + 01	5.64E + 03	2.67E + 00	1.32E + 01	3.00E + 02	2.83E + 02	2.64E + 02
	19th	2.37E + 01	1.24E + 02	1.93E + 01	2.34E + 04	3.34E + 00	1.33E + 01	3.00E + 02	2.95E + 02	2.76E + 02
	Worst	2.46E + 01	1.78E + 02	2.16E + 01	5.89E + 04	4.83E + 00	1.34E + 01	4.00E + 02	3.24E + 02	2.95E + 02
	Mean	2.30E + 01	1.09E + 02	1.77E + 01	1.84E + 04	3.01E + 00	1.30E + 01	2.80E + 02	2.31E + 02	2.73E + 02
	Std	1.14E + 00	4.02E + 01	2.71E + 00	2.17E + 04	1.02E + 00	4.27E − 01	7.48E + 01	1.17E + 02	1.21E + 01

Table 6.28 Error values achieved in 1.00E + 03, 1.00E + 04 and 1.00E + 05 and 3.00E + 05 function evaluations for the functions B18–B25 with dimension $D = 30$ [24]

$D = 30$		B18	B19	B20	B21	B22	B23	B24	B25
1.00E + 03	Best	1.20E + 03	1.17E + 03	1.20E + 03	1.14E + 03	1.43E + 03	1.23E + 03	2.03E + 02	2.22E + 02
	7th	1.29E + 03	1.21E + 03	1.27E + 03	1.20E + 03	1.58E + 03	1.25E + 03	2.03E + 02	2.43E + 02
	Median	1.32E + 03	1.23E + 03	1.31E + 03	1.23E + 03	1.62E + 03	1.26E + 03	2.04E + 02	2.73E + 02
	19th	1.35E + 03	1.26E + 03	1.35E + 03	1.25E + 03	1.74E + 03	1.26E + 03	2.05E + 02	2.83E + 02
	Worst	1.42E + 03	1.40E + 03	1.38E + 03	1.28E + 03	1.88E + 03	1.29E + 03	2.06E + 02	2.89E + 02
	Mean	1.32E + 03	1.25E + 03	1.30E + 03	1.22E + 03	1.65E + 03	1.26E + 03	2.04E + 02	2.62E + 02
	Std	7.23E + 01	7.86E + 01	6.31E + 01	4.77E + 01	1.52E + 02	1.94E + 01	1.17E + 00	2.55E + 01
1.00E + 04	Best	9.03E + 02	9.89E + 02	9.89E + 02	5.00E + 02	9.06E + 02	7.28E + 02	2.01E + 02	2.00E + 02
	7th	1.01E + 03	9.93E + 02	1.01E + 03	5.60E + 02	1.01E + 03	7.43E + 02	2.01E + 02	2.00E + 02
	Median	1.01E + 03	1.03E + 03	1.04E + 03	6.30E + 02	1.10E + 03	8.14E + 02	2.01E + 02	2.03E + 02
	19th	1.03E + 03	1.03E + 03	1.06E + 03	7.20E + 02	1.14E + 03	8.63E + 02	2.01E + 02	2.03E + 02
	Worst	1.03E + 03	1.05E + 03	1.07E + 03	8.60E + 02	1.18E + 03	8.87E + 02	2.01E + 02	2.05E + 02
	Mean	9.97E + 02	1.02E + 03	1.03E + 03	6.54E + 02	1.07E + 03	8.07E + 02	2.01E + 02	2.02E + 02
	Std	5.33E + 01	2.63E + 01	3.39E + 01	1.41E + 02	1.10E + 02	7.06E + 01	0.00E + 00	2.17E + 00
1.00E + 05	Best	9.06E + 02	9.06E + 02	9.06E + 02	5.00E + 02	8.91E + 02	5.34E + 02	2.01E + 02	2.00E + 02
	7th	9.06E + 02	9.58E + 02	9.06E + 02	5.00E + 02	9.12E + 02	5.43E + 02	2.01E + 02	2.00E + 02
	Median	9.07E + 02	9.63E + 02	9.08E + 02	5.00E + 02	9.19E + 02	5.56E + 02	2.01E + 02	2.00E + 02
	19th	9.07E + 02	1.01E + 03	9.57E + 02	5.60E + 02	9.27E + 02	5.63E + 02	2.01E + 02	2.00E + 02
	Worst	9.70E + 02	1.05E + 03	1.04E + 02	6.44E + 02	9.31E + 02	5.97E + 02	2.01E + 02	2.00E + 02
	Mean	9.19E + 02	9.77E + 02	7.56E + 02	5.41E + 02	9.16E + 02	5.59E + 02	2.01E + 02	2.00E + 02
	Std	2.54E + 01	4.90E + 01	3.27E + 02	5.66E + 01	1.41E + 01	2.17E + 01	3.89E − 14	0.00E + 00
3.00E + 05	Best	9.08E + 02	9.08E + 02	9.03E + 02	5.00E + 02	8.60E + 02	5.34E + 02	2.00E + 02	2.00E + 02
	7th	9.08E + 02	9.08E + 02	9.08E + 02	5.00E + 02	8.77E + 02	5.34E + 02	2.01E + 02	2.00E + 02
	Median	9.08E + 02	9.08E + 02	9.08E + 02	5.00E + 02	9.01E + 02	5.36E + 02	2.01E + 02	2.00E + 02
	19th	9.09E + 02	9.09E + 02	9.09E + 02	5.02E + 02	9.03E + 02	5.36E + 02	2.01E + 02	2.00E + 02
	Worst	9.09E + 02	9.10E + 02	9.09E + 02	5.05E + 02	9.04E + 02	5.56E + 02	2.01E + 02	2.00E + 02
	Mean	9.08E + 02	9.09E + 02	9.07E + 02	5.01E + 02	8.89E + 02	5.39E + 02	2.01E + 02	2.00E + 02
	Std	4.90E − 01	8.00E − 01	2.24E + 00	1.96E + 00	1.76E + 01	8.45E + 00	4.00E − 01	0.00E + 00

Table 6.29 Computational complexity of TLBO

	T0	*T1*	$T2_{(\text{mean})}$	Complexity
$D = 10$	0.4367	1,222.9	2,786.8	3,581.2
$D = 30$	0.4367	1,412.8	2,997.5	3,628.8
$D = 50$	0.4367	1,861.1	3,535.4	3,834.0

graphical representation is shown in Fig. 6.10. It is observed from the results that ECHT with TLBO has outperformed other constraint handling technique by performing 1.44, 1.3, 1.3 and 1.18 times better than SR, SF, SP and EC techniques respectively in searching the best solutions. Moreover, ECHT with TLBO has performed 6, 6, 4 and 2.4 times better than SF, EC, SR and SP techniques respectively in searching the mean solution. The performance of TLBO is also compared with differential evolution (DE) for all the 13 problems. It is observed from the results that performance of TLBO and DE with ECHT techniques is nearly same and have produced similar results for most of the benchmark functions except H02 and H09 for which both the algorithms have outperformed each other in searching the mean solution.

All the nature-inspired algorithms such as GA, PSO, ACO, ABC, HS, etc. require algorithm parameters to be set for their proper working. Proper selection of parameters is essential for the searching of the optimum solution by these algorithms. A change in the algorithm parameters influences the effectiveness of the algorithm. To avoid this difficulty an optimization method, TLBO, which is algorithm parameter free, is presented in this book. This method works on the effect of influence of a teacher on learners. Like other nature-inspired algorithms, TLBO is also a population-based method which uses a population of solutions to proceed to the global solution. As in PSO, TLBO uses the best solution of the iteration to change the existing solution in the population thereby increasing the convergence rate. TLBO does not divide the population like ABC and SFLA. Like GA which uses selection, crossover and mutation phase and ABC which uses employed, onlooker and scout bees phase, TLBO uses two different phases, 'teacher phase' and 'learner phase'. TLBO uses the mean value of the population to update the solution. TLBO implements greediness to accept the good solution like ABC. In teacher phase of TLBO, the update of solution from the old solution is considered as the exploration and the greedy selection which follows it is considered as the exploitation. Similarly, in the learner phase, the updating of the solution is the exploration and the greedy selection is the exploitation. So, TLBO incorporates both exploration and exploitation effectively in the balanced manner. Applications of TLBO can be found so far in the works of [20, 21, 22, 23, 24].

Table 6.30 Comparison of TLBO with other state of art optimization algorithms for functions B01–B08 with dimension $D = 10$ and 1.00E + 05 function evaluations [24]

Algorithm	B01	B02	B03	B04	B05	B06	B07	B08
1	2.50E − 14	1.77E − 13	9.68E − 02	2.47E − 07	2.09E − 07	9.57E − 01	5.73E − 02	2.00E + 01
2	0.00E + 00	1.30E − 13	7.01E − 09	1.89E − 03	1.14E − 06	6.89E − 08	4.52E − 02	2.00E + 01
3	8.90E − 09	9.63E − 09	1.08E + 05	9.38E − 09	9.15E − 09	1.89E + 01	8.26E − 02	2.10E + 01
4	0.00E + 00	0.00E + 00	1.94E − 06	9.09E − 15	0.00E + 00	1.59E + 01	1.46E − 01	2.04E + 01
5	0.00E + 00	1.05E − 13	1.67E − 05	1.42E − 05	1.23E − 02	1.20E − 08	1.99E − 02	2.00E + 01
6	5.20E − 09	4.70E − 09	5.60E − 09	5.02E − 09	6.58E − 09	4.87E − 09	3.31E − 09	2.00E + 01
7	4.89E − 17	4.81E − 17	2.50E + 03	1.50E − 16	5.82E + 01	3.31E + 00	2.52E − 01	2.03E + 01
8	0.00E + 00	0.00E + 00	2.86E + 03	0.00E + 00	0.00E + 00	4.10E − 02	1.26E − 02	2.00E + 01

Algorithm: 1: Jian [25]; 2: Liang et al. [26]; 3: Ballester et al. [27]; 4: Ronkkonen et al. [28]; 5: Qin and Suganthan [29]; 6: Auger and Hansen [30]; 7:Akay and Karaboga [4] ; 8: TLBO

Table 6.31 Comparison of TLBO with other state of art optimization algorithms for function B09–B17 with dimension $D = 10$ and 1.00E + 05 function evaluations [24]

Algorithm	B09	B10	B11	B12	B13	B14	B15	B16	B17
1	1.25E + 01	3.86E + 01	5.58E + 00	1.31E + 02	8.87E − 01	3.78E + 00	2.71E + 02	2.20E + 02	2.22E + 02
2	0.00E + 00	3.62E + 00	4.62E + 00	2.40E + 00	3.69E − 01	2.36E + 00	4.85E + 00	9.48E + 01	1.10E + 02
3	4.02E + 00	7.30E + 00	1.91E + 00	2.60E + 02	8.38E − 01	3.05E + 00	2.54E + 02	1.10E + 02	1.19E + 02
4	9.55E − 01	1.25E + 01	8.47E − 01	3.17E + 01	9.77E − 01	3.45E + 00	2.59E + 02	1.13E + 02	1.15E + 02
5	0.00E + 00	4.97E + 00	4.89E + 00	4.50E − 07	2.20E − 01	2.92E + 00	3.20E + 01	1.01E + 02	1.14E + 02
6	2.39E − 01	7.96E − 02	9.34E − 01	2.93E + 01	6.96E − 01	3.01E + 00	2.28E + 02	9.13E + 01	1.23E + 02
7	4.87E − 17	2.22E + 01	5.46E + 00	9.85E + 01	2.96E − 02	3.41E + 00	1.53E − 01	1.75E + 02	1.96E + 02
8	9.92E − 01	1.17E + 01	3.88E + 00	2.35E + 03	2.14E − 01	2.67E + 00	8.60E + 01	9.06E + 01	1.07E + 02

Algorithm: 1: Jian [25]; 2: Liang et al. [26]; 3: Ballester et al. [27]; 4: Ronkkonen et al. [28]; 5: Qin and Suganthan [29]; 6: Auger and Hansen [30]; 7: [4]; 8: TLBO

Table 6.32 Comparison of TLBO with other state of art optimization algorithms for function B18–B25 with dimension $D = 10$ and 1.00E + 05 function evaluations [24]

Algorithm	B18	B19	B20	B21	B22	B23	B24	B25
1	1.02E + 03	9.85E + 02	9.59E + 02	9.94E + 02	8.87E + 02	1.08E + 03	7.20E + 02	1.76E + 03
2	7.61E + 02	7.14E + 02	8.22E + 02	5.36E + 02	6.92E + 02	7.30E + 02	2.24E + 02	3.66E + 02
3	4.40E + 02	3.80E + 02	4.40E + 02	6.80E + 02	7.49E + 02	5.76E + 02	2.00E + 02	4.06E + 02
4	4.00E + 02	4.20E + 02	4.60E + 02	4.92E + 02	7.18E + 02	5.72E + 02	2.00E + 02	9.23E + 02
5	7.19E + 02	7.05E + 02	7.13E + 02	4.64E + 02	7.32E + 02	6.64E + 02	2.00E + 02	3.76E + 02
6	3.32E + 02	2.26E + 02	3.00E + 02	5.00E + 02	7.29E + 02	5.59E + 02	2.00E + 02	3.74E + 02
7	4.46E + 02	4.51E + 02	4.38E + 02	4.07E + 02	8.59E + 02	4.98E + 02	2.02E + 02	2.00E + 02
8	5.02E + 02	3.77E + 02	4.26E + 02	4.06E + 02	5.29E + 02	5.95E + 02	2.00E + 02	5.77E + 02

Algorithm: 1: Jian [25]; 2: Liang et al. [26]; 3: Ballester et al. [27]; 4: Ronkkonen et al. [28]; 5: Qin and Suganthan [29]; 6: Auger and Hansen [30]; 7: Akay and Karaboga [4]; 8: TLBO

Table 6.33 Results of constrained benchmark functions using TLBO with different constraint handling techniques

Function		SF-TLBO	SP-TLBO	EC-TLBO	SR-TLBO	ECHT-TLBO
G01						
−15.00000						
	B	−15.00000	−15.00000	−15.00000	−15.00000	−15.00000
	MD	−14.99990	−14.99990	−14.99990	−14.99990	−15.00000
	MN	−14.99990	−14.99990	−14.99940	−14.99930	−15.00000
	W	−14.99980	−14.99990	−14.99960	−14.99860	−15.00000
	SD	0.00008	0.00005	0.00028	0.00065	0.00000
G02						
−0.80362						
	B	−0.80362	−0.80361	−0.80361	−0.80362	−0.80362
	MD	−0.80359	−0.79854	−0.79923	−0.79653	−0.80362
	MN	−0.80237	−0.79723	−0.79865	−0.79548	−0.80359
	W	−0.79362	−0.79127	−0.79248	−0.78294	−0.80117
	SD	0.00482	0.00507	0.00458	0.00860	0.00122
G03						
−1.00050						
	B	−1.00050	−1.00050	−1.00050	−1.00050	−1.00050
	MD	−1.00050	−1.00050	−1.00050	−1.00050	−1.00050
	MN	−1.00050	−1.00050	−1.00050	−1.00050	−1.00050
	W	−1.00050	−1.00050	−1.00050	−1.00050	−1.00050
	SD	0.00000	0.00000	0.00000	0.00000	0.00000
G04						
−30,665.53870						
	B	−30,665.53870	−30,665.53870	−30,665.53870	−30,665.53870	−30,665.53870
	MD	−30,665.53870	−30,665.53870	−30,665.53870	−30,665.53870	−30,665.53870
	MN	−30,665.53870	−30,665.53870	−30,665.53870	−30,665.53870	−30,665.53870
	W	−30,665.53870	−30,665.53870	−30,665.53870	−30,665.53870	−30,665.53870
	SD	0.00000	0.00000	0.00000	0.00000	0.00000

(Continued)

Table 6.33 (continued)

G05						
5,126.49670						
	B	5,126.49690	5,126.49670	5,126.49670	5,126.49690	5,126.49670
	MD	5,131.71470	5,126.52110	5,126.49680	5,130.01310	5,126.49670
	MN	5,161.53880	5,127.71820	5,126.50580	5,158.33170	5,126.49670
	W	5,485.18000	5,134.67510	5,126.60480	5,329.38660	5,126.49720
	SD	173.32040	3.92320	0.05269	96.60985	0.00025
G06						
−6,961.81390						
	B	−6,961.81390	−6,961.81390	−6,961.81390	−6,961.81390	−6,961.81390
	MD	−6,961.81390	−6,961.81390	−6,961.81390	−6,961.81390	−6,961.81390
	MN	−6,961.81390	−6,961.81390	−6,961.81390	−6,961.81390	−6,961.81390
	W	−6,961.81390	−6,961.81390	−6,961.81390	−6,961.81390	−6,961.81390
	SD	0.00000	0.00000	0.00000	0.00000	0.00000
G07						
24.30620						
	B	24.30620	24.30620	24.30630	24.30630	24.30620
	MD	24.30630	24.30660	24.30860	24.30770	24.30620
	MN	24.30830	24.30740	24.30920	24.30820	24.30630
	W	24.31230	24.31170	24.31230	24.32420	24.30660
	SD	0.00285	0.00253	0.00247	0.00844	0.00019
G08						
−0.09583						
	B	−0.09583	−0.09583	−0.09583	−0.09583	−0.09583
	MD	−0.09583	−0.09583	−0.09583	−0.09583	−0.09583
	MN	−0.09583	−0.09583	−0.09583	−0.09583	−0.09583
	W	−0.09583	−0.09583	−0.09583	−0.09583	−0.09583
	SD	0.00000	0.00000	0.00000	0.00000	0.00000

(Continued)

Table 6.33 (continued)

G09						
680.63006						
	B	680.63006	680.63006	680.63006	680.63006	680.63006
	MD	680.63006	680.63006	680.63006	680.63006	680.63006
	MN	680.63006	680.63006	680.63006	680.63006	680.63006
	W	680.63006	680.63006	680.63006	680.63006	680.63006
	SD	0.00000	0.00000	0.00000	0.00000	0.00000
G10						
7,049.24800						
	B	7,049.24950	7,049.24870	7,049.24890	7,049.24900	7,049.24870
	MD	7,102.45801	7,049.89430	7,093.26899	7,105.63774	7,049.24860
	MN	7,094.94359	7,050.78570	7,064.75955	7,145.59676	7,049.24880
	W	7,223.64669	7,142.54660	7,191.25121	7,405.22466	7,050.24900
	SD	74.52132	46.28947	63.74196	157.56928	0.50015
G11						
0.74990						
	B	0.74990	0.74990	0.74990	0.74990	0.74990
	MD	0.74990	0.74990	0.74990	0.74990	0.74990
	MN	0.74990	0.74990	0.74990	0.74990	0.74990
	W	0.74990	0.74990	0.74990	0.74990	0.74990
	SD	0.00000	0.00000	0.00000	0.00000	0.00000
G12						
−1.00000						
	B	−1.00000	−1.00000	−1.00000	−1.00000	−1.00000
	MD	−1.00000	−1.00000	−1.00000	−1.00000	−1.00000
	MN	−1.00000	−1.00000	−1.00000	−1.00000	−1.00000
	W	−1.00000	−1.00000	−1.00000	−1.00000	−1.00000
	SD	0.00000	0.00000	0.00000	0.00000	0.00000

(Continued)

Table 6.33 (continued)

G13						
0.05394						
	B	0.05394	0.05394	0.05394	0.05394	0.05394
	MD	0.05394	0.05394	0.05394	0.05394	0.05394
	MN	0.05394	0.05394	0.05394	0.05394	0.05394
	W	0.05394	0.05394	0.05394	0.05394	0.05394
	SD	0.00000	0.00000	0.00000	0.00000	0.00000
G14						
−47.76490						
	B	−47.23530	−47.75870	−47.65460	−47.76440	−47.76490
	MD	−45.31976	−47.55407	−46.38745	−44.83807	−47.76490
	MN	−45.21888	−47.60297	−46.36839	−45.01175	−47.76480
	W	−43.56421	−46.84574	−45.35002	−43.13874	−47.76480
	SD	1.50117	0.40591	0.94361	1.91406	0.00006
G15						
961.71502						
	B	961.71502	961.71502	961.71502	961.71502	961.71502
	MD	961.71502	961.71502	961.71502	961.71502	961.71502
	MN	961.71502	961.71502	961.71502	961.71502	961.71502
	W	961.71502	961.71502	961.71502	961.71502	961.71502
	SD	0.00000	0.00000	0.00000	0.00000	0.00000
G16						
−1.90516						
	B	−1.90516	−1.90516	−1.90516	−1.90516	−1.90516
	MD	−1.90516	−1.90516	−1.90516	−1.90516	−1.90516
	MN	−1.90516	−1.90516	−1.90516	−1.90516	−1.90516
	W	−1.90516	−1.90516	−1.90515	−1.90516	−1.90516
	SD	0.00000	0.00000	0.00000	0.00000	0.00000

(Continued)

Table 6.33 (continued)

G17						
8,853.53970						
	B	8,853.87660	8,862.34250	8,853.53970	8,853.88430	8,853.53970
	MD	8,938.15755	8,909.51570	8,853.68450	8,935.38418	8,853.53970
	MN	8,927.05966	8,877.20021	8,853.74350	8,934.44236	8,853.53970
	W	8,945.51610	8,920.84473	8,853.98434	9,275.02030	8,853.53970
	SD	42.20465	27.30393	0.18520	187.49095	0.00000
G18						
−0.86603						
	B	−0.86595	−0.86600	−0.86596	−0.86600	−0.86603
	MD	−0.86569	−0.86528	−0.86575	−0.86600	−0.86603
	MN	−0.86586	−0.86445	−0.86576	−0.86582	−0.86603
	W	−0.86594	−0.85861	−0.86592	−0.86594	−0.86603
	SD	0.00012	0.00338	0.00011	0.00008	0.00000
G19						
32.65560						
	B	32.89237	32.69850	32.80446	32.79010	32.65608
	MD	33.23746	32.79345	33.23009	33.29771	32.65944
	MN	33.27372	32.83084	33.21009	33.33337	32.65106
	W	33.93610	33.14444	34.04003	33.74028	32.65305
	SD	0.43608	0.19329	0.51781	0.38927	0.00366
G21						
193.72450						
	B	193.73700	195.36281	193.72547	193.71960	193.72450
	MD	198.73734	232.39366	240.14259	193.73840	193.72460
	MN	225.37104	241.79494	234.72651	206.11650	193.73250
	W	294.00432	330.41344	275.86452	261.96352	193.74470
	SD	46.16540	57.23883	33.64048	32.58027	0.00952

(Continued)

Table 6.33 (continued)

G23						
−400.05510						
	B	−380.36797	−324.67697	−385.13745	−371.23539	−396.95605
	MD	−353.17245	−314.81888	−354.39886	−365.46312	−386.72738
	MN	−332.78161	−289.69716	−352.32407	−342.73427	−377.06648
	W	−321.61719	−248.59725	−292.52528	−322.89408	−332.73754
	SD	25.80678	33.92946	38.73751	22.18271	28.28063
G24						
−5.50800	B	−5.50800	−5.50800	−5.50800	−5.50800	−5.50800
	MD	−5.50800	−5.50800	−5.50800	−5.50800	−5.50800
	MN	−5.50800	−5.50800	−5.50800	−5.50800	−5.50800
	W	−5.50800	−5.50800	−5.50800	−5.50800	−5.50800
	SD	0.00000	0.00000	0.00000	0.00000	0.00000

B Best; *MD* Median; *MN* Mean; *W* Worst; *SD* Standard deviation

Table 6.34 Comparison of TLBO with other optimization techniques for the constrained benchmark functions

		(ES + SR)	SMES	ATMES	Multiobjective	ISR	ECHT-EP2	ECHT-TLBO
G01								
−15.0000								
	B	−15.0000	−15.0000	−15.0000	−15.0000	−15.0000	−15.0000	-15.00000
	MD	−15.0000	−15.0000	−15.0000	−15.0000	−15.0000	−15.0000	-15.00000
	MN	−15.0000	−15.0000	−15.0000	−15.0000	−15.0000	−15.0000	-15.00000
	W	−15.0000	−15.0000	−15.0000	−14.9999	−15.0000	−15.0000	-15.00000
	SD	0.00000	0.00000	1.60E + 14	4.297E − 07	5.80E − 14	0.00000	0.00000
G02								
−0.803619								
	B	−0.803515	−0.803601	−0.803339	−0.803550	−0.803619	−0.8036191	-0.80362
	MD	−0.785800	−0.792549	−0.792420	−0.794591	−0.793082	−0.8033239	-0.80362
	MN	−0.781975	−0.785238	−0.790148	−0.792610	−0.782715	−0.7998220	-0.80359
	W	−0.726288	−0.751322	−0.756986	−0.756938	−0.723591	−0.7851820	-0.80117
	SD	2.0E − 02	1.67E − 02	1.3E − 02	1.0E − 02	2.20E − 02	6.29E − 03	0.00122
G03								
−1.0005								
	B	−1.000	−1.000	−1.000	−1.000	−1.001	−1.0005	-1.00050
	MD	−1.000	−1.000	−1.000	−1.000	−1.001	−1.0005	-1.00050
	MN	−1.000	−1.000	−1.000	−1.000	−1.001	−1.0005	-1.00050
	W	−1.000	−1.000	−1.000	−1.000	−1.001	−1.0005	-1.00050
	SD	1.9E − 04	2.09E − 04	5.9E − 05	1.304E − 12	8.20E − 09	0.00000	0.00000
G04								
−30,665.5387								
	B	−30,665.5387	−30,665.5387	−30,665.5387	−30,665.5387	−30,665.5387	−30,665.5387	−30,665.5387
	MD	−30,665.5387	−30,665.5387	−30,665.5387	−30,665.5387	−30,665.5387	−30,665.5387	−30,665.5387
	MN	−30,665.5387	−30,665.5387	−30,665.5387	−30,665.5387	−30,665.5387	−30,665.5387	−30,665.5387

(continued)

Table 6.34 (continued)

		(ES + SR)	SMES	ATMES	Multiobjective	ISR	ECHT-EP2	ECHT-TLBO
	W	−30,665.5387	−30,665.5387	−30,665.5387	−30,665.5387	−30,665.5387	−30,665.5387	−30,665.5387
	SD	2.0E − 05	0.00000	7.4E − 12	5.404E − 07	1.10E − 11	0.00000	0.00000
G05								
5,126.49670	B	5,126.49700	5,126.59900	5,126.49890	5,126.49810	5,126.49700	5,126.49670	5,126.49670
	MD	5,127.37200	5,160.19800	5,126.77600	5,126.49810	5,126.49700	5,126.49670	5,126.49670
	MN	5,128.88100	5,174.49200	5,127.64800	5,126.49810	5,126.49700	5,126.49670	5,126.49670
	W	5,142.47200	5,160.19800	5,135.25600	5,126.49840	5,126.49700	5,126.49670	5,126.49720
	SD	3.50000	50.06000	1.80000	1.727E − 07	7.20E − 13	0.00000	0.00025
G06								
−6,961.8139	B	−6,961.814	−6,961.814	−6,961.814	−6,961.81388	−6,961.814	−6,961.8139	−6,961.81390
	MD	−6,961.814	−6,961.814	−6,961.814	−6,961.81388	−6,961.814	−6,961.8139	−6,961.81390
	MN	−6,875.940	−6,961.284	−6,961.814	−6,961.81388	−6,961.814	−6,961.8139	−6,961.81390
	W	−6,350.262	−6,952.482	−6,961.814	−6,961.81388	−6,961.814	−6,961.8139	−6,961.81390
	SD	160.00000	1.85000	4.6E − 12	8.507E − 12	1.90E − 12	0.00000	0.00000
G07								
24.30620	B	24.30700	24.32700	24.30600	24.30646	24.30600	24.30620	24.30620
	MD	24.35700	24.42600	24.31300	24.30731	24.30600	24.30630	24.30620
	MN	24.37400	24.47500	24.31600	24.30740	24.30600	24.30630	24.30630
	W	24.64200	24.84300	24.35900	24.30924	24.30600	24.30630	24.30660
	SD	6.6E − 02	1.32E − 01	1.1E − 02	7.118E − 04	6.30E − 05	3.19E − 05	0.00019
G08								
−0.095825	B	−0.095825	−0.095825	−0.095825	−0.095825	−0.095825	−0.09582504	−0.09583
	MD	−0.095825	−0.095825	−0.095825	−0.095825	−0.095825	−0.09582504	−0.09583
	MN	−0.095825	−0.095825	−0.095825	−0.095825	−0.095825	−0.09582504	−0.09583
	W	−0.095825	−0.095825	−0.095825	−0.095825	−0.095825	−0.09582504	−0.09583
	SD	2.6E − 17	0.00000	2.8E − 17	2.417E − 17	2.70E − 13	0.00000	0.00000

(continued)

Table 6.34 (continued)

		(ES + SR)	SMES	ATMES	Multiobjective	ISR	ECHT-EP2	ECHT-TLBO
G09								
680.63006	B	680.63000	680.63200	680,630.00000	680.63006	680.63000	680.63006	680.63006
	MD	680.64100	680.64200	680.63300	680.63006	680.63000	680.63006	680.63006
	MN	680.65600	680.64300	680.63900	680.63006	680.63000	680.63006	680.63006
	W	680.76300	680.71900	680.67300	680.63006	680.63000	680.63006	680.63006
	SD	3.4E − 02	1.55E − 02	1.0E − 12	9.411E − 08	3.20E − 13	2.61E − 08	0.00000
G10								
7,049.24800	B	7,054.31600	7,051.90300	7,052.25300	7,049.28660	7,049.24800	7,049.24830	7,049.24870
	MD	7,372.61300	7,253.60300	7,215.35700	7,049.48615	7,049.24800	7,049.24880	7,049.24860
	MN	7,559.19200	7,253.04700	7,250.43700	7,049.52544	7,049.25000	7,049.24900	7,049.24880
	W	8,835.65500	7,638.36600	7,560.22400	7,049.98421	7,049.27000	7,049.25010	7,050.24900
	SD	530.00000	136.00000	120.00000	1.502E − 01	3.20E − 03	6.60E − 04	0.50015
G11								
0.74990	B	0.75000	0.75000	0.75000	0.75000	0.75000	0.74990	0.74990
	MD	0.75000	0.75000	0.75000	0.75000	0.75000	0.74990	0.74990
	MN	0.75000	0.75000	0.75000	0.75000	0.75000	0.74990	0.74990
	W	0.75000	0.75000	0.75000	0.75000	0.75000	0.74990	0.74990
	SD	8.0E − 05	1.52E − 04	340.00000	1.546E − 12	1.10E − 16	0.00000	0.00000
G12								
−1.000	B	−1.0000	−1.0000	−1.0000	−1.0000	−1.0000	−1.0000	−1.00000
	MD	−1.0000	−1.0000	−1.0000	−1.0000	−1.0000	−1.0000	−1.00000
	MN	−1.0000	−1.0000	−1.0000	−1.0000	−1.0000	−1.0000	−1.00000
	W	−1.0000	−1.0000	−1.0000	−1.0000	−1.0000	−1.0000	−1.00000
	SD	0.00000	0.00000	1.0E − 03	0.00000	1.20E − 09	0.00000	0.00000

(continued)

Table 6.34 (continued)

		(ES + SR)	SMES	ATMES	Multiobjective	ISR	ECHT-EP2	ECHT-TLBO
G13								
0.05394	B	0.05396	0.05399	0.05395	0.05395	0.05394	0.05394	0.05394
	MD	0.05701	0.06187	0.05395	0.05395	0.05394	0.05394	0.05394
	MN	0.06754	0.16639	0.05396	0.05395	0.06677	0.05394	0.05394
	W	0.21692	0.46829	0.05400	0.05395	0.43880	0.05394	0.05394
	SD	3.1E − 02	1.77E − 01	1.3E − 05	8.678E − 08	7.09E − 02	1.00E − 12	0.00000

B Best; *MD* Median; *MN* Mean; *W* Worst; *SD* Standard deviation

Table 6.35 Comparison of TLBO with DE for constrained benchmark functions H01–H13

	PROB	SF-TLBO	SP-TLBO	SR-TLBO	EC-TLBO	ECHT-TLBO	SF-DE	SP-DE	SR-DE	EC-DE	ECHT-DE
H01	B	0.00E + 00	0.00E + 00	0.00E + 00	0.00E + 00	0.00E + 00	5.59E − 11	1.24E − 85	7.57E − 81	5.59E − 11	8.29E − 83
	MD	3.35E + 03	0.00E + 00	0.00E + 00	3.67E + 03	0.00E + 00	3.33E + 03	3.42E − 84	3.26E − 78	3.33E + 03	2.19E − 80
	MN	5.67E + 03	0.00E + 00	0.00E + 00	5.78E + 03	0.00E + 00	4.58E + 03	2.95E − 83	1.37E − 74	4.58E + 03	2.66E − 78
	W	2.97E + 04	0.00E + 00	0.00E + 00	1.77E + 04	0.00E + 00	1.67E + 04	3.47E − 82	4.04E − 37	1.67E + 04	7.41E − 77
	SD	1.35E + 04	0.00E + 00	0.00E + 00	7.65E + 03	0.00E + 00	5.63E + 03	6.92E − 83	7.37E − 74	5.63E + 03	1.35E − 77
H02	B	−9.56E − 01	−2.27E + 00	5.36E − 01	−9.18E − 01	−2.28E + 00	−1.01E + 00	−2.27E + 00	7.52E − 02	−1.01E + 00	−2.28E + 00
	MD	1.06E + 00	−1.95E + 00	4.25E + 00	6.39E − 01	−2.24E + 00	5.91E − 01	−1.98E + 00	2.11E + 00	5.91E − 01	−2.26E + 00
	MN	8.72E − 01	−8.80E − 01	5.95E + 00	9.72E − 01	−2.23E + 00	7.73E − 01	−1.42E + 00	2.17E + 00	7.73E − 01	−2.25E + 00
	W	4.27E + 00	4.37E + 00	1.07E + 01	4.00E + 00	−2.17E + 00	2.90E + 00	1.36E + 00	5.12E + 00	2.90E + 00	−2.22E + 00
	SD	2.17E + 00	3.09E + 00	4.20E + 00	2.06E + 00	4.71E − 02	1.30E + 00	1.05E + 00	1.43E + 00	1.30E + 00	3.13E − 02
H03	B	0.00E + 00	0.00E + 00	0.00E + 00	0.00E + 00	0.00E + 00	1 08E − 81	2.95E − 85	6.25E − 80	1.08E − 81	1.19E − 83
	MD	0.00E + 00	0.00E + 00	0.00E + 00	0.00E + 00	0.00E + 00	4.59E − 80	2.15E − 83	3.40E − 78	4.59E − 80	1.14E − 81
	MN	0.00E + 00	0.00E + 00	0.00E + 00	0.00E + 00	0.00E + 00	2.05E − 78	8.31E − 83	2.20E − 77	2.05E − 78	6.90E − 81
	W	0.00E + 00	0.00E + 00	0.00E + 00	0.00E + 00	0.00E + 00	3.39E − 77	7.15E − 82	2.20E − 76	3.39E − 77	3.68E − 80
	SD	0.00E + 00	0.00E + 00	0.00E + 00	0.00E + 00	0.00E + 00	6.53E − 78	1.58E − 82	5.02E − 77	6.53E − 78	1.12E − 80
H04	B	0.00E + 00	0.00E + 00	0.00E + 00	0.00E + 00	0.00E + 00	3.89E − 93	4.17E − 94	2.50E − 94	3.89E − 93	4.91E − 95
	MD	0.00E + 00	0.00E + 00	0.00E + 00	0.00E + 00	0.00E + 00	4.01E − 91	6.93E − 93	1.99E − 92	4.01E − 91	3.24E − 93
	MN	0.00E + 00	0.00E + 00	0.00E + 00	1.72E + 02	0.00E + 00	1.12E + 03	5.81E − 92	3.41E − 91	1.12E + 03	1.01E − 92
	W	2.18E + 04	0.00E + 00	0.00E + 00	1.53E + 04	0.00E + 00	1.24E + 04	7.01E − 91	4.69E − 90	1.24E + 04	7.98E − 92
	SD	1.09E + 04	0.00E + 00	0.00E + 00	7.64E + 03	0.00E + 00	3.21E + 03	1.41E − 91	9.57E − 91	3.21E + 03	1.85E − 92
H05	B	−2.01E + 01	−2.01E + 01	−1.91E + 01	−2.01E + 01	−2.01E + 01	−2.01E + 01	−2.01E + 01	−1.90E + 01	−2.01E + 01	−2.01E + 01
	MD	−1.96E + 01	−2.00E + 01	−1.68E + 01	−1.94E + 01	−2.01E + 01	−1.95E + 01	−2.01E + 01	−1.65E + 01	−1.90E + 01	−2.01E + 01
	MN	−1.98E + 01	−1.99E + 01	−1.52E + 01	−1.98E + 01	−2.01E + 01	−1.95E + 01	−1.99E + 01	−1.67E + 01	−1.94E + 01	−2.01E + 01
	W	−1.96E + 01	−1.89E + 01	−1.23E + 01	−1.81E + 01	−2.01E + 01	−1.89E + 01	−1.88E + 01	−1.49E + 01	−1.80E + 01	−2.01E + 01
	SD	2.14E − 01	5.64E − 01	2.84E + 00	8.82E − 01	9.05E − 03	5.56E − 01	3.98E − 01	9.70E − 01	6.00E − 01	3.30E − 03
H06	B	−8.38E + 00	−8.38E + 00	−8.38E + 00	−8.38E + 00	−8.38E + 00	−8.38E + 00	−8.38E + 00	−8.38E + 00	−8.38E + 00	−8.38E + 00
	MD	−2.55E + 00	−8.19E + 00	−8.39E + 00	−2.38E + 00	−8.38E + 00	−2.55E + 00	−8.38E + 00	−8.38E + 00	−2.55E + 00	−8.38E + 00
	MN	−2.41E + 00	−7.65E + 00	−8.04E + 00	−1.21E + 00	−8.38E + 00	−4.10E + 00	−8.19E + 00	−8.17E + 00	−2.98E + 00	−8.38E + 00
	W	−2.35E + 00	−2.68E + 00	−1.15E + 00	−6.93E − 01	−8.38E + 00	−2.55E + 00	−2.55E + 00	−2.55E + 00	−2.55E + 00	−8.38E + 00
	SD	2.97E + 00	2.72E + 00	3.56E + 00	3.55E + 00	0.00E + 00	2.62E + 00	1.07E + 00	1.12E + 00	1.35E + 00	3.76E − 15

(continued)

Table 6.35 (continued)

	PROB	SF-TLBO	SP-TLBO	SR-TLBO	EC-TLBO	ECHT-TLBO	SF-DE	SP-DE	SR-DE	EC-DE	ECHT-DE
H07	B	−7.62E + 00	−7.62E + 00	−7.62E + 00	−7.62E + 00	−7.62E + 00	−7.62E + 00	−7.62E + 00	−7.62E + 00	−6.02E + 00	−7.62E + 00
	MD	−7.23E + 00	−7.53E + 00	−7.60E + 00	−6.39E + 00	−7.62E + 00	−4.42E + 00	−7.53E + 00	−7.62E + 00	−1.42E + 00	−7.62E + 00
	MN	−6.99E + 00	−6.83E + 00	−6.42E + 00	−4.48E + 00	−7.62E + 00	−4.50E + 00	−6.33E + 00	−6.82E + 00	−1.86E + 00	−7.62E + 00
	W	−1.66E + 00	−2.55E + 00	3.23E + 00	−1.74E + 00	−7.62E + 00	−1.42E + 00	−1.42E + 00	−1.42E + 00	−1.25E + 00	−7.62E + 00
	SD	2.82E + 00	2.41E + 00	5.25E + 00	2.56E + 00	0.00E + 00	3.25E + 00	2.59E + 00	1.96E + 00	1.46E + 00	4.26E − 10
H08	B	−1.55E + 02	−4.80E + 02	−4.39E + 02	1.01E + 03	−4.84E + 02	−1.62E + 02	−4.82E + 02	−4.72E + 02	5.00E + 02	−4.84E + 02
	MD	2.45E + 02	−4.60E + 02	−4.39E + 02	5.61E + 02	−4.84E + 02	1.92E + 02	−4.81E + 02	−4.69E + 02	5.00E + 02	−4.84E + 02
	MN	2.69E + 02	−4.40E + 02	−3.98E + 02	6.22E + 02	−4.84E + 02	1.80E + 02	−4.80E + 02	−4.67E + 02	5.00E + 02	−4.84E + 02
	W	5.26E + 02	−3.41E + 02	−2.75E + 02	6.81E + 02	−4.84E + 02	5.00E + 02	−4.78E + 02	−4.55E + 02	5.00E + 02	−4.84E + 02
	SD	2.81E + 02	6.20E + 01	7.79E + 01	1.99E + 02	0.00E + 00	3.70E + 02	1.13E + 00	6.41E + 00	0.00E + 00	0.00E + 00
H09	B	−6.84E + 01	−4.50E + 01	−5.12E + 01	−6.84E + 01	−6.84E + 01	−6.84E + 01	−5.15E + 01	−5.94E + 01	−6.84E + 01	−6.84E + 01
	MD	−7.15E + 01	−4.03E + 01	−4.77E + 01	−7.22E + 01	−6.84E + 01	−6.84E + 01	−4.49E + 01	−5.00E + 01	−6.84E + 01	−6.84E + 01
	MN	−6.88E + 01	−3.70E + 01	−5.03E + 01	−6.92E + 01	−6.81E + 01	−6.75E + 01	−4.47E + 01	−5.04E + 01	−6.76E + 01	−6.79E + 01
	W	−6.44E + 01	−2.20E + 01	−3.82E + 01	−5.33E + 01	−6.45E + 01	−6.35E + 01	−3.86E + 01	−4.43E + 01	−5.35E + 01	−6.35E + 01
	SD	2.91E + 00	9.91E + 00	5.94E + 00	8.45E + 00	1.90E + 00	1.71E + 00	2.84E + 00	3.21E + 00	1.86E + 00	1.09E + 00
H10	B	0.00E + 00	0.00E + 00	0.00E + 00	0.00E + 00	0.00E + 00	0.00E + 00	0.00E + 00	0.00E + 00	0.00E + 00	0.00E + 00
	MD	0.00E + 00	0.00E + 00	0.00E + 00	0.00E + 00	0.00E + 00	0.00E + 00	0.00E + 00	8.99E + 00	0.00E + 00	0.00E + 00
	MN	4.05E + 00	3.76E + 00	5.63E + 00	3.30E + 00	5.84E − 02	4.20E + 00	3.90E + 00	5.39E + 00	4.20E + 00	5.99E − 01
	W	8.99E + 00	8.99E + 00	8.99E + 00	8.99E + 00	8.99E + 00	8.99E + 00	8.99E + 00	8.99E + 00	8.99E + 00	8.99E + 00
	SD	4.27E + 00	4.26E + 00	4.44E + 00	4.24E + 00	4.49E + 00	4.56E + 00	4.53E + 00	4.48E + 00	4.56E + 00	2.28E + 00
H11	B	5.81E + 02	5.81E + 02	5.81E + 02	5.81E + 02	5.81E + 02	5.81E + 02	5.81E + 02	8.94E + 02	5.81E + 02	5.81E + 02
	MD	5.81E + 02	5.81E + 02	5.81E + 02	5.81E + 02	5.81E + 02	5.81E + 02	5.81E + 02	9.68E + 02	5.81E + 02	5.81E + 02
	MN	6.18E + 02	5.81E + 02	1.01E + 03	6.01E + 02	5.81E + 02	6.01E + 02	5.81E + 02	9.74E + 02	6.01E + 02	5.81E + 02
	W	6.78E + 02	5.81E + 02	1.28E + 03	7.04E + 02	5.81E + 02	6.55E + 02	5.81E + 02	1.11E + 03	6.55E + 02	5.81E + 02
	SD	4.57E + 01	2.89E − 04	3.43E + 02	5.91E + 01	0.00E + 00	2.70E + 01	2.11E − 14	4.91E + 01	2.70E + 01	1.32E − 11
H12	B	0.00E + 00	0.00E + 00	0.00E + 00	0.00E + 00	0.00E + 00	1.14E − 29	1.25E − 30	3.18E + 01	3.20E − 29	1.54E − 32
	MD	6.79E + 00	0.00E + 00	6.86E + 00	7.46E + 00	0.00E + 00	5.00E + 01	1.55E − 28	5.00E + 01	5.00E + 01	2.41E − 31
	MN	8.57E + 00	0.00E + 00	7.92E + 00	8.19E + 00	0.00E + 00	3.99E + 01	7.97E − 02	4.80E + 01	4.28E + 01	4.55E − 31
	W	9.81E + 00	1.43E + 00	9.63E + 00	1.14E + 01	0.00E + 00	5.00E + 01	1.29E + 00	5.00E + 01	5.00E + 01	1.75E − 30
	SD	4.37E + 00	7.13E − 01	4.23E + 00	4.81E + 00	0.00E + 00	2.03E + 01	3.04E − 01	4.81E + 00	1.72E + 01	4.61E − 31

(continued)

Table 6.35 (continued)

	PROB	SF-TLBO	SP-TLBO	SR-TLBO	EC-TLBO	ECHT-TLBO	SF-DE	SP-DE	SR-DE	EC-DE	ECHT-DE
H13	B	−4.64E + 01	−4.64E + 01	−4.64E + 01	−4.64E + 01	−4.64E + 01	−4.64E + 01	−4.64E + 01	−4.64E + 01	−4.64E + 01	−4.64E + 01
	MD	−4.64E + 01	−4.64E + 01	−4.64E + 01	−4.64E + 01	−4.64E + 01	−4.64E + 01	−4.64E + 01	−4.64E + 01	−4.64E + 01	−4.64E + 01
	MN	−4.63E + 01	−4.64E + 01	−4.60E + 01	−4.63E + 01	−4.64E + 01	−4.63E + 01	−4.63E + 01	−4.63E + 01	−4.61E + 01	−4.64E + 01
	W	−4.36E + 01	−4.36E + 01	−4.36E + 01	−4.51E + 01	−4.64E + 01	−4.36E + 01	−4.36E + 01	−4.36E + 01	−4.36E + 01	−4.64E + 01
	SD	1.35E + 00	1.40E + 00	1.36E + 00	6.16E − 01	0.00E + 00	5.13E − 01	5.13E − 01	5.12E − 01	8.57E − 01	9.47E − 15

References

1. Ahrari A, Atai A (2010) Grenade explosion method-a novel tool for optimization of multimodal functions. Appl Soft Comput 10(4):1132–1140
2. Shu SFK, Erwie Z (2007) A hybrid simplex search and particle swarm optimization for unconstrained optimization. Eur J Oper Res 181:527–548
3. Karaboga D, Akay B (2009) Artificial bee colony (ABC), harmony search and bees algorithms on numerical optimization. In: IPROMS-2009, Innovative production machines and systems virtual conference, Cardiff, UK
4. Akay B, Karaboga D (2010) Artificial bee colony algorithm for large-scale problems and engineering design optimization. J Intell Manuf. doi: 10.1007/s10845-010-0393-4
5. Liu H, Cai Z, Wang Y (2010) Hybridizing particle swarm optimization with differential evolution for constrained numerical and engineering optimization. Appl Soft Comput 10:629–640
6. Amirjanov A (2006) The development of a changing range genetic algorithm. Comput Methods Appl Mech Eng 195:2495–2508
7. Tessema B, Yen G (2006) A self adaptive penalty function based algorithm for constrained optimization. Proceedings of 2006 IEEE congress on evolutionary computation, pp 246–253
8. Huang FA, Wang L, He Q (2007) An effective co-evolutionary differential evolution for constrained optimization. Appl Math Comput 186(1):340–356
9. Becerra RL, Coello CAC (2006) Cultured differential evolution for constrained optimization. Comput Methods Appl Mech Eng 195:4303–4322
10. Krohling RA, Coelho LS (2006) Coevolutionary particle swarm optimization using Gaussian distribution for solving constrained optimization problems. IEEE Trans Syst Man Cybern Part B Cybern 36(6):1407–1416
11. Montes E, Coello CAC (2005) A simple multi-membered evolution strategy to solve constrained optimization problems. IEEE Trans Evol Comput 9(1):1–17
12. Parsopoulos K, Vrahatis M (2005) Unified particle swarm optimization for solving constrained engineering optimization problems. In: Proceedings of advances in natural computation, LNCS 3612, Springer-Verlag, Berlin, pp 582–591
13. He Q, Wang L (2007) An effective co-evolutionary particle swarm optimization for constrained engineering design problems. Eng Appl Artif Intell 20:89–99
14. Suganthan PN, Hansen N, Liang JJ, Deb K, Chen A, Auger YP, Tiwari S (2005) Problem definitions and evaluation criteria for the CEC 2005 special session on real-parameter optimization. Technical report, Nanyang Technological University, Singapore. <http://www.ntu.edu.sg/home/EPNSugan>
15. Deb K (2000) An efficient constraint handling method for genetic algorithms. Comput Methods Appl Mech Eng 186:311–338
16. Takahama T, Sakai S (2006) Constrained optimization by the constrained differential evolution with gradient-based mutation and feasible elites. Proceedings of IEEE congress on evolutionary computation, Vancouver, BC, Canada, pp 1–8
17. Runarsson TP, Yao X (2005) Search biases in constrained evolutionary optimization. IEEE Trans Syst Man Cybern 35:233–243
18. Mallipeddi R, Suganthan PN (2010) Ensemble of constraint handling techniques. IEEE Trans Evol Comput 14:561–579
19. Wang Y, Cai Z, Guo G, Zhou Y (2007) Multiobjective optimization and hybrid evolutionary algorithm to solve constrained optimization problems. IEEE Trans Syst Man Cybern 37:560–575
20. Rao RV, Savsani VJ, and Vakharia DP (2011b) Teaching–learning-based optimization: an optimization method for continuous non-linear large scale problems. Inf Sci. doi: 10.1016/j.ins.2011.08.006

21. Rao RV, Savsani VJ, Vakharia DP (2011) Teaching–learning-based optimization: a novel method for constrained mechanical design optimization problems. Comput Aided Des 43:303–315
22. Rao RV, Kalyankar VD (2011) Parameter optimization of machining processes using a new optimization algorithm. Mat Manuf Process (in press)
23. Rao RV, Patel V (2011) Multi-objective optimization of combined Brayton and inverse Brayton cycle using advanced optimization algorithms. Engg Opt (in press)
24. Rao RV, Savsani VJ, Balic J (2012) Teaching-learning-based optimization algorithm for constrained and unconstrained real parameter optimization problems. Engg Opt (in press)
25. Jian MC (2006) Introducing recombination with dynamic linkage discovery to particle swarm optimization. Technical report NCL-TR-2006006, Department of Computer Science, National Chiao Tung University, Taiwan
26. Liang JJ, Runarsson TP, Montes EM, Clerc M, Suganthan PN, Coello CAC, Deb K (2006) Problem definitions and evolution criteria for the CEC 2006 special session on constrained real-parameter optimization. Technical report, Nanyang Technological University, Singapore. <http://www.ntu.edu.sg/home/EPNSugan>
27. Ballester PJ, Stephenson J, Carter JN, Gallagher K (2005) Real-parameter optimization performance study on the CEC-2005 benchmark with SPC-PNX. In: The 2005 IEEE congress on evolutionary computation, vol 1, pp 498–505
28. Ronkkonen J, Kukkonen S, Price KV (2005) Real-parameter optimization with differential evolution. In: The 2005 IEEE congress on evolutionary computation, vol 1, pp 506–513
29. Qin AK, Suganthan PN (2005) Self-adaptive differential evolution algorithm for numerical optimization. In: The 2005 IEEE congress on evolutionary computation, vol 2, pp 1785–1791
30. Auger A, Hansen N (2005) A restart CMA evolution strategy with increasing population size. In: The 2005 IEEE congress on evolutionary computation, vol 2, pp 1769–1776

Chapter 7
Design Optimization of Selected Thermal Equipment Using Advanced Optimization Techniques

7.1 Design Optimization of Thermoelectric Cooler

The application of thermoelectric coolers (TECs) has grown appreciably because of the need for a steady, low-temperature, environment friendly operating environment for various applications such as aerospace, military, medicine, biology and other electronic devices. However, the cooling capacity and coefficient of performance (COP) of TECs are low compared with traditional devices such as vapor compression system and vapor absorption system. Therefore, performance improvement of the TECs is an important issue in their applications [1, 2].

With the help of one-stage TEC, maximum 70 K temperature difference is produced when its hot end is maintained at room temperature. So, when a large temperature difference is required, two-stage TECs should be employed [3]. Usually two-stage TECs are commercially arranged in cascade; the cold stage is attached to the heat source and the hot stage pumps total heat to the environment. Moreover the two-stage TECs are arranged in two different design configurations as shown in Fig. 7.1. In such two-stage TECs, the determination of the number of thermoelectric (TE) modules in hot stage and cold stage as well as the supply current to the hot stage and the cold stage are important for improving the COP and cooling capacity of TECs. Moreover, the consideration of temperature-dependent material properties and existence of thermal and electric contact resistance between the contact surfaces of TECs make the determination of these parameters complex ([4, 6]).

Several investigators had used different methodologies considering different objective functions to optimize the TECs design. Chen et al. [5] carried out the optimal performance comparison of single and two-stage TE refrigeration systems. The authors had calculated the maximum COP and rate of refrigeration and optimized the internal structure parameter of the TE device. Xuan et al. [6] carried out the optimization of a two-stage TEC with two design configurations. The authors had found out the optimum ratio of the number of TE modules between the stages and optimum ratio of the electric current between stages for

R. V. Rao and V. J. Savsani, *Mechanical Design Optimization Using Advanced Optimization Techniques*, Springer Series in Advanced Manufacturing,
DOI: 10.1007/978-1-4471-2748-2_7,

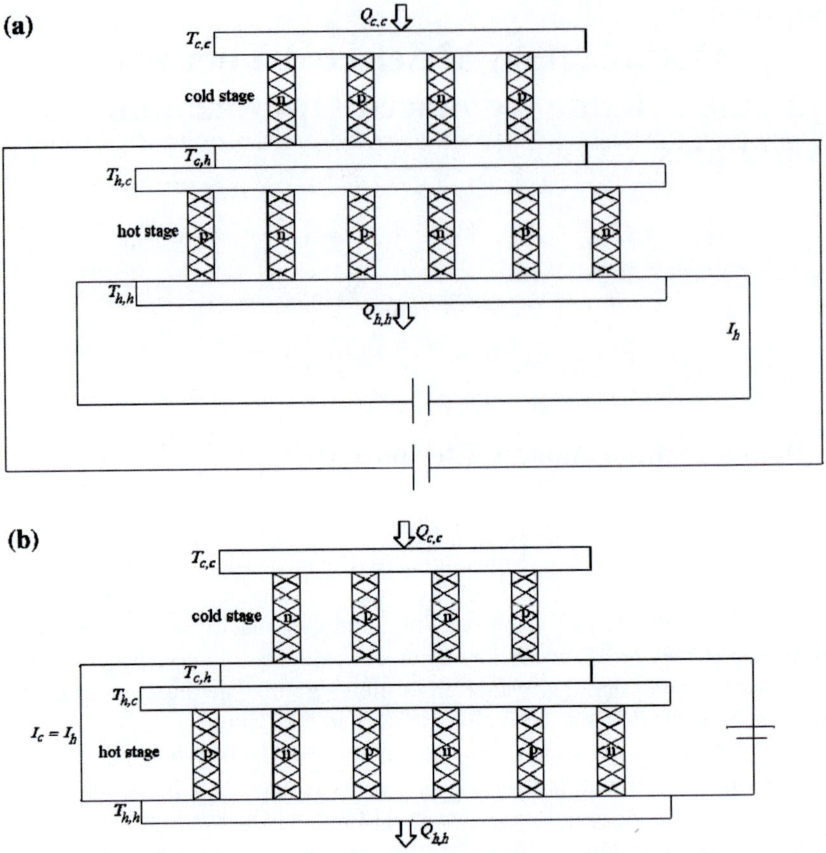

Fig. 7.1 Two-stage TEC. **a** Electrically separated. **b** Electrically connected in series (from [4], reprinted with permission from Elsevier)

maximization of cooling capacity and COP of TEC. Xuan [7] and Xuan et al. [8] carried out the performance analysis of a two-stage TEC with three design configurations. The authors had considered the maximum cooling capacity, maximum COP and the maximum temperature difference of the two-stage TEC. Chen et al. [9] carried out the parametric optimum design of a TE device. The authors had calculated the maximum COP and rate of refrigeration of the system and determined the optimal operating region of the system. Pan et al. [2] carried out the performance analysis and parametric optimization of a multi-couple TE refrigerator. The authors had determined the optimal operating-state of the COP for a TE refrigeration device.

Cheng and Lin [1] used genetic algorithm for geometric optimization of TEC. The authors had considered maximization of cooling capacity as an objective function and determined the optimum value of structure parameter of TE modules. Cheng and Shih [4] used GA for maximizing the cooling capacity and COP of a two-stage TEC. The authors had considered the effect of thermal resistance and determined the optimum value of input current and number of TE modules for two different design configurations of TEC. Abramzon [10] used multi start adaptive random search method for the numerical optimization of the TEC. The author had considered maximization of total cooling rate of the TEC as an objective function. Yu et al. [11] analyzed the optimum configuration of two-stage TE modules. The authors had investigated the influence of different parameters on the cooling performance of the TE modules. Chen et al. [12] analyzed the performance of a two-stage TE heat pump system driven by a two-stage TE generator. The authors had optimized the allocations of the TE element pairs among the two TE generators and the two TE heat pumps for maximizing the heating load and COP respectively. Several other researchers [13–16] investigated the two-stage TECs for optimization of COP or for optimum allocation of TE module.

So far, only GA is used for the optimization of TECs. Moreover, only single objective optimization of TECs was carried out by previous researchers. Considering this fact, the main objectives of this work are: (1) multi-objective optimization of the influential parameters of a two-stage TEC using the TLBO algorithm and (2) to demonstrate the effectiveness of the TLBO algorithm for multi-objective optimization of the TEC. The optimization results obtained by using TLBO are compared with those obtained by using GA for the same example considered by previous researchers.

7.1.1 Thermal Modeling of Two-Stage TECs

Based on the work of Cheng and Shih [4], thermal model of the two-stage TECs is formulated as described below (from Cheng and Shih [4]; reprinted with permission from Elsevier).

The cascade two-stage TECs are stacked one on the top of the other (as shown if Fig. 1). Here in this arrangement the top stage is the cold stage and the bottom stage is the hot stage. The COP of the two-stage TECs is given by,

$$\mathrm{COP} = \frac{Q_{c,c}}{Q_{h,h} - Q_{c,c}} \tag{7.1}$$

where, $Q_{c,c}$ and $Q_{h,h}$ are the cooling capacity of the cold side of the cold stage and the heat rejected at the hot side of hot stage, respectively and are obtained by heat balance at relevant junction of TECs.

$$Q_{c,c} = \frac{N_t}{r+1}\left[\alpha_c I_c T_{c,c} - \frac{1}{2} I_c^2 R_c - K_c (T_{c,h} - T_{c,c})\right] \tag{7.2}$$

$$Q_{h,h} = \frac{N_t r}{r+1}\left[\alpha_h I_h T_{h,h} + \frac{1}{2} I_h^2 R_h - K_h (T_{h,h} - T_{h,c})\right] \tag{7.3}$$

where, N_t is the total number of TE modules of two-stages and r is the ratio of the number of TE modules between the hot stage (N_h) to the cold stage (N_c). I_c and I_h are the input current to the cold stage and the hot stage, respectively. T is the temperature of the TEC as shown in Fig. 7.1. α, R and K are the Seebeck coefficient, electrical resistance and thermal conductance of the cold stage and the hot stage, respectively and their relation to TE material properties is given by,

$$\alpha_i = \left(\alpha_{i,p} - \alpha_{i,n}\right)_{T_{i,\mathrm{ave}}} \tag{7.4}$$

$$R_i = \frac{\left[\rho_{i,p} + \rho_{i,n}\right]_{T_{i,\mathrm{ave}}}}{G} \tag{7.5}$$

$$K_i = \left[k_{i,p} + k_{i,n}\right]_{T_{i,\mathrm{ave}}} G \tag{7.6}$$

where, subscript i stands for the cold side (c) and the hot side (h) of TEC; subscript *ave* indicates the average value and subscripts p and n indicate the properties of p- and n-type TE modules. G is the structure parameter of the TE modules and indicates the ratio of cross-section area to the length of TE modules. ρ and k are the electric resistivity and thermal conductivity of the TE material, respectively. As the material properties are considered to be dependent on the average temperature of the cold side and hot side of each stage, their values are calculated by the following correlation [4]

$$\alpha_{i,p} = -\alpha_{i,n} = \left(22224 + 9300.6\, T_{i,\mathrm{ave}} - 0.9905\, T^2_{i,\mathrm{ave}}\right) 10^{-9} \tag{7.7}$$

$$\rho_{i,p} = \rho_{i,n} = \left(5112 + 163.4\, T_{i,\mathrm{ave}} + 0.6279\, T^2_{i,\mathrm{ave}}\right) 10^{-10} \tag{7.8}$$

$$k_{i,p} = k_{i,n} = \left(62605 - 277.7\, T_{i,\mathrm{ave}} + 0.4131\, T^2_{i,\mathrm{ave}}\right) 10^{-4} \tag{7.9}$$

The total thermal resistance (RS_t) existing between the interface of the TECs is given by,

$$\mathrm{RS}_t = \mathrm{RS}_{\mathrm{sprd}} + \mathrm{RS}_{\mathrm{cont}} \tag{7.10}$$

where, $\mathrm{RS}_{\mathrm{sprd}}$ and $\mathrm{RS}_{\mathrm{cont}}$ are the spreading resistance and contact resistance between the interface of the two TECs, respectively.

Based on the work of Lee et al. [17] and Cheng and Shih [4], the spreading resistances between the interface of the two TECs are calculated from the following equation,

$$\mathrm{RS}_{\mathrm{sprd}} = \frac{\psi_{\max}}{k_{h,s}\, rad_{c,s} \sqrt{\pi}} \tag{7.11}$$

where, $rad_{c,s}$ is the equilibrium radius of the substrates of the cold stage and $k_{h,s}$ is the thermal conductivity of the substrate of the hot stage. The detailed explanation related to the equilibrium radius is available in the work of Lee et al. [17]. However, it is calculated by the following equation.

$$rad_{c,s} = \sqrt{\frac{(2a\,N_t/r + 1)}{\pi}} \quad (7.12)$$

where, factor $2a$ represents the linear relationship between the cross-sectional area of the substrate and the TE modules [4].

The dimensionless parameter $\psi_{\max}$ of the Eq. 7.11 is given by,

$$\psi_{\max} = \frac{\varepsilon\,\tau}{\sqrt{\pi}} + \frac{1}{\sqrt{\pi}}(1 - \varepsilon)\varphi \quad (7.13)$$

where, ε and τ are the dimensionless parameters and are calculated by,

$$\varepsilon = \frac{rad_{c,s}}{rad_{h,s}} = \sqrt{\frac{1}{r}} \quad (7.14)$$

$$\tau = \frac{s_{h,s}}{rad_{h,s}} \quad (7.15)$$

where, $rad_{h,s}$ is the equilibrium radius of the substrate of the hot stage and $S_{h,s}$ is the substrate thickness of the hot stage, respectively and given by,

$$rad_{h,s} = \sqrt{\frac{(2a\,N_t r/r + 1)}{\pi}} \quad (7.16)$$

The dimensionless parameter φ of the Eq. 7.13 is given by,

$$\varphi = \frac{\tanh(\lambda \times \tau) + \frac{\lambda}{Bi}}{1 + \frac{\lambda}{Bi}\tanh(\lambda \times \tau)} \quad (7.17)$$

where, *Bi* is the *Biot* number and its value is infinity i.e. ($Bi = \infty$) for isothermal cold side of the hot stage.

The dimensionless parameter λ of the Eq. 7.17 is given by [4],

$$\lambda = \pi + \frac{1}{\varepsilon\sqrt{\pi}} \quad (7.18)$$

The contact thermal resistance (RS_{cont}) at the interface of the two TECs is calculated by,

$$RS_{cont} = \frac{RS_j}{\left(2a\,N_t/_r + 1\right)} \quad (7.19)$$

where, Rs_j is the joint resistance at the interface of two TECs.

The heat rejected at the hot side of the cold stage ($Q_{c,h}$) and cooling capacity at the cold side of the hot stage ($Q_{h,c}$) is obtained by considering the heat balance at the interface of TECs.

$$Q_{c,h} = \frac{N_t}{r+1}\left[\alpha_c I_c T_{c,h} + \frac{1}{2} I_c^2 R_c - K_c (T_{c,h} - T_{c,c})\right] \tag{7.20}$$

$$Q_{h,c} = \frac{N_t r}{r+1}\left[\alpha_h I_h T_{h,h} - \frac{1}{2} I_h^2 R_h - K_h (T_{h,h} - T_{h,c})\right] \tag{7.21}$$

As the hot side of the cold stage and cold side of the hot stage are at the interface $Q_{c,h} = Q_{h,c}$, but due to the thermal resistance at the interface, the temperature of both sides is not same. The relation between both these temperatures is given by [4],

$$T_{h,c} = T_{c,h} + \mathrm{RS}_t \, Q_{c,h} \tag{7.22}$$

The next section describes the objective function formulation based on this thermal model of two-stage TECs.

7.1.2 *Multi-Objective Optimization and Formulation of Objective Functions*

Multi-objective optimization has been defined as finding a vector of decision variables while optimizing (i.e. minimizing or maximizing) several objectives simultaneously, with a given set of constraints. In the present work, two such objectives namely maximizing the cooling capacity and maximizing the COP of the two-stage TECs are considered simultaneously for multi-objective optimization.

The first objective is to maximize the cooling capacity of a two-stage TEC as given by the Eq. 7.23.

$$\begin{gathered} Z_1 = Maximize\ Q_{c,c}(X), \quad X = [x_1, x_2, \ldots\ldots x_{D_n}], \\ x_{i,\min} \le x_i \le x_{i,\max}, \quad i = 1, 2, \ldots\ldots D_n \end{gathered} \tag{7.23}$$

Subject to the set of constraints (m),

$$g_j(X) \le 0, \quad j = 1, 2. \ldots\ldots m \tag{7.24}$$

The second objective is to maximize the COP of a two-stage TEC as given by the Eq. 7.25.

$$\begin{gathered} Z_2 = Maximize\ \mathrm{COP}(Y), \quad Y = [y_1, y_2, \ldots\ldots y_{D_n}], \\ y_{i,\min} \le y_i \le y_{i,\max}, \quad i = 1, 2, \ldots\ldots D_n \end{gathered} \tag{7.25}$$

Subject to a set of constraints (m)

$$g_j(Y) \leq 0, \quad j = 1, 2\ldots\ldots m \tag{7.26}$$

The above-mentioned single objective functions are put together for multi-objective optimization. The normalized multi-objective function (Z) is formulated considering different weight factors to both the objectives and is given by the following equation:

$$Maximize\ Z = w_1\left({Z_1}/{Z_{1,\max}}\right) + (1 - w_1)\left({Z_2}/{Z_{2,\max}}\right) + \sum_{j=1}^{m} R1\left(g_j(X)\right)^2 + \sum_{j=1}^{m} R1\left(g_j(Y)\right)^2 \tag{7.27}$$

where, w_1 is weight factor for the first objective function. $Z_{1,\max}$ and $Z_{2,\max}$ are the maximum values of the objective functions Z_1 and Z_2, respectively when these objectives are considered independently. The last two terms in Eq. 7.27 takes into account the constraints violation. $R1$ is the penalty parameter having a large value. The value of weight factor w_1 can be decided by the designer. The result is a set of optimum solutions, called Pareto solutions, each of which is a trade off between the considered objective functions. The designer can choose any set of optimal solutions by selecting desired value of w_1 between 0 and 1.

Now an example is considered to demonstrate the effectiveness of the TLBO algorithm for the optimization of two-stage TECs.

7.1.3 Application Example of a Two-Stage TEC

The effectiveness of the TLBO algorithm is assessed by analyzing an example of two-stage TECs which was earlier analyzed by Cheng and Shih [4] using GA. A Two-stage TEC used to produce temperature of 210 K at the cold stage when its hot stage is maintained at a temperature of 300 K is needed to be optimized for maximum cooling capacity and maximum COP. The total number of TE modules of the two stages is 50 and the ratio of cross-sectional area to the length of TE modules is 0.0018 m. Thermal resistance exists at the interface of TEC. Alumina having thermal conductivity 30 W/m K is acting as a substrate to take into account the spreading resistance. The thickness of the substrate is 1 mm. To take into account the contact resistance between the two-stages, the joint resistance is varied between 0.02 and 2 cm^2 K/W. The property values of TE material are considered to be temperature dependent. Moreover, the two-stage TECs, electrically separated and electrically connected in series as shown in Fig. 7.1 are considered for the optimization.

Following inequality constraints which are bound by lower and upper limits of the design variables are considered in the present work of TECs optimization.

Table 7.1 Comparison of the two-stage TEC (electrically separated)

	GA [4] TLBO			
	Max. $Q_{c,c}$	Max. COP	Max. $Q_{c,c}$	Max.COP
RS_j = 0.02 cm² K/W				
I_h (A)	8.613	6.611	9.3077	6.7299
I_c (A)	7.529	7.592	7.7146	7.581
r	5.25	6.143	5.25	6.143
N_c	8	7	8	7
$Q_{c,c}$ (W)	0.755	–	0.784	0.5968
COP	–	0.019	0.015	0.0192
RS_j = 0.2 cm² K/W				
I_h (A)	8.652	6.769	9.3278	6.5338
I_c (A)	7.805	7.465	8.0121	7.8165
r	5.25	6.143	5.25	6.143
N_c	8	7	8	7
$Q_{c,c}$ (W)	0.838	–	0.8826	0.6544
COP	–	0.021	0.0168	0.0219
RS_j = 2 cm² K/W				
I_h (A)	9.29	5.204	9.609	4.4163
I_c (A)	9.41	9.889	11	10.722
r	4.556	5.25	4.556	7.333
N_c	9	8	9	6
$Q_{c,c}$ (W)	2.103	–	2.254	1.201
COP	–	0.061	0.0393	0.0654

$$4 \leq I_h \leq 11 \tag{7.28}$$

$$4 \leq I_c \leq 11 \tag{7.29}$$

$$2 \leq r \leq 7 \tag{7.30}$$

7.1.3.1 Single Objective Consideration

Table 7.1 shows the optimized parameters of the considered example obtained by using the TLBO approach for maximum cooling capacity as well as maximum COP when the considered two-stage TEC is electrically separated and its comparison with the optimized parameters obtained by Cheng and Shih [4] using the GA approach. When the joint resistance is 0.02 cm^2 K/W, present approach using the TLBO results in such combination of input current and TE module which increases the cooling capacity by 3.84% as compared to the GA approach suggested by Cheng and Shih [4]. Also as the joint resistance increases from 0.02 to 0.2 and then 2 cm^2 K/W, the increment in cooling capacity is 5.32 and 7.18%, respectively as compared to the GA approach. Similarly, for the maximum COP consideration, the present approaches yield 1.1, 4.29 and 7.21% higher COP as

Table 7.2 Comparison of the two-stage TEC (electrically connected in series)

	GA [4] TLBO				
	Max. $Q_{c,c}$	Max. COP		Max. $Q_{c,c}$	Max.COP
RS_j = 0.02 cm^2 K/W					
I_h (A)	8.415	7.27		8.5737	7.1558
I_c (A)	8.415	7.27		8.5737	7.1558
r	6.143	5.25		6.143	5.25
N_c	7	8		7	8
$Q_{c,c}$ (W)	0.73	–	0.7479	0.6405	
COP	–	0.019		0.0159	0.0191
RS_j = 0.2 cm^2 K/W					
I_h (A)	8.663	7.135		8.7375	7.1681
I_c (A)	8.663	7.135		8.7375	7.1681
r	6.143	5.25		6.143	6.143
N_c	7	8		7	8
$Q_{c,c}$ (W)	0.818	–	0.8838	0.7098	
COP	–	0.02		0.0172	0.0215
RS_j = 2 cm^2 K/W					
I_h (A)	9.482	7.133		10.387	7.305
I_c (A)	9.482	7.133		10.387	7.305
r	4	4.555		4.556	3.546
N_c	10	9		9	11
$Q_{c,c}$ (W)	2.123	–	2.276	1.6947	
COP	–	0.048		0.0354	0.0506

compared to the GA approach when the joint resistance is 0.02, 0.2 and 2 cm^2 K/W, respectively.

Table 7.2 shows the comparison of the optimized parameters for a two-stage TEC electrically connected in series. In this case also, the increment in cooling capacity is corresponding to 2.45, 8.1 and 7.2% when the joint resistance is 0.02, 0.2 and 2 cm^2 K/W, respectively as compared to the GA approach considered by Cheng and Shih [4]. Similarly, for the maximum COP consideration the present approaches results in 0.5, 7.5 and 5.41% higher COP as compared to the GA approach when the joint resistance is 0.02, 0.2 and 2 cm^2 K/W, respectively.

7.1.3.2 Multi-Objective Consideration

The results of single objective optimization for maximum cooling capacity and maximum COP reveal that higher cooling capacity accompanies the lower COP and vice versa which reflects the necessity of multi-objective optimization for two-stage TECs. Equation 7.27 represents the normalized objective function for multi-objective optimization. Figure. 7.2 shows the Pareto-optimal curve obtained by using the odified TLBO algorithm for multi-objective optimization when the considered two-stage TECs are electrically separated. As seen from Fig. 7.2a–c

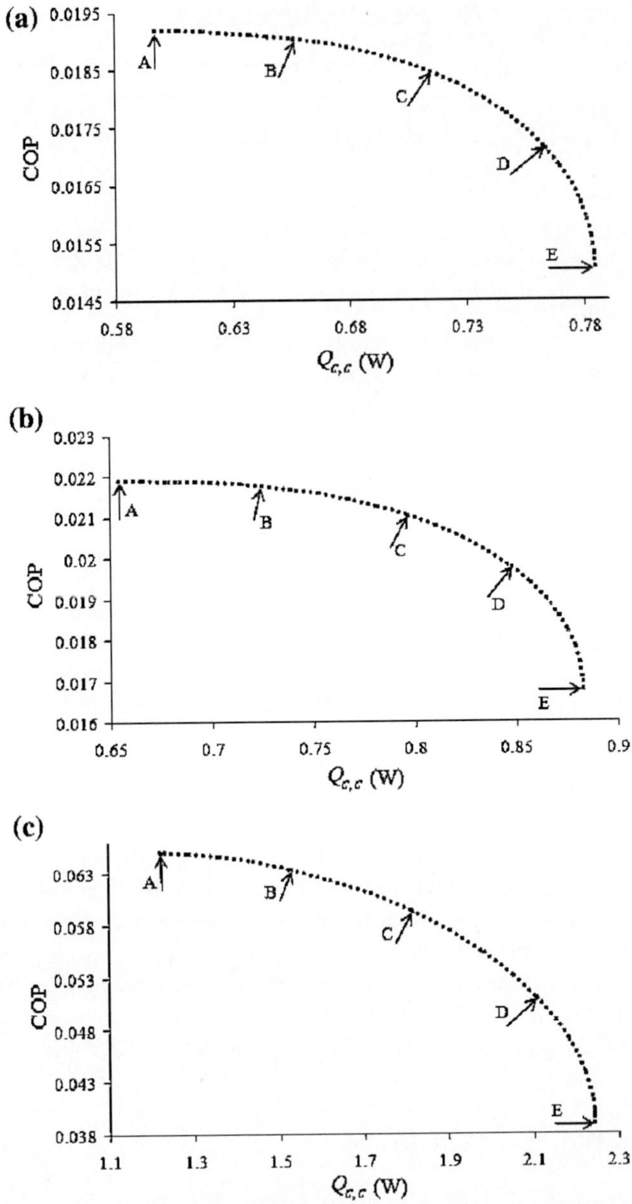

Fig. 7.2 The distribution of Pareto-optimal points solutions for electrically separated TEC using the TLBO algorithm. **a** $Rs_j = 0.2\text{cm}^2\text{K/W}$. **b** $Rs_j = 2\text{cm}^2\text{K/W}$. **c** $Rs_j = 0.2\text{cm}^2\text{K/W}$

that for different values of joint resistance, the maximum cooling capacity exists at design point E where the COP is minimum. On the other hand, the maximum COP occurs at design point A where the cooling capacity has minimum value.

Table 7.3 Optimal output variables for A to E Pareto-optimal front shown in Fig. 7.2

Output variable	Design point				
	A	B	C	D	E
RS_j = 0.02 cm^2 K/W					
I_h (A)	6.7299	7.4285	8.0476	8.7347	9.3077
I_c (A)	7.581	7.4018	7.5229	7.6351	7.7146
r	6.143	5.25	5.25	5.25	5.25
N_c	7	8	8	8	8
$Q_{c,c}$ (W)	0.5968	0.6788	0.7375	0.7745	0.784
COP	0.0192	0.0189	0.018	0.0165	0.015
RS_j = 0.2 cm^2 K/W					
I_h (A)	6.5338	7.0084	7.5076	8.0907	9.3278
I_c (A)	7.8165	7.5756	7.6925	7.8118	8.0121
r	6.143	5.25	5.25	5.25	5.25
N_c	7	8	8	8	8
$Q_{c,c}$ (W)	0.6544	0.717	0.782	0.8368	0.8826
COP	0.0219	0.0217	0.0212	0.0201	0.0168
RS_j = 2 cm^2 K/W					
I_h (A)	4.4163	5.5156	6.9828	7.9011	9.609
I_c (A)	10.722	10.759	10.866	10.581	11
r	7.333	6.143	5.25	4.556	4.556
N_c	6	7	8	9	9
$Q_{c,c}$ (W)	1.201	1.5826	1.9754	2.1289	2.254
COP	0.0654	0.0631	0.0559	0.0506	0.0393

Specifications of five sample design points A–E in Pareto-optimal fronts for different values of joint resistance are listed in Table 7.3. It is observed from the Fig. 7.2a–c and Table 7.3 that by properly modulating the input current of hot stage and cold stage as well as TE module of each stage, the cooling capacity and COP of the two-stage TEC increases with the increase in joint resistance.

Figure 7.3 represents the Pareto-optimal curve obtained by using the TLBO algorithm for the two-stage TECs electrically connected in series. Looking at the Pareto front obtained for different values of joint resistance it is found that the maximum cooling capacity exists at design point E where the COP is lowest. On the other hand, the maximum COP occurs at design point A where the cooling capacity has minimum value. Table 7.4 shows the specifications of sample design points A–E in Pareto-optimal fronts for different values of the joint resistance.

In the present work, TLBO algorithm is applied successfully to the multi-objective optimization of a two-stage TEC considering two conflicting objectives: cooling capacity and COP. Two different configuration of TECs, electrically separated and electrically connected in series are investigated for the optimization. Moreover, the contact and spreading resistance of TEC are also considered. The ability of the TLBO algorithm is demonstrated and the performance of the TLBO algorithm is compared with the performance of GA. The proposed algorithm can

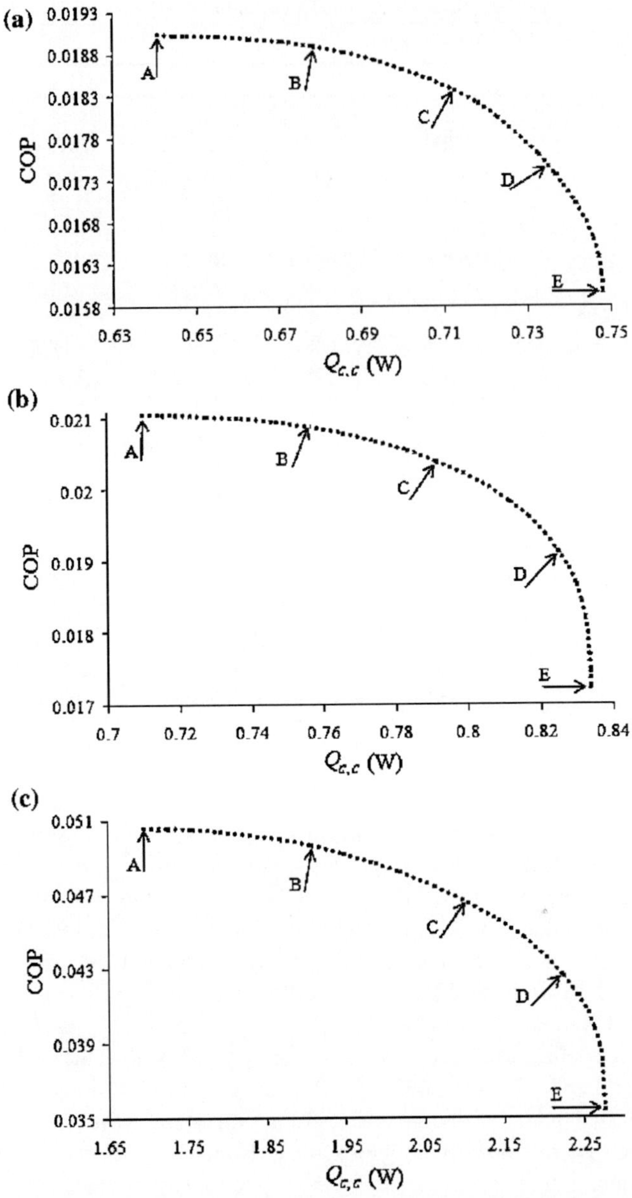

Fig. 7.3 The distribution of Pareto-optimal points solutions for electrically connected TEC using the TLBO algorithm. **a** $Rs_j = 0.2$ cm^2 K/W. **b** $Rs_j = 2$ cm^2 K/W. **c** $Rs_j = 0.2$ cm^2 K/W

be easily customized to suit the optimization of other types of thermal systems involving large number of variables and objectives. These features boost up the applicability of the proposed algorithm for the thermal systems optimization.

Table 7.4 Optimal output variables for A to E Pareto-optimal front shown in Fig. 7.3

Output variable	Design point				
	A	B	C	D	E
RS_j = 0.02 cm^2 K/W					
I_h (A)	7.1558	7.4227	7.7519	8.002	8.5737
I_c (A)	7.1558	7.4227	7.7519	7.6351	8.5737
r	5.25	5.25	5.25	5.25	6.143
N_c	8	8	8	8	7
$Q_{c,c}$ (W)	0.6405	0.6785	0.7127	0.7294	0.7479
COP	0.0191	0.0189	0.0184	0.0178	0.0159
RS_j = 0.2 cm^2 K/W					
I_h (A)	7.1681	7.4634	7.7568	8.223	8.7375
I_c (A)	7.1681	7.4634	7.7568	8.223	8.7375
r	6.143	5.25	5.25	5.25	6.143
N_c	8	8	8	8	7
$Q_{c,c}$ (W)	0.7098	0.7563	0.7915	0.825	0.8838
COP	0.0215	0.0209	0.0204	0.0191	0.0172
RS_j = 2 cm^2 K/W					
I_h (A)	7.305	7.77	8.285	9.32	10.387
I_c (A)	7.305	7.77	8.285	9.32	10.387
r	3.546	3.546	3.546	4	4.556
N_c	11	11	11	10	9
$Q_{c,c}$ (W)	1.6947	1.868	2.02	2.2258	2.276
COP	0.0506	0.0499	0.0481	0.0426	0.0354

7.2 Design Optimization of Shell and Tube Heat Exchanger Using Shuffled Frog Leaping Algorithm

Heat Exchanger is a process equipment designed for the effective transfer of heat energy between two or more fluids; a hot fluid and a coolant. The purpose may be either to remove heat from a fluid or to add heat to a fluid. In heat exchangers, there are usually no external heat and work interactions. Typical applications involve heating or cooling of a fluid stream of concern and evaporation or condensation of single or multicomponent fluid streams. The heat transferred in the heat exchanger may be in the form of latent heat (e.g. in boilers and condensers), or sensible heat (in heaters and coolers). Some examples are:

- Boilers, Evaporators, super heaters and condensers of a power plant.
- Automobile radiators and oil coolers of heat engines.
- Evaporator of an ice plant and milk chillers of pasteurizing plant.
- Condensers and evaporators in refrigeration units.
- Water and air heaters or coolers.

In other applications, the objective may be to recover or reject heat, or sterilize, pasteurize, fractionate, distil, concentrate, crystallize or control a process fluid.

Shell and tube heat exchangers consist of a series of tubes. One set of these tubes contains the fluid that must be either heated or cooled. The second fluid runs over the tubes that are being heated or cooled so that it can either provide the heat or absorb the heat required. A set of tubes is called the tube bundle and can be made up of several types of tubes: plain, longitudinally finned, etc. Shell and Tube heat exchangers are typically used for high pressure applications (with pressures greater than 30 bar and temperatures greater than 260°C). This is because the shell and tube heat exchangers are robust due to their shape.

A variety of different internal constructions are used in shell and tube exchangers, depending on the desired heat transfer and pressure drop performance and the methods employed to reduce thermal stresses, to prevent leakages, to provide for ease of cleaning, to contain operating pressures and temperatures, to control corrosion, to accommodate highly asymmetric flows and so on. Shell and tube exchangers are classified and constructed in accordance with the widely used TEMA (Tubular Exchanger Manufacturers Association) standards, other standards in Europe and elsewhere and ASME (American Society of Mechanical Engineers) boiler and pressure vessel codes [18].

There are several thermal design features that are to be taken into account when designing the tubes in the shell and tube heat exchangers. The optimum thermal design of a shell and tube heat exchanger involves the consideration of many interacting design parameters which can be summarized as follows:

- Process fluid assignments to shell side or tube side.
- Selection of stream temperature specifications.
- Setting shell side and tube side pressure drop design limits.
- Setting shell side and tube side velocity limits.
- Selection of heat transfer models and fouling coefficients for shell and tube side.
- Selection of heat exchanger TEMA layout and number of passes.
- Specification of tube parameters—size, layout, pitch and material.
- Setting upper and lower design limits on tube length.
- Specifications of shell side parameters materials baffle cut, baffle spacing and clearances.
- Setting upper and lower design limits on shell diameter, baffle cut and baffles spacing.

There are many previous studies on the optimization of heat exchanger. Several investigators had used different optimization techniques considering different objective functions to optimize heat exchanger design. Ravagnani et al. [19] proposed a new methodology to include features such as pressure drop and fouling effects which were usually neglected in grassroots as in retrofit designs. Pinch analysis was used to obtain the heat exchangers network with the maximum energy recovery, and a new systematic procedure was proposed to the identification and loop breaking. Bell–Delaware method for the shell side was used to design the heat exchangers. Results of an example show differences between heat exchangers, with and without the detailed design, relative to heat transfer area, fouling and pressure drop. The great contribution of this work was that Individual and global heat transfer coefficients were always calculated, which is generally assumed in the design step. The methodology proposed to the heat exchangers design assured the minor heat exchanger according to TEMA standards, contributing to the minimization of the heat exchanger network global annual cost. This method considered pressure drops and fouling effects, hence presents values more realistic then those neglecting the equipment detailed design.

Pariyani et al. [20] presented randomized algorithm with stream splitting for design of heat exchanger networks in this work. The algorithm had provisions for splitting any one of the process streams. Three benchmark problems taken from the literature were studied. The results obtained from study indicated the strength of the randomization method in finding a cost-effective network.

Babu and Munawar [21] applied Differential evolution (DE) and its various strategies for the optimal design of shell and tube heat exchangers. Minimum heat transfer area was main objective in heat exchanger design. In the presented study DE, an improved version of genetic algorithms (GAs) had been successfully applied with different strategies for many design configurations using Bell's method to find the heat transfer area. In the application of DE, 9,680 combinations of the key parameters have been considered. For this optimal design problem, it has been found that DE, an exceptionally simple evolution strategy, is significantly faster compared to GA and yields the global optimum for a wide range of the key parameters. Ravagnani and Caballero [22] presented an optimization model for the synbook of heat exchanger networks (HEN) including the detailed design of the equipments formulated as a decomposition method. The optimization model was based on area, energy and pumping costs. The algorithm combined two distinct models, in a decomposition method, a mixed integer nonlinear programing (MINLP) superstructure simultaneous optimization model for the heat exchanger network synbook considering stream splitting and a MINLP model for the detailed equipment design, following rigorously the standards of the TEMA. Two examples were used to test the algorithm developed, and the results confirmed the achievement of the optimum HEN configuration with the detailed heat exchangers design, following the TEMA standards.

Fakheri [23] proposed methodology for Optimization of Shell and Tube Heat Exchangers in Series. For a given total rate of heat transfer and the known inlet and exit temperatures of the hot and cold fluids, the total area of the heat exchanger network was minimized. In the proposed methodology, the heat exchangers were assumed to be different. This is a generalization compared to the traditional approach where all the heat exchangers are taken to have the same area and the same LMTD (Log Mean Temperature Difference) correction factor. In the traditional approach the minimum number of identical shells, for which a feasible solution exists and meets the design criteria, was used as the optimum solution. The proposed optimization approach shows that using larger number of smaller heat exchangers results in less overall heat exchanger area due to the more efficient operation of the individual heat exchangers.

Gholap and Khan [24] proposed a detailed thermodynamic model for a refrigerator based on an irreversible Carnot cycle is developed with the focus on forced air heat exchangers. A multi-objective optimization procedure was implemented to find optimal design values for design variables. Minimizations of energy consumption and material cost were the two objectives considered. Since these objectives were conflicting, no single design will satisfy both simultaneously. The result of this research was a set of multiple optimum solutions, which were called 'Pareto optimal solutions'. Air and refrigerant side correlations were combined with an elemental approach to model the heat exchangers. This paper has presented a detailed design model development. A limited validation is presented with experimental test data obtained from a typical household refrigerator. An optimization algorithm requires several evaluations of such models. Response surface based metamodels for objective functions were used to save computational effort. A genetic algorithm based optimization tool was used for multi-criteria optimization.

Caputo et al. [25] proposed a procedure for optimal design of shell and tube heat exchangers, which utilized a genetic algorithm to minimize the total cost of the equipment including capital investment and the sum of discounted annual energy expenditures related to pumping. In order to verify the capability of the proposed method, three case studies were also presented showing that significant cost reductions were feasible with respect to traditionally designed exchangers. In particular, in the examined cases a reduction of total costs up to more than 50% was observed. Soltan et al. [26] proposed a computer program that enables designers to determine the optimum baffle spacing for segmentally baffled shell and tube condensers. Total costs of heat transfer area and pumping power were involved to perform objective function, using a weight factor, which depends on the economical conditions of the desired location. As a result, a set of correlation was presented to determine the optimum baffle spacing, which could be considered as a complementary to HEDH recommendations.

Costa and Queiroz [27] formulated problem on design optimization of shell and tube heat exchangers which consists of the minimization of the thermal surface area for a certain service, involving discrete decision variables. Additional constraints represented geometrical features and velocity conditions were complied in order to reach a more realistic solution for the process task. The optimization algorithm was based on a search along the tube count table where the established constraints and the investigated design candidates were employed to eliminate non-optimal alternatives, thus reducing the number of rating runs executed. The obtained results illustrated the capacity of the proposed approach to direct the optimization toward more effective designs, considering important limitations usually ignored in the literature.

Thirumarimurugan et al. [28] investigated on comparative heat transfer study on a solvent and solutions were made using 1–1 Shell and Tube Heat Exchanger. Steam is the hot fluid; whereas Water and Acetic acid Water miscible solution serves as cold fluid. A series of runs were made between steam and water, steam and Acetic acid solution. In addition to, the volume fraction of Acetic acid was varied and the experiment was held. The flow rate of the cold fluid is maintained from 120 to 720 lph and the volume fraction of Acetic acid was varied from 10 to 50%. Experimental results such as exchanger effectiveness, overall heat transfer coefficients were calculated. A mathematical model was developed for the outlet temperatures of both the Shell and Tube side fluids and was simulated using MATLAB program. The model was compared with the experimental findings and found to be valid.

Ponce et al. [29] presented an approach based on GAs for the optimal design of shell and tube heat exchangers. The approach used the Bell–Delaware method for the description of the shell side flow with no simplifications. The optimization procedure involved the selection of the major geometric parameters such as the number of tube-passes, standard internal and external tube diameters, tube layout and pitch, type of head, fluids allocation, number of sealing strips, inlet and outlet baffle spacing, and shell side and tube side pressure drops. The methodology took into account the geometric and operational constraints typically recommended by design codes. The examples analyzed showed that GAs provide a valuable tool for the optimal design of heat exchangers. The use of GA together with the Bell–Delaware method allows several design factors, typically specified from experience and later subject to a rating test, to be calculated as part of the optimum solution. Genetic algorithms provide better expectations to detect global optimum solutions than gradient methods, in addition to being more robust for the solution of non-convex problems.

Guo et al. [30] applied the field synergy principle to the optimization design of the shell and tube heat exchanger with segmental baffles. The field synergy number which is defined as the indicator of the synergy between the velocity field and the heat flow was taken as the objective function. The genetic algorithm was employed to solve the heat exchanger optimization problems with multiple design variables. The field synergy number maximization approach for heat exchanger optimization design was thus formulated. In comparison with the initial design, the optimal design leads to a significant cost cut on the one hand and an improvement of the heat exchanger performance on the other hand. The comparison with the traditional heat exchanger optimization design approach with the total cost as the objective function showed that the field synergy number maximization approach was more advantageous. From the work he has concluded that the field synergy number maximization approach was more attractive in the sense that the reduction of water consumption or the heat exchanger effectiveness improvement can lead to much more profit than the total cost cut achieved by the traditional heat exchanger optimization design approach with the total cost as the objective function.

Sanaye and Hajabdollahi [31] proposed optimization of shell and tube heat exchanger. The effectiveness and cost are two important parameters in heat exchanger design. The total cost includes the capital investment for equipment (heat exchanger surface area) and operating cost (for energy expenditures related to pumping). Tube arrangement, tube diameter, tube pitch ratio, tube length, tube number, baffle spacing ratio as well as baffle cut ratio were considered as seven design parameters. For optimal design of a shell and tube heat exchanger, it was first thermally modeled using NTU method while Belle Delaware procedure was applied to estimate its shell side heat transfer coefficient and pressure drop. Fast and elitist non-dominated sorting genetic algorithm (NSGA-II) with continuous and discrete variables was applied to obtain the maximum effectiveness (heat recovery) and the minimum total cost as two objective functions. The results of optimal designs were a set of multiple optimum solutions, called 'Pareto optimal solutions'. The sensitivity analysis of change in optimum effectiveness and total cost with change in design parameters of the shell and tube heat exchanger was also performed and the results are reported. In this study the effectiveness and total cost were considered as two objective functions. Total cost included the investment cost of heat transfer surface area as well as the operating cost for the pumping power.

To maximize the effectiveness value and to minimize the total cost, seven design parameters including, tube arrangement, tube diameter, tube pitch ratio, tube length, tube number, baffle spacing ratio as well as baffle cut ratio were selected. Design parameters (decision variables) and the range of their variations calculated. The number of iterations for finding the global extremum in the whole searching domain was about 8.2×10^{15}. The genetic algorithm optimization was performed for 200 generations, using a search population size of 100 individuals, crossover probability of 0.9, gene mutation probability of 0.035 and controlled elitism value of 0.65.

Hosseini and Ceylan [32] obtained the heat transfer coefficient and pressure drop on the shell side of a shell and tube heat exchanger experimentally for three different types of copper tubes (smooth, corrugated and with micro-fins). Also, experimental data has been compared with theoretical data available. Correlations have been suggested for both pressure drop and Nusselt number for the three tube types. A shell and tube heat exchanger of an oil cooler used in a power transformer has been modeled and built for this experimental work in order to investigate the effect of surface configuration on the shell side heat transfer as well as the pressure drop of the three types of tube bundles. The bundles with the same geometry, configuration, number of baffles and length, but with different external tube surfaces inside the same shell were used for the experiment. Corrugated and micro-fin tubes have shown degradation of performance at a Reynolds number below a certain value ($Re < 400$). At a higher Reynolds number the performance of the heat exchanger greatly improved for micro-finned tubes. Patel and Rao [33] proposed the use of PSO for design optimization of shell and tube heat exchangers from economic view point. Minimization of total annual cost was considered as an objective function. Three design variables such as shell internal diameter, outer tube diameter and baffle spacing were considered for optimization. Four different case studies were presented to demonstrate the effectiveness and accuracy of the proposed algorithm. The results of optimization using PSO technique were compared with those obtained by using genetic algorithm.

From this literature survey it is clear that shell and tube heat exchanger optimization was attempted by many non-traditional optimization algorithms like GA, PSO, DE and ACO in the past. Now an attempt is made to implement the shuffled frog leaping algorithm to achieve shell and tube heat exchanger optimization.

In the present study the fluid inlet and outlet temperatures and flow rates are considered as design specifications while shell inside diameter (D_s), tube outside diameter (d_o) and baffle spacing (B) are considered as design variables. The following nomenclature is used:

A	heat exchanger surface area (m^2)
a_1	numerical constant (€)
a_2	numerical constant (€/m^2)
a_3	numerical constant
a_s	shell side pass area (m^2)
B	baffles spacing (m)
C	numerical constant
C_e	energy cost (€/kW h)
C_i	capital investment (€)
cl	clearance (m)
C_o	annual operating cost (€/yr)
C_{od}	total discounted operating cost (€)
C_p	specific heat (J/kg K)
C_{tot}	total annual cost (€)
d_e	equivalent shell diameter (m)
d_i	tube inside diameter (m)
d_o	tube outside diameter (m)
D_s	shell inside diameter (m)
F	temperature difference correction factor
f_s	shell side friction coefficient
f_t	tube side friction coefficient
H	annual operating time (h/yr)
h_s	shell side convective coefficient (W/m^2 K)
h_t	tube side convective coefficient (W/m^2 K)
i	annual discount rate (%)
k	Thermal conductivity (W/m K)
L	tubes length (m)
$LMTD$	Logarithmic mean temperature difference (K)
m_s	shell side mass flow rate (kg/s)
m_t	tube side mass flow rate (kg/s)
N_t	number of tubes
n	number of tube-passes
n_1	numerical constant
ny	equipment life (yr)
P	pumping power (W)
Pr_s	shell side Prandtl number

Pr_t	tube side Prandtl number
Q	heat duty (W)
Re_s	shell side Reynolds number
Re_t	tube side Reynolds number
R_{fs}	shell side fouling resistance (m^2 K/W)
R_{ft}	tube side fouling resistance (m^2 K/W)
S_t	tube pitch (m)
T_{ci}	cold fluid inlet temperature (K)
T_{co}	cold fluid outlet temperature (K)
T_{hi}	hot fluid inlet temperature (K)
T_{ho}	hot fluid outlet temperature (K)
U	overall heat transfer coefficient (W/m^2 K)
v_s	shell side fluid velocity (m/s)
v_t	tube side fluid velocity (m/s)
Δh	heat transfer difference (W/m^2 K)
ΔP	Pressure drop (Pa)
$\Delta P_{\text{tube elbow}}$	Tube elbow pressure drop
$\Delta P_{\text{tube length}}$	Tube length pressure drop (Pa)

Greek letters

μ	Dynamic viscosity (Pa s)
ρ	density (kg/m^3)
η	overall pumping efficiency

Subscript

c	cold stream
e	equivalent
h	hot stream
i	inlet
o	outlet
s	shell side
t	tube side
wt	wall

7.2.1 Mathematical Model

7.2.1.1 Heat Transfer

According to flow regime, the tube side heat transfer coefficient (h_t) is computed from following correlation (from Caputo et al. [25]; reprinted with permission from Elsevier),

$$h_t = \frac{k_t}{d_i}\left[3.657 + \frac{0.0677\left(\mathrm{Re_t\,Pr_t}\,\frac{\mathrm{d_i}}{\mathrm{L}}\right)^{1.33^{1/3}}}{1 + 0.1\,\mathrm{Pr}_t\left(\mathrm{Re}_t\,\frac{d_i}{L}\right)^{0.3}}\right] \quad (7.31)$$

(If $Re_t < 2{,}300$)

$$h_t = \frac{k_t}{d_i}\left[\frac{\frac{f_t}{8}\left(\mathrm{Re_t} - 1000\right)\mathrm{Pr}_t}{1 + 12.7\left(\frac{f_t}{8}\right)^{1/2}\left(\mathrm{Pr}_t^{2/3} - 1\right)}\left(1 + \frac{d_i}{L}\right)^{0.67}\right] \quad (7.32)$$

(If 2,300 < $\mathrm{Re_t}$ < 10,000, [10])

$$h_t = 0.027\frac{k_t}{d_o}\mathrm{Re}_t^{0.8}\,\mathrm{Pr}_t^{1/3}\left(\frac{\mu_t}{\mu_{\mathrm{wt}}}\right)^{0.14} \quad (7.33)$$

(For $\mathrm{Re_t} > 10{,}000$) where, f_t is the Darcy friction factor given as,

$$f_t = \left(1.82\log 10^{\mathrm{Re_t}} - 1.64\right)^{-2} \quad (7.34)$$

Re_t is the tube side Reynolds Number and given by,

$$\mathrm{Re}_t = \frac{\rho_t v_t d_i}{\mu_t} \quad (7.35)$$

Flow velocity for tube side is found by,

$$v_t = \frac{m_t}{\left(\frac{\pi}{4}\right)d_t^2\rho_t}\left(\frac{n}{N_t}\right) \quad (7.36)$$

N_t is number of tubes and n is the number of tube-passes which can be found approximately from the following equation,

$$N_t = C\left(\frac{D_s}{d_o}\right)^{n_1} \quad (7.37)$$

C and n_1 are coefficients taking values according to flow arrangement and number of passes. Pr_t is the tube side prandtl number and given by,

$$\mathrm{Pr}_t = \frac{\mu_t C_{\mathrm{pt}}}{k_t} \quad (7.38)$$

Also, $d_i = 0.8\ d_o$

Kern's formulation for segmental baffle shell and tube exchanger is used for computing shell side heat transfer coefficient h_s,

$$h_s = 0.36\frac{k_t}{d_e}\mathrm{Re}_s^{0.55}\,\mathrm{Pr}_s^{1/3}\left(\frac{\mu_s}{\mu_{\mathrm{wts}}}\right)^{0.14} \quad (7.39)$$

where, d_e is the shell hydraulic diameter and computed as,

$$d_e = \frac{4\left(S_t^2 - \frac{\pi d_o^2}{4}\right)}{\pi d_o} \tag{7.40}$$

(For Square pitch)

$$d_e = \frac{4\left(0.43 S_t^2 - \frac{0.5\pi d_o^2}{4}\right)}{0.5\pi d_o} \tag{7.41}$$

(For Triangular pitch) Cross-section area normal to flow direction is determined by,

$$A_s = D_s B\left(1 - \frac{d_o}{S_t}\right) \tag{7.42}$$

Flow velocity for the shell side can be obtained from,

$$v_s = \frac{m_s}{\rho_s A_s} \tag{7.43}$$

Reynolds number for shell side follows,

$$\text{Re}_s = \frac{m_s d_e}{A_s \mu_s} \tag{7.44}$$

Prandtl number for shell side follows,

$$\Pr_s = \frac{\mu_s C_{ps}}{k_s} \tag{7.45}$$

The overall heat transfer coefficient (U) depends on both the tube side and shell side heat transfer coefficient and fouling resistances are given by,

$$U = \frac{1}{\left(\frac{1}{h_s}\right) + R_{fs} + \frac{d_o}{d_i}\left(R_{ft} + \frac{1}{h_t}\right)} \tag{7.46}$$

Considering the cross flow between adjacent baffle, the logarithmic mean temperature difference ($LMTD$) is determined by,

$$LMTD = \frac{(T_{hi} - T_{co}) - (T_{ho} - T_{ci})}{\ln\left(\frac{T_{hi} - T_{co}}{T_{ho} - T_{ci}}\right)} \tag{7.47}$$

The correction factor F for the flow configuration involved is found as a function of dimensionless temperature ratio for most flow configuration of interest.

$$F = \sqrt{\frac{R^2+1}{R-1}} \frac{\ln\left(\frac{1-P}{1-PR}\right)}{\ln\left(\frac{2-P\left(R+1-\sqrt{R^2+1}\right)}{2-P\left(R+1+\sqrt{R^2+1}\right)}\right)} \tag{7.48}$$

where R is the correction coefficient given by,

$$R = \frac{(T_{\text{hi}} - T_{\text{ho}})}{(T_{\text{co}} - T_{\text{ci}})} \tag{7.49}$$

P is the efficiency given by,

$$P = \frac{(T_{\text{co}} - T_{\text{ci}})}{(T_{\text{hi}} - T_{\text{ci}})} \tag{7.50}$$

Considering overall heat transfer coefficient, the heat exchanger surface area (A) is computed by,

$$A = \frac{Q}{U\,F\,LMTD} \tag{7.51}$$

For sensible heat transfer, the heat transfer rate is given by,

$$Q = m_h C_{\text{ph}}(T_{\text{hi}} - T_{\text{ho}}) = m_c C_{\text{pc}}(T_{\text{co}} - T_{\text{ci}}) \tag{7.52}$$

Based on total heat exchanger surface area (A), the necessary tube length (L) is,

$$L = \frac{A}{\pi\, d_o\, N_t} \tag{7.53}$$

7.2.1.2 Pressure Drop

The pressure drop allowance in heat exchanger is the static fluid pressure which may be expended to drive the fluid through the exchanger. In all heat exchanger there is close physical and economical affinity between heat transfer and pressure drop. For a constant heat capacity in the heat exchanger that is to be designed, increasing the flow velocity will cause a rise of heat transfer coefficient which result in compact exchanger design and lower investment cost. However increase of flow velocity will cause more pressure drop in heat exchanger which results in additional running cost. For this reason when designing a heat exchanger pressure drop must be considered with heat transfer and best solution for the system must be found.

Tube side pressure drop include distributed pressure drop along the tube length and concentrated pressure losses in elbows and in the inlet and out let nozzle.

$$\Delta P_t = \Delta P_{\text{tube length}} + \Delta P_{\text{tube elbow}} \tag{7.54}$$

$$\Delta P_t = \frac{\rho_t v_t^2}{2}\left(\frac{L}{d_i} f_t + p\right) n \tag{7.55}$$

Different value of constant p is considered by different authors. Kern [34] assumed $p = 4$, while Sinnot [35] assumed $p = 2.5$. The shell side pressure drop is,

$$\Delta P_s = f_s \left(\frac{\rho_s v_s^2}{2}\right)\left(\frac{L}{B}\right)\left(\frac{D_s}{d_e}\right) \tag{7.56}$$

where,

$$f_s = 2\, b_o\, \mathrm{Re}_s^{-0.15} \tag{7.57}$$

And $b_o = 0.72$ valid for $Re_s < 40{,}000$.

Considering pumping efficiency (η), pumping power computed by,

$$P = \frac{1}{\eta}\left(\frac{m_t}{\sigma_t}\Delta P_t + \frac{m_s}{\sigma_s}\Delta P_s\right) \tag{7.58}$$

7.2.1.3 Objective Function

Total cost C_{tot} is taken as the objective function, which includes capital investment (C_i), energy cost (C_e), annual operating cost (C_o) and total discounted operating cost (C_{od}).

$$C_{\mathrm{tot}} = C_i + C_{\mathrm{od}} \tag{7.59}$$

Adopting Hall's correlation, the capital investment C_i is computed as a function of the exchanger surface area.

$$C_i = a_1 + a_2 A^{a/3} \tag{7.60}$$

where, $a_1 = 8{,}000$, $a_2 = 259.2$ and $a_3 = 0.93$ for exchanger made with stainless steel for both shell and tubes.

The total discounted operating cost related to pumping power to overcome friction losses is computed from the following equation,

$$C_o = P\, C_e H \tag{7.61}$$

$$C_{\mathrm{od}} = \sum_{x=1}^{\mathrm{ny}} \frac{C_o}{(1+i)^x} \tag{7.62}$$

Table 7.5 Process parameters and physical properties for case study 7.2.2.1

Fluid allocation	Shell side	Tube side
Fluid	Methanol	Sea water
Mass flow rate (kg/s)	27.80	68.90
Inlet temperature (°C)	95.00	25.00
Outlet temperature (°C)	40.00	40.00
Heat capacity (kJ/kg K)	750	995
Density (kg/m^3)	2.84	4.2
Viscosity (Pa s)	0.00034	0.0008
Thermal conductivity(W/m K)	0.19	0.59
Fouling factor (m^2 K/W)	0.00033	0.0002

Based on all above calculations, total cost is computed. The procedure is repeated computing new value of exchanger area (A), exchanger length (L), total cost (C_{tot}) and a corresponding exchanger architecture meeting the specifications. Each time the optimization algorithm changes the values of the design variables d_o, D_s and B in an attempt to minimize the objective function.

Following case studies have been considered for demonstration and validation of the Shuffled Frog Leaping Algorithm.

7.2.2 Case Study

7.2.2.1 (4.34 MW Duty) Methanol–Brackish Water Exchanger

Table 7.5 shows input parameters and physical properties for this case study taken from Sinnot [35]. It is a 4.34 MW duty, Methanol-Brackish water heat exchanger. Sea water is allocated to tube side as the mass flow rate of sea water is much higher compared to methanol. Also it is easy to clean tubes from sludge by chemical wash. Procedure explained in Sect. 7.2.1 is used for calculating other geometric parameters, pressure drops on both shell and tube side and overall heat transfer coefficient of heat exchanger. The shell side inside diameter is not more than 1.5 m, the tubes outer diameter ranges from 0.015 to 0.15 m and the baffle spacing should not be more than 0.5 m. The pressure drops and calculated heat transfer is used to find total annual operating and overhead costs. Tube side and shell side velocities and the ratio of baffle spacing to shell side inner diameter are considered as constraints. Tube side velocities for water and similar fluids ranges from 0.5 to 2.5 m/s, shell side velocities generally ranges from 0.2 to 1.5 m/s and the ratio of baffle spacing to shell side inner diameter ranges from 0.2 to 1. Other design parameters are based on the design variables considered. All the values of discounted operating costs are computed with ny = 10 years. Annual discount rate (i) = 10%. Energy cost (Ce) = 0.12 €/kW h. An annual work hours H = 7,000 h/yr. This problem had been solved by Caputo et al. [25] using GA and Patel and Rao [33] using PSO technique. Minimization of total annual cost is considered as objective. The process input and physical properties for this case study are as follows:

Table 7.6 Optimal heat exchanger geometry using different optimization methods for case study 7.2.2.1

	Literature (Sinnot [35])	GA [25]	PSO [33]	SFLA
L(m)	4.83	3.379	3.115	18.35
d_o(m)	0.02	0.016	0.015	0.0747
B(m)	0.356	0.5	0.424	0.4283
D_s(m)	0.894	0.83	0.81	0.7954
N_t	918	1,567	1,658	47
v_t(m/s)	0.75	0.69	0.67	1.072
Re_t	14,925	10,936	10,503	79,678
Pr_t	5.7	5.7	5.7	5.6949
$h_t(w/m^2)$	3,812	3,762	3,721	3318.7
ΔP_t	6,251	4,298	4,171	9483.3
$A_s(m^2)$	0.032	0.0831	0.0687	0.0681
D_e(m)	0.014	0.011	0.0107	0.0532
v_s(m/s)	0.58	0.44	0.53	0.544
Re_s	18,381	11,075	12,678	63,807
Pr_s	5.1	5.1	5.1	5.0821
$h_s(w/m^2)$	1,573	1,740	1950.8	999.99
ΔP_s	35,789	13,267	20,551	19,487
$U(w/m^2)$	615	660	713.9	483.90
$A(m^2)$	278.6	262.8	243.3	198.28
C_i(€)	51,507	49,259	46,453	43,491
C_o(€/yr)	2,111	947	1038.7	1362.8
C_{od}(€)	12,937	5,818	6778.2	8373.7
C_{tot}(€)	64,480	55,077	53523.1	**51,865**

The following parameters of optimization are selected after various trials for this case study.

Number of memeplexes: 10; number of frogs in each memeplexes: 10 and number of iterations: 100. The best results obtained by SFLA in this work for this case study are compared with the literature results of Sinnot et al. (1996) and with the best results of Caputo et al. [25] and Patel and Rao [33] and are presented in Table 7.6. The best solution by SFLA is 19.56% superior to the solution previously reported in the literature. In this case, reduction in heat transfer area is observed and as a result capital investment is reduced by 6.37% as compared to the PSO approach considered by Patel and Rao [33]. However, increment in tube side pressure losses results in increase in operating cost. Therefore, a cumulative effect of reduction in capital investment and increment in operating expense led to a reduction of the total cost of about 3.09% compared to PSO approach considered by Patel and Rao [33].

7.2.2.2 (0.46 MW Duty) Distilled Water-Raw Water Exchanger

The case study is based on the design problem discussed in Sinnot [35]. The shell side inside diameter is not more than 1.5 m, the tubes outer diameter ranges from 0.015 to 0.61 m and the baffle spacing should not be more than 0.5 m. Tube side

Table 7.7 Process parameters and physical properties for case study 7.2.2.2

Fluid location	Shell side	Tube side
Fluid	Distilled water	Raw water
Mass flow rate (kg/s)	22.07	35.31
Inlet temperature (°C)	33.90	23.90
Outlet temperature (°C)	29.40	26.70
Density (kg/m^3)	995	999
Heat capacity (kJ/kg K)	4.18	4.18
Viscosity (Pa s)	0.0008	0.00092
Thermal conductivity(W/m K)	0.62	0.62
Fouling factor (mK/W)	0.00017	0.00017

and shell side velocities and the ratio of baffle spacing to shell side inner diameter are considered as constraints. Tube side velocities for water and similar fluids ranges from 0.5 to 2.5 m/s, shell side velocities generally ranges from 0.2 to 1.5 m/s and and the ratio of baffle spacing to shell side inner diameter ranges from 0.2 to 1. Other design parameters are based on the design variables considered. All the values of discounted operating costs are computed with ny = 10 years. Annual discount rate (i) = 10%. Energy cost (Ce) = 0.12 €/kW h. An annual work hours H = 7,000 h/yr. This problem had been solved by Caputo et al. [25] using GA and Patel and Rao [33] using PSO technique. Minimization of total annual cost is considered as objective. The process input and physical properties for this case study are given in Table 7.7.

The following parameters of optimization are selected after various trials for this case study. Number of memeplexes: 10; number of frogs in each memeplexes: 50 and number of iterations: 100. The best results obtained by SFLA in this work for this case study are compared with the literature results of Sinnot [35] and with the best results of Caputo et al. [25] and Patel and Rao [33] and are presented in Table 7.8.

In this case decrease in heat transfer area is observed by 22.01% as a result capital investment is decreased by 7.89% as compared to PSO approach Patel and Rao [33]. The shell side and tube side pressure losses are increased as a result the operating expenses are increased. A cumulative effect of decrease in capital investment and increase in operating expense lead to a reduction of the total cost of about 2.70% compared to the PSO approach used by Patel and Rao [33].

7.2.2.3 (1.44 MW Duty) Kerosene-Crude Oil Heat Exchanger

The case study is based on the design problem discussed in Sinnot [35]. The shell side inside diameter is not more than 1.5 m, the tubes outer diameter ranges from 0.051 to 0.081 m and the baffle spacing should not be more than 0.5 m. Tube side and shell side velocities and the ratio of baffle spacing to shell side inner diameter are considered as constraints. Tube side velocities for water and similar fluids

Table 7.8 Optimal heat exchanger geometry using different optimization methods for case study 7.2.2.2

	Literature (Sinnot [35])	GA [25]	PSO [33]	SFLA
L(m)	4.88	1.548	1.45	11.84
d_o(m)	0.013	0.016	0.0145	0.0533
B(m)	0.305	0.44	0.423	0.1494
D_s(m)	0.387	0.62	0.59	0.4199
S_t	0.023	0.02	0.0181	0.0666
N_t	160	803	894	24
v_t(m/s)	1.76	0.68	0.74	1.1171
Re_t	36,400	9,487	9,424	13,239
Pr_t	6.2	6.2	6.2	5.64
$h_t(w/m^2)$	6,558	6,043	5,618	641.09
f_t	0.023	0.031	0.0314	0.0291
ΔP_t	62,812	3,673	4,474	12,964
$A_s(m^2)$	0.0236	0.0541	0.059	0.0125
D_e(m)	0.013	0.011	0.0103	0.0379
v_s(m/s)	0.94	0.41	0.375	0.5176
Re_s	16,200	8,039	4,814	41,728
Pr_s	5.4	5.4	5.4	7.6
$h_s(w/m^2)$	5,735	3,476	4088.3	799.12
f_s	0.337	0.374	0.403	0.2919
ΔP_s	6768.4	4,365	4,721	28,651
$U(w/m^2)$	1,471	1,121	1,177	218.643
$A(m^2)$	46.6	62.5	59.15	46.13
C_i(€)	16,549	19,163	18,614	17,144
C_o(€/yr)	4,466	272	276	425.94
C_{od}(€)	27,440	1,671	1,696	2617.3
C_{tot}(€)	43,989	20,834	20,310	**19,761**

ranges from 0.5 to 2.5 m/s, shell side velocities generally ranges from 0.2 to 1.5 m/s and R_{bs} ranges from 0.2 to 1. Other design parameters are based on the design variables considered. All the values of discounted operating costs are computed with ny = 10 years. Annual discount rate (i) = 10%. Energy cost (Ce) = 0.12 €/kW h. Annual work hours H = 7,000 h/yr. This problem had been solved by Caputo et al. [25] using GA and Patel and Rao [33] using PSO technique. Minimization of total annual cost is considered as objective. The process input and physical properties for this case study are given in Table 7.9.

The following parameters of optimization are selected after various trials for this case study. Number of memeplexes: 10; number of frogs in each memeplexes: 50 and number of iterations: 100. The best results obtained by SFLA in this work for this case study are compared with the literature results of Sinnot [35] and with the best results of Caputo et al. [25] and Patel and Rao [33] and are presented in Table 7.10. In this case, decrease in heat transfer area is observed by 8.21% and as a result the capital investment is decreased by 5.81% as compared to PSO approach used by Patel and Rao [33]. The shell side and tube side pressure losses

Table 7.9 Process parameters and physical properties for case study 7.2.2.3

Fluid location	Shell side	Tube side
Fluid	Kerosene	Crude oil
Mass flow rate (kg/s)	5.52	18.80
Inlet temperature (°C)	199.00	37.80
Outlet temperature (°C)	93.30	76.70
Density (kg/m^3)	850	995
Heat capacity (kJ/kg K)	2.47	2.05
Viscosity (Pa s)	0.0004	0.00358
Thermal conductivity(W/m K)	0.13	0.13
Fouling factor (mK/W)	0.00061	0.00061

are decreased and as a result the operating expenses are also decreased. A cumulative effect of decrease in capital investment and decrease in operating expense lead to a reduction of the total cost of about 5.64% compared to results by PSO presented by Patel and Rao [33].

7.3 Design Optimization of Heat Pipe Using Grenade Explosion Algorithm

A heat pipe is a simple device that can quickly transfer heat from one point to another. They are also called as the "superconductors" of heat as they possess an extra ordinary heat transfer capacity and rate [36]. Heat pipes are passive devices that transport heat from a heat source (evaporator) to a heat sink (condenser) over relatively long distances via the latent heat of vaporization of a working fluid. Heat pipes are offer high effective thermal conductivities (5,000–200,000 W/mK), energy-efficiency, light weight, low cost and the flexibility of many different size and shape options. As passive heat transfer systems, heat pipes offer simple and reliable operation, with high effective thermal conductivity, no moving parts, ability to transport heat over long distances and quiet vibration-free operation [37].

A heat pipe is similar to a thermosyphon. Thermosyphon refers to a method of passive heat exchange based on natural convection which circulates liquid without the necessity of a mechanical pump. It differs from a thermosyphon by virtue of its ability to transport heat against gravity by an evaporation–condensation cycle with the help of porous capillaries that form the wick [38]. A heat pipe generally has three sections: an evaporator section, an adiabatic (or transport) section and a condenser section shown in Fig. 7.4. The major components of a heat pipe are a sealed container, a wick structure and a working fluid. The wick structure is placed on the inner surface of the heat pipe wall and is saturated with the liquid working fluid and provides the structure to develop the capillary action for liquid returning from the condenser to the evaporator section [37].

Table 7.10 Optimal heat exchanger geometry using different optimization methods for case study 7.2.2.3

	Literature (Sinnot [35])	GA [25]	PSO [33]	SFLA
d_o(m)	0.025	0.02	0.015	0.6430
B(m)	0.127	0.12	0.112	0.3061
D_s(m)	0.539	0.63	0.59	0.7351
S_t	0.031	0.025	0.0187	0.8038
v_t(m/s)	1.44	0.87	0.93	1.0168
Re_t	8,287	4,068	3,283	56,794
Pr_t	55.2	55.2	55.2	6.2026
$h_t(w/m^2)$	619	1,168	1,205	2,045
f_t	0.033	0.041	0.044	0.0128
ΔP_t	49,245	14,009	16,926	4,243
$A_s(m^2)$	0.0137	0.0148	0.0131	0.0450
D_e(m)	0.025	0.019	0.0149	0.4577
v_s(m/s)	0.47	0.43	0.495	0.4929
Re_s	25,281	18,327	15,844	28,056
Pr_s	7.5	7.5	7.5	5.3935
$h_s(w/m^2)$	920	1,034	1,288	886.66
f_s	0.315	0.331	0.337	0.2194
ΔP_s	24,909	15,717	12,745	8979.3
$U(w/m^2)$	317	376	409.3	471.35
$A(m^2)$	61.5	52.9	47.5	43.60
C_i (€)	19,007	17,599	17,707	16,678
C_o (€/yr)	1,304	440	523.3	345.03
C_{od} (€)	8,012	2,704	3215.6	2120.3
C_{tot} (€)	27,020	20,303	19922.6	**18,798**

With evaporator heat addition, the working fluid is evaporated as it absorbs an amount of heat equivalent to the latent heat of vaporization; while in the condenser section, the working fluid vapour is condensed. The mass addition in the vapour core of the evaporator section and mass rejection in the condenser end results in a pressure gradient along the vapour channel which drives the corresponding vapour flow. Return of the liquid to the evaporator from the condenser is provided by the wick structure. As vaporization occurs in the evaporator, the liquid meniscus recedes correspondingly into the wick structure. The difference between the capillary radii in the evaporator and condenser ends of the wick structure results in a net pressure difference in the liquid- saturated wick. This pressure difference drives the liquid from the condenser through the wick structure to the evaporator region, thus allowing the overall process to be continuous [36].

Theoretically, heat pipe operation is possible at any temperature between the triple state and the critical point of the working fluid utilized. Each heat pipe application has a temperature range in which the heat pipe is intended to operate. Therefore, the working fluid must be chosen to take into account this operating temperature. Heat pipes can be built in almost any size and shape. Inside the

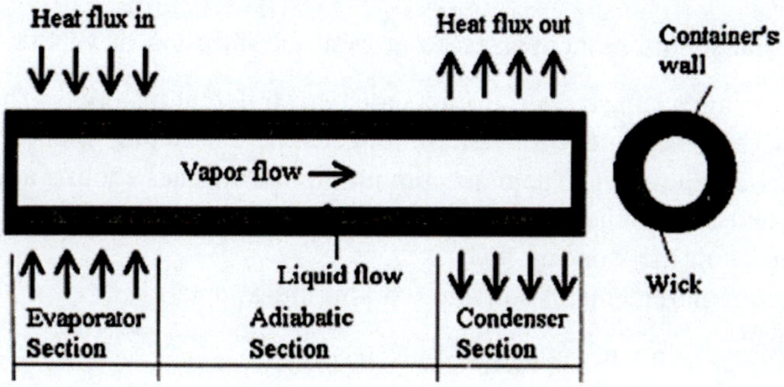

Fig. 7.4 Conceptual drawing of a conventional heat pipe in operation (from Sousa et al. [39]; reprinted with permission from Elsevier)

container of the heat pipe is a liquid under its own pressure that enters the pores of the capillary material, wetting all internal surfaces. Applying heat at any point along the surface of the heat pipe causes the liquid at that point to boil and enter a vapour state. When that happens, the liquid picks up the latent heat of vaporization. The gas which then has a higher pressure moves inside the sealed container to a colder location where it condenses. Thus, the gas gives up the latent heat of vaporization and moves heat from the input to the output end of the heat pipe. Heat pipes have an effective thermal conductivity many thousands of times that of copper. The heat transfer or transport capacity of a heat pipe is specified by its axial power rating (APR). It is the energy moving axially along the pipe. The larger the heat pipe diameter, greater is the APR. Similarly, longer the heat pipe lesser is the APR.

The three basic components of a heat pipe are: (1) Container, (2) Working fluid and (3) Wick or capillary structure. They all work together to transfer heat more efficiently and evenly. The wick structure lines the inner surface of the heat pipe shell and is saturated with the working fluid. The wick provides the structure to develop the capillary action for the liquid returning from the condenser (heat output/sink) to the evaporator (heat input/source). Since the heat pipe contains a vacuum, the working fluid will boil and take up latent heat at well below its boiling point at atmospheric pressure. Water, for instance, will boil at just above 273 K (0°C) and start to effectively transfer latent heat at this low temperature.

Selection of the container material depends on many factors [36]. These are given below:

- Compatibility (both with working fluid and external environment)
- Strength to weight ratio
- Thermal conductivity
- Ease of fabrication, including welding, machinability and ductility
- Porosity

- Wet ability
- The material should be nonporous to prevent the diffusion of vapour.

Heat pipes can be constructed from a variety of different materials. Thermacore has constructed heat pipes from aluminum, copper, titanium, Monel, stainless steel, inconel and tungsten. The most common for electronics cooling applications is copper. The choice of heat pipe containment material is largely dependent on the compatibility with the working fluid.

The prime requirements of suitable working fluids are:

- Compatibility with wick and wall materials
- Good thermal stability
- Wetability of wick and wall materials
- Vapour pressure not too high or low over the operating temperature range
- High latent heat
- High thermal conductivity
- Low liquid and vapour viscosities
- High surface tension
- Acceptable freezing or pour point

The selection of the working fluid must also be based on thermodynamic considerations which are concerned with the various limitations to heat flow occurring within the heat pipe line, viscous, sonic, capillary, entrainment and nucleate boiling levels. In heat pipe design, a high value of surface tension is desirable in order to enable the heat pipe to operate against gravity and to generate a high capillary driving force. In addition to high surface tension, it is necessary for the working fluid to wet the wick and the container material that is contact angle should be zero or very small. The vapour pressure over the operating temperature range must be sufficiently great to avoid high vapour velocities, which tend to setup large temperature gradient and cause flow instabilities.

A high latent heat of vaporization is desirable in order to transfer large amounts of heat with minimum fluid flow, and hence to maintain low pressure drops within the heat pipe. The thermal conductivity of the working fluid should preferably be high in order to minimize the radial temperature gradient and to reduce the possibility of nucleate boiling at the wick or wall surface. The resistance to fluid flow will be minimized by choosing fluids with low values of vapour and liquid viscosities.

7.3.1 Case Study

This problem was proposed by Sousa et al. [39]. In its basic form, the heat pipe is a hermetically sealed tube-type container with a wick structure placed on its internal walls and filled with a working fluid. In Fig. 7.4, a drawing of the heat pipe concept is shown. Methanol is used as working fluid and stainless steel (SS 304) is

Table 7.11 Optimal results with decision variable by using GEM

Q	°C	N_m	d (m) e-5	d_v (m)	t_w (m) e-5	L_e (m)	L_c (m)	t_t (m) e-4	ε	M_{total} (kg)
25	−15	314.87	3.12	0.0270	7.80	0.0554	0.0807	1.11	0.9663	0.0522
50	−15	316.32	4.01	0.0318	8.75	0.0546	0.0501	1.02	0.8180	0.0605
75	−15	319.42	2.85	0.0363	6.14	0.0795	0.0619	1.02	0.9049	0.0663
100	−15	319.16	2.76	0.0392	7.37	0.1295	0.1353	1.01	0.9998	0.0812
25	0	316.14	2.62	0.0231	8.43	0.0528	0.0511	1.02	0.9656	0.0396
50	0	314.43	2.57	0.0306	5.87	0.0738	0.0527	1.06	0.9706	0.0546
75	0	315.48	3.48	0.0340	8.19	0.1224	0.0916	1.01	0.9447	0.0688
100	0	314.53	4.12	0.0375	8.67	0.1157	0.0880	1.00	0.8072	0.0819
25	15	316.60	3.89	0.0227	9.32	0.0754	0.0675	1.01	0.8657	0.0446
50	15	315.92	4.39	0.0286	9.30	0.0733	0.0770	1.17	0.9350	0.0617
75	15	315.51	3.10	0.0372	6.78	0.0956	0.1419	1.01	0.9340	0.0764
100	15	319.29	2.66	0.0397	5.63	0.1879	0.0740	1.20	0.7522	0.1045
25	30	316.87	2.78	0.0234	5.88	0.0990	0.0688	1.02	0.9821	0.0428
50	30	317.40	3.72	0.0283	7.89	0.0788	0.1147	1.15	0.8988	0.0639
75	30	316.38	4.33	0.0326	9.22	0.0869	0.0541	1.71	0.9592	0.0966
100	30	317.85	3.14	0.0397	8.04	0.0940	0.0875	1.77	0.7048	0.1400

used as the material of the container since it is compatible. The wick is of the mesh type and is made of SS 304.

The objective function to be minimized is the total mass of the heat pipe (m_{total}). The design variables are the wick‘s mesh number (N_m), the diameter of the vapour core (d_v), the thickness of wick (t_w), the thickness of the container‘s wall (t_t), the length of the evaporator section (L_e) and the length of the condenser section (L_c). The length of the adiabatic section (L_a) is dependent on the application and here was fixed equal to 0.5 m. The details of the optimization problem with the objective function and the constraints are available in Sousa et al. [39].

Following are the upper and lower bounds of design variables imposed for this case study:

- Mesh number of wick (N_m) ranging from 314 to 15,000;
- Diameter of wick wire (d) ranging from 0.025 to 1 mm;
- Diameter of vapour core (d_v) ranging from 5 to 80 mm;
- Thickness of wick (t_w) ranging from 0.05 to 10 mm;
- Length of evaporator section (L_e) ranging from 50 to 400 mm;
- Length of condenser section (L_c) ranging from 50 to 400 mm;
- Thickness heat pipe tube (t_t) ranging from 0.3 to 3 mm;
- Porosity (ε) ranging from 0.0001 to 0.9999.

Table 7.11 shows the minimum mass of heat pipe for different heat inputs and different sink temperatures with decision variables by using the grenade explosion algorithm (GEM).

The best results produced by GEM are compared with the generalized extremal optimization (GEO) algorithm used by Sousa et al. [39]. It is observed that, using the GEM, there is a reduction of up to 30.32% in the mass of heat pipe (M_{total}) with varying heat transfer rates (Q) at different temperatures.

The TLBO technique presented in this book can also be attempted for the design optimization problems of shell and tube heat exchangers and the heat pipe. However, this has not been taken up in the present work.

References

1. Cheng YH, Lin WK (2005) Geometric optimization of thermoelectric coolers in a confined volume using genetic algorithms. Appl Therm Eng 25:2983–2997
2. Pan Y, Lin B, Chen J (2007) Performance analysis and parametric optimal design of an irreversible multi-couple thermoelectric refrigerator under various operating conditions. Appl Energy 84:882–892
3. Rowe DM (1996) CRC handbook of thermoelectric. CRC Press, London
4. Cheng YH, Shih C (2006) Maximizing the cooling capacity and COP of two-stage thermoelectric coolers through genetic algorithm. Appl Therm Eng 26:937–947
5. Chen J, Zhou Y, Wang H, Wang JT (2002) Comparison of the optimal performance of single- and two-stage thermoelectric refrigeration systems. Appl Energy 73:285–298
6. Xuan XC, Ng KC, Yap C, Chua HT (2002) Optimization of two stage thermoelectric coolers with two design configurations. Energy Convers Manage 43:2041–2052
7. Xuan XC (2002) Analyses of the performance and polar characteristics of two-stage thermoelectric coolers. Semicond Sci Tech 17:414–420
8. Xuan XC, Ng KC, Yap C, Chua HT (2002) The maximum temperature difference and polar characteristic of two-stage thermoelectric coolers. Cryog 42:273–278
9. Chen X, Lin B, Chen J (2006) The parametric optimum design of a new combined system of semiconductor thermoelectric devices. Appl Energy 83:681–686
10. Abramzon B (2007) Numerical optimization of the thermoelectric cooling devices. J Electron Packag 129(3):339–347
11. Yu J, Zhao H, Xie K (2007) Analysis of optimum configuration of two-stage thermoelectric modules. Cryog 47(2):89–93
12. Chen L, Meng F, Sun F (2009) A novel configuration and performance for a two-stage thermoelectric heat pump system driven by a two-stage thermoelectric generator. P I Mech Eng A-J Pow 223(4):329–339
13. Lai H, Pan Y, Chen J (2004) Optimum design on the performance parameters of a two-stage combined semiconductor thermoelectric heat pump. Semicond Sci Tech 19:17–22
14. Chen L, Li J, Sun F, Wu C (2005) Effect of heat transfer on the performance of two-stage semiconductor thermoelectric refrigerators. J Appl Phys 98(3):1–7
15. Chen L, Li J, Sun F, Wu C (2008) Performance optimization for a two-stage thermoelectric heat-pump with internal and external irreversibilities. Appl Energy 85(7):641–649
16. Meng F, Chen L, Sun F (2009) Performance optimization for two-stage thermoelectric refrigerator system driven by two-stage thermoelectric generator. Cryog 49(2):57–65
17. Lee S, Song S, Au V, Moran KP (1995) Constriction/spreading resistance model for electronic packaging. In: Proceedings of ASME/JSME thermal engineering joint conference, vol 4. Maui, Hawaii, pp 199–206

18. Edwards JE (2008) Design and rating of shell and tube heat exchangers. P and I Design Ltd, Teesside
19. Ravagnani M, Silva A, Andrade AL (2003) Detailed equipment design in heat exchanger networks synthesis and optimisation. Appl Therm Eng 23:141–151
20. Pariyani A, Gupta A, Ghosh P (2006) Design of heat exchanger networks using randomized algorithm. Comput Chem Eng 30:1046–1053
21. Babu BV, Munawar SA (2007) Differential evolution strategies for optimal design of shell and tube heat exchangers. Chem Eng Sci 14:3720–3739
22. Ravagnani J, Caballero A (2007) Optimal heat exchanger network synthesis with the detailed heat transfer equipment design. Comput Chem Eng 31(11):1432–1448
23. Fakheri A (2007) Profile design optimization of heat exchangers. Appl Therm Eng 24:1–9
24. Gholap AK, Khan JA (2007) Design and multi-objective optimization of heat exchangers for refrigerators. Appl Therm Eng 84:1226–1239
25. Caputo AC, Pelagagge PM, Salini P (2008) Heat exchanger design based on economic optimization. Appl Therm Eng 10:1151–1159
26. Soltan BK, Saffar-Avval M, Damangir E (2004) Minimizing capital and operating costs of shell and tube condensers using optimum baffle spacing. Appl Therm Eng 24:2801–2810
27. Costa LH, Queiroz M (2008) Design optimization of shell-and-tube heat exchangers. Appl Therm Eng 28:1798–1805
28. Thirumarimurugan M, Kannadasan T, Ramasamy E (2008) Performance analysis of shell and tube heat exchanger using miscible system. Am J Appl Sci 5(5):548–552
29. Ponce OJM, Serna GM, Jimenez GA (2009) Use of genetic algorithms for the optimal design of shell-and-tube heat exchangers. Appl Therm Eng 29:203–209
30. Guo J, Xu M, Cheng L (2009) The application of field synergy number in shell-and-tube heat exchanger optimization design. Appl Energy 86:2079–2087
31. Sanaye S, Hajabdollahi H (2009) Multi-objective optimizationMulti-objective optimization of shell and tube heat exchangers. Appl Therm Eng 29:1–9
32. Hosseini R, Ceylan H (2009) A new solution algorithm for improving performance of ant colony optimization. Appl Math Computat 211:75–84
33. Patel VK, Rao RV (2010) Design optimization of shell-and-tube heat exchanger using particle swarm optimization technique. Appl Therm Eng 30:1–9
34. Kern DQ (1950) Process Heat Transfer. McGraw-Hill, New York
35. Sinnot RK (1996) Coulson and Richarson's chemical engineering-chemical engineering design. Butterworth-Heinemann, Oxford
36. Bejan A, Kraus AD (2003) Heat transfer handbook. John Wiley, New York, pp 1181–1230
37. Reay D, Kew P (2006) Heat pipes: theory design and applications, butterworth-heinemann. Elsevier, Netherlands
38. Faghri A (1995) Heat pipe science and technology. Taylor and Francis, New York
39. Sousa FL, Vlassov V, Ramos FM (2004) Generalized external optimization: an application in heat pipe design. Appl Math Model 28:911–931

Chapter 8
Conclusions

Mechanical elements such as gears, bearings, clutches, springs, power screws, hydraulic cylinders, etc., are widely used in machine tools, transmission systems, material handling equipments, automobiles, etc. Design of these mechanical elements includes an optimization process in which the designers consider certain objectives such as strength, deflection, weight, wear, corrosion, etc. depending on the requirements. It is required to optimize such mechanical elements to make the whole assembly efficient and cost effective. Although, traditional optimization techniques had been employed in the past to solve optimization problems in mechanical design, these techniques have their own drawbacks. Considering the drawbacks of the traditional optimization techniques, this book deals with the applications of advanced optimization techniques, modifications in existing advanced optimization techniques, hybridization of existing advanced optimization techniques and development of a new optimization technique for the design optimization of different mechanical elements.

Advanced optimization techniques such as PSO, ABC, BBO, DE and AIA are applied to the mechanical design problems of gear train, radial ball bearing, Belleville spring, multi-plate disc clutch brake, robot gripper, hydrostatic thrust bearing and 4-stage gear train. The advanced optimization techniques have shown better results than the methods (like GA, PSO or both) which were applied to the same problems by the previous researchers.

The existing algorithms such as PSO, HEA and ABC are modified in this book by incorporating changes in their basic search mechanisms. The modified algorithms are tested on 13 unconstrained and 24 constrained benchmark functions, and also on 20 mechanical design problems. It is observed from the results that modification in PSO is effective in comparison to the basic PSO for all the considered problems in finding the best and the mean solutions. Modification in ABC is effective for the unconstrained benchmark functions and mechanical design problems, but it is slightly inferior to basic ABC for the constrained benchmark

R. V. Rao and V. J. Savsani, *Mechanical Design Optimization Using Advanced Optimization Techniques*, Springer Series in Advanced Manufacturing,
DOI: 10.1007/978-1-4471-2748-2_8, © Springer-Verlag London 2012

functions. Modification in HEA is also effective over basic HEA. The results of modified PSO are slightly inferior to the basic ABC in finding the best solutions but it is equivalent in finding the mean solutions. HEA and modified HEA are inferior to the different variants of PSO and ABC in finding best and the mean solutions.

Different hybrid algorithms are developed in this book by keeping ABC as the base algorithm. Four different hybrid algorithms are developed viz. HPABC (Hybrid Particle swarm based Artificial Bee Colony), HBABC (Hybrid Biogeography based Artificial Bee Colony), HDABC (Hybrid Differential evolution based Artificial Bee Colony) and HGABC (Hybrid Genetic algorithm based Artificial bee colony). All the algorithms are tested on unconstrained and constrained benchmark functions and also on mechanical design problems. The hybridization of ABC with BBO and DE is very effective than the basic ABC in finding the best solutions for the considered problems. Hybridization of ABC with PSO and GA is not so effective than the basic ABC. Hybridization of ABC with PSO is effective than all the variants of PSO. In searching the mean solution, only hybridization of ABC with DE is effective than all other algorithms.

All the nature-inspired algorithms such as GA, PSO, BBO, ABC, DE, etc. require algorithm-specific parameters to be set for their proper working in addition to the common control parameters of population size, number of generations, and number of elites. Proper selection of parameters is essential for the searching of the optimum solution by these algorithms. A change in the algorithm parameters changes the effectiveness of the algorithms. To avoid this difficulty an optimization method, TLBO, which is algorithm-specific parameter free, is presented in this book. This method works on the effect of influence of a teacher on learners. Like other nature-inspired algorithms, TLBO is also a population-based method which uses a population of solutions to proceed to the global solution. For TLBO, the population is considered as a group of learners or a class of learners. The process of working of TLBO is divided into two parts. The first part consists of 'Teacher Phase' and the second part consists of 'Learner Phase'. The 'Teacher Phase' means learning from the teacher and the 'Learner Phase' means learning through the interaction between learners.

Performance of the TLBO method is checked with the other optimization techniques available in the literature for different unconstrained and constrained benchmark functions. The TLBO method has outperformed the results of other algorithms from the literature for the unconstrained and constrained benchmark functions. Moreover the TLBO method requires less number of function evaluations than that of the other optimization algorithms to find the global optimum. Convergence of the TLBO method is compared with the convergence of the ABC algorithm for the unconstrained benchmark problems. The TLBO method has shown better convergence than the ABC algorithm.

The TLBO method is applied to different mechanical element design optimization problems available in the literature. The results of the TLBO method are compared with the results given by the other researchers for the mechanical element design optimization problems. The TLBO method requires less number of function evaluations than that required by the other algorithms to find the optimum results. The TLBO method is also applied successfully to the design optimization of a two-stage thermoelectric cooler. This paves the way for the application of TLBO for the design optimization of thermal equipment and devices.

The proposed TLBO method is also compared with the hybrid algorithms developed in this book by considering all the unconstrained and the constrained benchmark functions and also the mechanical element design optimization problems. It is observed from the comparison that the TLBO method has shown superior results than the hybrid algorithms. Convergence plots are obtained for the comparison of HDABC with TLBO for the problems having same mean solutions. It is observed from the convergence graphs that the TLBO method has better convergence than the HDABC hybrid method.

The performance of the proposed TLBO method is checked with the recent and well-known optimization algorithms such as GA, ABC, PSO, HS, DE, Hybrid-PSO, etc. by experimenting with different benchmark problems and mechanical element design optimization problems with different characteristics. The effectiveness of TLBO method is also checked for different performance criteria, like success rate, mean solution, average function evaluations required, convergence rate, etc. The results show better performance of TLBO method over other natured-inspired optimization methods for the considered benchmark functions and mechanical element design optimization problems. Also, the TLBO method shows better performance with less computational efforts for the large scale problems, i.e. problems with high dimensions. This method can be further extended for the other engineering applications to check its suitability and robustness.

Appendix 1
Additional Demonstrative Examples Solved by TLBO Algorithm

Now few benchmark functions are considered for demonstrating the steps of the proposed TLBO algorithm. Population size of ten, number of generations of 3 and two design variables are considered for demonstration. The results are shown only for 3 generations. Actually, the final solutions for these benchmark functions can be obtained after a sufficient number of generations.

A1.1 Example 1: Sphere Function

$$f(x) = \sum_{i=1}^{n} x_i^2$$
$$-5 \leq x_i \leq 5, \quad \min(f) = f(0, 0\ldots0) = 0$$

Generation: 1

Initial population		Initial population cost
1.3638	−0.8495	2.5815
−0.7540	−2.0739	4.8695
1.8853	2.6906	10.7936
−3.7511	−0.0789	14.0766
−3.9645	0.9684	16.6554
−3.2782	−2.6811	17.9349
1.6293	−4.2761	20.9400
−4.5295	−1.2920	22.1860
−4.5299	2.2808	25.7218
−3.8581	−4.4127	34.3565

R. V. Rao and V. J. Savsani, *Mechanical Design Optimization Using Advanced Optimization Techniques*, Springer Series in Advanced Manufacturing,
DOI: 10.1007/978-1-4471-2748-2, © Springer-Verlag London 2012

Teacher Phase: TF = 2; Teacher = 1.3638 −0.8495

Population after teacher phase		Population after checking feasibility		Population cost after checking feasibility	Population after greedy selection		Population cost after greedy selection
2.6080	−0.6697	2.6080	−0.6697	7.2504	1.3638	−0.8495	2.5815
0.4903	−1.8941	0.4903	−1.8941	3.8280	0.4903	−1.8941	3.8280
3.1296	2.8704	3.1296	2.8704	18.0334	1.8853	2.6906	10.7936
−2.5068	0.1009	−2.5068	0.1009	6.2941	−2.5068	0.1009	6.2941
−2.7203	1.1482	−2.7203	1.1482	8.7182	−2.7203	1.1482	8.7182
−2.0340	−2.5013	−2.0340	−2.5013	10.3933	−2.0340	−2.5013	10.3933
2.8736	−4.0964	2.8736	−4.0964	25.0375	1.6293	−4.2761	20.9400
−3.2853	−1.1122	−3.2853	−1.1122	12.0300	−3.2853	−1.1122	12.0300
−3.2856	2.4606	−3.2856	2.4606	16.8496	−3.2856	2.4606	16.8496
−2.6138	−4.2329	−2.6138	−4.2329	24.7493	−2.6138	−4.2329	24.7493

Learner phase

Population after learner phase		Population after checking feasibility		Population cost after checking feasibility	Population after greedy selection		Population cost after greedy selection
2.5782	−1.4436	2.5782	−1.4436	8.7309	1.3638	−0.8495	2.5815
2.1657	−1.4911	2.1657	−1.4911	6.9137	0.4903	−1.8941	3.8280
−0.6155	−0.6222	−0.6155	−0.6222	0.7660	−0.6155	−0.6222	0.7660
−2.8892	2.2057	−2.8892	2.2057	13.2125	−2.5068	0.1009	6.2941
−5.1436	4.1704	−5.0000	4.1704	42.3920	−2.7203	1.1482	8.7182
−2.5611	0.3017	−2.5611	0.3017	6.6500	−2.5611	0.3017	6.6500
1.7730	−4.2776	1.7730	−4.2776	21.4413	1.6293	−4.2761	20.9400
−5.9137	0.5800	−5.0000	0.5800	25.3364	−3.2853	−1.1122	12.0300
−3.2854	0.1005	−3.2854	0.1005	10.8038	−3.2854	0.1005	10.8038
−2.4336	−3.6947	−2.4336	−3.6947	19.5737	−2.4336	−3.6947	19.5737

Generation: 2

Initial population		Initial population cost
−0.6155	−0.6222	0.7660
1.3638	−0.8495	2.5815
−2.6575	−0.8495	2.5815
0.4903	−1.8941	3.8280
−2.5068	0.1009	6.2941
−2.5611	0.3017	6.6500
−2.7203	1.1482	8.7182
−3.2854	0.1005	10.8038
−3.2853	−1.1122	12.0300
−2.4336	−3.6947	19.5737

Teacher Phase: TF = 1; Teacher = − 0.6155 −0.6222

Population after teacher phase		Population after checking feasibility		Population cost after checking feasibility	Population after greedy selection		Population cost after greedy selection
0.5575	−0.5766	0.5575	−0.5766	0.6432	0.5575	−0.5766	0.6432
2.5367	−0.8039	2.5367	−0.8039	7.0811	1.3638	−0.8495	2.5815
−1.4845	−0.8039	−1.4845	−0.8039	2.8501	−2.6575	−0.8495	2.5815
1.6632	−1.8485	1.6632	−1.8485	6.1832	0.4903	−1.8941	3.8280
−1.3338	0.1466	−1.3338	0.1466	1.8006	−1.3338	0.1466	1.8006
−1.3881	0.3473	−1.3881	0.3473	2.0475	−1.3881	0.3473	2.0475
−1.5473	1.1939	−1.5473	1.1939	3.8195	−1.5473	1.1939	3.8195
−2.1125	0.1461	−2.1125	0.1461	4.4838	−2.1125	0.1461	4.4838
−2.1123	−1.0666	−2.1123	−1.0666	5.5996	−2.1123	−1.0666	5.5996
−1.2607	−3.6491	−1.2607	−3.6491	14.9055	−1.2607	−3.6491	14.9055

Learner phase

Population after learner phase		Population after checking feasibility		Population cost after checking feasibility	Population after greedy selection		Population cost after greedy selection
2.1365	−0.2868	2.1365	−0.2868	4.6471	0.5575	−0.5766	0.6432
4.1181	−1.6384	4.1181	−1.6384	19.6435	1.3638	−0.8495	2.5815
−2.4764	−0.7133	−2.4764	−0.7133	6.6414	−2.6575	−0.8495	2.5815
0.3368	−1.6614	0.3368	−1.6614	2.8738	0.3368	−1.6614	2.8738
−0.9028	0.1468	−0.9028	0.1468	0.8367	−0.9028	0.1468	0.8367
−0.1484	−0.2414	−0.1484	−0.2414	0.0803	−0.1484	−0.2414	0.0803
−1.5243	1.2365	−1.5243	1.2365	3.8525	−1.5473	1.1939	3.8195
−1.6329	0.1464	−1.6329	0.1464	2.6877	−1.6329	0.1464	2.6877
−1.8683	−0.5902	−1.8683	−0.5902	3.8390	−1.8683	−0.5902	3.8390
−2.5849	−0.9949	−2.5849	−0.9949	7.6716	−2.5849	−0.9949	7.6716

Generation: 3

Initial population		Initial population cost
−0.1484	−0.2414	0.0803
0.5575	−0.5766	0.6432
−0.6155	−0.6222	0.7660
−0.9028	0.1468	0.8367
1.3638	−0.8495	2.5815
−1.6329	0.1464	2.6877
0.3368	−1.6614	2.8738
−1.5473	1.1939	3.8195
−1.8683	−0.5902	3.8390
−2.5849	−0.9949	7.6716

Teacher phase: TF = 2; Teacher = − 0.1484 −0.2414

Population after teacher phase		Population after checking feasibility		Population cost after checking feasibility	Population after greedy selection		Population cost after greedy selection
0.6927	−0.1171	0.6927	−0.1171	0.4936	−0.1484	−0.2414	0.0803
1.3986	−0.4523	1.3986	−0.4523	2.1607	0.5575	−0.5766	0.6432
0.2257	−0.4979	0.2257	−0.4979	0.2989	0.2257	−0.4979	0.2989
−0.0617	0.2711	−0.0617	0.2711	0.0773	−0.0617	0.2711	0.0773
2.2049	−0.7252	2.2049	−0.7252	5.3876	1.3638	−0.8495	2.5815
−0.7917	0.2707	−0.7917	0.2707	0.7001	−0.7917	0.2707	0.7001
1.1779	−1.5371	1.1779	−1.5371	3.7503	0.3368	−1.6614	2.8738
−0.7062	1.3181	−0.7062	1.3181	2.2362	−0.7062	1.3181	2.2362
−1.0272	−0.4659	−1.0272	−0.4659	1.2722	−1.0272	−0.4659	1.2722
−1.7437	−0.8706	−1.7437	−0.8706	3.7987	−1.7437	−0.8706	3.7987

Learner phase

Population after learner phase		Population after checking feasibility		Population cost after checking feasibility	Population after greedy selection		Population cost after greedy selection
−0.6314	1.1720	−0.6314	1.1720	1.7721	−0.1484	−0.2414	0.0803
0.7420	−0.6925	0.7420	−0.6925	1.0302	0.5575	−0.5766	0.6432
0.1923	−0.4085	0.1923	−0.4085	0.2038	0.1923	−0.4085	0.2038
0.4241	0.2714	0.4241	0.2714	0.2535	−0.0617	0.2711	0.0773
0.9183	−0.6987	0.9183	−0.6987	1.3315	0.9183	−0.6987	1.3315
−0.8064	0.0913	−0.8064	0.0913	0.6586	−0.8064	0.0913	0.6586
2.0102	−2.2975	2.0102	−2.2975	9.3194	0.3368	−1.6614	2.8738
−0.7583	0.6796	−0.7583	0.6796	1.0369	−0.7583	0.6796	1.0369
−0.4104	0.0049	−0.4104	0.0049	0.1685	−0.4104	0.0049	0.1685
−1.5116	−0.7395	−1.5116	−0.7395	2.8318	−1.5116	−0.7395	2.8318

A1.2 Example 2: Schwefel 2.22 Function

$$f(x) = \sum_{i=1}^{n} |x_i| + \prod_{i=1}^{n} |x_i|$$

$$-100 \le x_i \le 100, \quad \min(f) = f(0, 0\ldots0) = 0$$

Generation: 1

Initial population		Initial population cost
−1.0264	−1.0271	3.1077
3.3883	0.2132	4.3237
−0.2588	−4.7546	6.2438
7.6304	0.0446	8.0154
−0.8446	−4.5691	9.2730
2.4142	4.5386	17.9099
−4.9259	3.4846	25.5753
5.7342	−3.6804	30.5190
3.6019	−6.7873	34.8361
−9.6584	4.1030	53.3903

Teacher phase: TF = 2; Teacher = − 1.0264 −1.0271

Population after teacher phase		Population after checking feasibility		Population cost after checking feasibility	Population after greedy selection		Population cost after greedy selection
−2.0713	−0.9968	−2.0713	−0.9968	5.1328	−1.0264	−1.0271	3.1077
2.3434	0.2434	2.3434	0.2434	3.1573	2.3434	0.2434	3.1573
−1.3037	−4.7243	−1.3037	−4.7243	12.1869	−0.2588	−4.7546	6.2438
6.5855	0.0749	6.5855	0.0749	7.1535	6.5855	0.0749	7.1535
−1.8895	−4.5389	−1.8895	−4.5389	15.0047	−0.8446	−4.5691	9.2730
1.3693	4.5689	1.3693	4.5689	12.1944	1.3693	4.5689	12.1944
−5.9708	3.5149	−5.9708	3.5149	30.4721	−4.9259	3.4846	25.5753
4.6893	−3.6502	4.6893	−3.6502	25.4563	4.6893	−3.6502	25.4563
2.5570	−6.7570	2.5570	−6.7570	26.5917	2.5570	−6.7570	26.5917
−10.7033	4.1333	−10.0000	4.1333	55.4664	−9.6584	4.1030	53.3903

Learner phase

Population after learner phase		Population after checking feasibility		Population cost after checking feasibility	Population after greedy selection		Population cost after greedy selection
6.4764	−5.4861	6.4764	−5.4861	47.4924	−1.0264	−1.0271	3.1077
2.3023	1.5901	2.3023	1.5901	7.5532	2.3434	0.2434	3.1573
−0.2635	−4.7317	−0.2635	−4.7317	6.2420	−0.2635	−4.7317	6.2420
7.5028	1.8768	7.5028	1.8768	23.4612	6.5855	0.0749	7.1535
−0.5765	−4.1644	−0.5765	−4.1644	7.1419	−0.5765	−4.1644	7.1419
1.6883	3.1525	1.6883	3.1525	10.1629	1.6883	3.1525	10.1629
−0.5785	−4.1901	−0.5785	−4.1901	7.1927	−0.5785	−4.1901	7.1927
3.5026	−1.6806	3.5026	−1.6806	11.0696	3.5026	−1.6806	11.0696
2.6711	−6.5635	2.6711	−6.5635	26.7664	2.5570	−6.7570	26.5917
0.9506	−5.3288	0.9506	−5.3288	11.3450	0.9506	−5.3288	11.3450

Generation: 2

Initial population		Initial population cost
−1.0264	−1.0271	3.1077
7.6341	−1.0271	3.1077
2.3434	0.2434	3.1573
−0.2635	−4.7317	6.2420
−0.5765	−4.1644	7.1419
6.5855	0.0749	7.1535
−0.5785	−4.1901	7.1927
1.6883	3.1525	10.1629
3.5026	−1.6806	11.0696
0.9506	−5.3288	11.3450

Teacher phase: TF = 1; Teacher = − 1.0264 −1.0271

Population after teacher phase		Population after checking feasibility		Population cost after checking feasibility	Population after greedy selection		Population cost after greedy selection
−1.3634	−0.7118	−1.3634	−0.7118	3.0458	−1.3634	−0.7118	3.0458
7.2971	−0.7118	7.2971	−0.7118	13.2032	7.6341	−1.0271	3.1077
2.0064	0.5587	2.0064	0.5587	3.6859	2.3434	0.2434	3.1573
−0.6005	−4.4165	−0.6005	−4.4165	7.6690	−0.2635	−4.7317	6.2420
−0.9135	−3.8492	−0.9135	−3.8492	8.2790	−0.5765	−4.1644	7.1419
6.2486	0.3901	6.2486	0.3901	9.0762	6.5855	0.0749	7.1535
−0.9155	−3.8749	−0.9155	−3.8749	8.3379	−0.5785	−4.1901	7.1927
1.3513	3.4677	1.3513	3.4677	9.5048	1.3513	3.4677	9.5048
3.1656	−1.3653	3.1656	−1.3653	8.8532	3.1656	−1.3653	8.8532
0.6136	−5.0136	0.6136	−5.0136	8.7036	0.6136	−5.0136	8.7036

Learner phase

Population after learner phase		Population after checking feasibility		Population cost after checking feasibility	Population after greedy selection		Population cost after greedy selection
−2.0594	2.3724	−2.0594	2.3724	9.3176	−1.3634	−0.7118	3.0458
8.2026	−1.6245	8.2026	−1.6245	23.1519	7.6341	−1.0271	3.1077
4.4724	4.3066	4.4724	4.3066	28.0396	2.3434	0.2434	3.1573
−0.9285	−2.3012	−0.9285	−2.3012	5.3665	−0.9285	−2.3012	5.3665
−1.1636	−6.4886	−1.1636	−6.4886	15.2023	−0.5765	−4.1644	7.1419
6.6034	0.0561	6.6034	0.0561	7.0299	6.6034	0.0561	7.0299
−0.3157	−3.7913	−0.3157	−3.7913	5.3038	−0.3157	−3.7913	5.3038
0.8566	−2.2202	0.8566	−2.2202	4.9786	0.8566	−2.2202	4.9786
2.8941	−1.7535	2.8941	−1.7535	9.7225	3.1656	−1.3653	8.8532
0.1071	−4.6522	0.1071	−4.6522	5.2574	0.1071	0−4.6522	5.2574

Generation: 3

Initial population		Initial population cost
−1.3634	−0.7118	3.0458
−1.0264	−1.0271	3.1077
2.3434	0.2434	3.1573
0.8566	−2.2202	4.9786
0.1071	−4.6522	5.2574
−0.3157	−3.7913	5.3038
−0.9285	−2.3012	5.3665
6.6034	0.0561	7.0299
−0.5765	−4.1644	7.1419
3.1656	−1.3653	8.8532

Teacher phase: TF = 2; Teacher = − 1.3634 −0.7118

Population after teacher phase		Population after checking feasibility		Population cost after checking feasibility	Population after greedy selection		Population cost after greedy selection
−2.6427	0.7969	−2.6427	0.7969	5.5456	−1.3634	−0.7118	3.0458
−2.3058	0.4817	−2.3058	0.4817	3.8980	−1.0264	−1.0271	3.1077
1.0641	1.7521	1.0641	1.7521	4.6806	2.3434	0.2434	3.1573
−0.4227	−0.7115	−0.4227	−0.7115	1.4350	−0.4227	−0.7115	1.4350
−1.1722	−3.1435	−1.1722	−3.1435	8.0006	0.1071	−4.6522	5.2574
−1.5950	−2.2826	−1.5950	−2.2826	7.5183	−0.3157	−3.7913	5.3038
−2.2078	−0.7925	−2.2078	−0.7925	4.7502	−2.2078	−0.7925	4.7502
5.3241	1.5648	5.3241	1.5648	15.2200	6.6034	0.0561	7.0299
−1.8558	−2.6557	−1.8558	−2.6557	9.4401	−0.5765	−4.1644	7.1419
1.8863	0.1434	1.8863	0.1434	2.3001	1.8863	0.1434	2.3001

Learner phase

Population after learner phase		Population after checking feasibility		Population cost after checking feasibility	Population after greedy selection		Population cost after greedy selection
−1.6841	−0.4118	−1.6841	−0.4118	2.7895	−1.6841	−0.4118	2.7895
0.0980	−1.2503	0.0980	−1.2503	1.4709	0.0980	−1.2503	1.4709
0.8996	−0.2550	0.8996	−0.2550	1.3840	0.8996	−0.2550	1.3840
0.0438	−0.6903	0.0438	−0.6903	0.7644	0.0438	−0.6903	0.7644
−5.3532	−8.6095	−5.3532	−8.6095	60.0515	0.1071	−4.6522	5.2574
−0.3749	−3.8242	−0.3749	−3.8242	5.6329	−0.3157	−3.7913	5.3038
−3.3775	1.6251	−3.3775	1.6251	10.4912	−2.2078	−0.7925	4.7502
1.2595	−0.7025	1.2595	−0.7025	2.8470	1.2595	−0.7025	2.8470
−0.5266	−4.0930	−0.5266	−4.0930	6.7752	−0.5266	−4.0930	6.7752
−0.1810	0.1816	−0.1810	0.1816	0.3955	−0.1810	0.1816	0.3955

A1.3 Example 3: Schwefel 1.2 Function

$$f(x) = \sum_{i=1}^{n}\left(\sum_{j=1}^{i} x_j\right)^2$$

$$-100 \le x_i \le 100, \quad \min(f) = f(0, 0\ldots0) = 0$$

Generation: 1

Initial population		Initial population cost (1e3)
43.2047	−33.3457	1.9639
−34.8956	6.9403	1.9992
43.2156	−31.2374	2.0111
37.0539	−2.6692	2.5553
−51.2416	54.5154	2.6364
−52.2736	47.9308	2.7514
−36.1370	−5.8343	3.0675
55.9135	−10.6063	5.1791
−49.5863	−20.0747	7.3115
38.4306	42.9056	8.0925

Population after teacher phase		Population after checking feasibility		Population cost after checking feasibility (1e4)	Population after greedy selection		Population cost after greedy selection (1e3)
77.4899	−42.5756	65.5360	−42.5756	0.4822	43.2047	−33.3457	1.9639
−0.6104	−2.2897	−0.6104	−2.2897	0.0009	−0.6104	−2.2897	0.0088
77.5007	−40.4674	65.5360	−40.4674	0.4923	43.2156	−31.2374	2.0111
71.3391	−11.8992	65.5360	−11.8992	0.7172	37.0539	−2.6692	2.5553
−16.9564	45.2854	−16.9564	45.2854	0.109	−16.9564	45.2854	1.09
−17.9884	38.7008	−17.9884	38.7008	0.0753	−17.9884	38.7008	0.7526
−1.8518	−15.0643	−1.8518	−15.0643	0.029	−1.8518	−15.0643	0.2896
90.1986	−19.8363	65.5360	−19.8363	0.6383	55.9135	−10.6063	5.1791
−15.3011	−29.3047	−15.3011	−29.3047	0.2224	−15.3011	−29.3047	2.2238
72.7158	33.6756	65.5360	33.6756	1.4138	38.4306	42.9056	8.0925

Population after learner phase		Population after checking feasibility		Population cost after checking feasibility (1e3)	Population after greedy selection		Population cost after greedy selection (1e3)
−10.0974	36.3206	−10.0974	36.3206	0.7896	−10.0974	36.3206	0.7896
−0.4343	−0.4773	−0.4343	−0.4773	0.001	−0.4343	−0.4773	0.001
35.3720	−26.0566	35.3720	−26.0566	1.338	35.3720	−26.0566	1.338
21.6050	3.8326	21.6050	3.8326	1.1138	21.6050	3.8326	1.1138
−6.3926	3.0782	−6.3926	3.0782	0.0519	−6.3926	3.0782	0.0519
−29.9437	37.8098	−29.9437	37.8098	0.9585	−17.9884	38.7008	0.7526
−40.3814	−1.2373	−40.3814	−1.2373	3.3628	−1.8518	−15.0643	0.2896
−7.9103	38.3470	−7.9103	38.3470	0.989	−7.9103	38.3470	0.989
−15.3737	−26.0349	−15.3737	−26.0349	1.951	−15.3737	−26.0349	1.951
16.6245	11.5246	16.6245	11.5246	1.0687	16.6245	11.5246	1.0687

Generation: 2

Initial population		Initial population cost (1e3)
−0.4343	−0.4773	0.001
−6.3926	3.0782	0.0519
−1.8518	−15.0643	0.2896
−17.9884	38.7008	0.7526
−10.0974	36.3206	0.7896
−7.9103	38.3470	0.989
16.6245	11.5246	1.0687
21.6050	3.8326	1.1138
35.3720	−26.0566	1.338
43.2047	−33.3457	1.9639

Population after teacher phase		Population after checking feasibility		Population cost after checking feasibility (1e3)	Population after greedy selection		Population cost after greedy selection (1e3)
−9.7903	−1.1687	−9.7903	−1.1687	0.2159	−0.4343	−0.4773	0.001
−15.7486	2.3869	−15.7486	2.3869	0.4266	−6.3926	3.0782	0.0519
−11.2078	−15.7557	−11.2078	−15.7557	0.8526	−1.8518	−15.0643	0.2896
−27.3444	38.0094	−27.3444	38.0094	0.8615	−17.9884	38.7008	0.7526
−19.4534	35.6293	−19.4534	35.6293	0.6401	−19.4534	35.6293	0.6401
−17.2663	37.6556	−17.2663	37.6556	0.7138	−17.2663	37.6556	0.7138
7.2685	10.8333	7.2685	10.8333	0.3805	7.2685	10.8333	0.3805
12.2490	3.1412	12.2490	3.1412	0.3869	12.2490	3.1412	0.3869
26.0160	−26.7480	26.0160	−26.7480	0.6774	26.0160	−26.7480	0.6774
33.8487	−34.0370	33.8487	−34.0370	1.1458	33.8487	−34.0370	1.1458

Population after learner phase		Population after checking feasibility		Population cost after checking feasibility (1e3)	Population after greedy selection		Population cost after greedy selection (1e3)
9.5768	−22.8203	9.5768	−22.8203	0.2671	−0.4343	−0.4773	1.0195
−35.6012	30.0177	−35.6012	30.0177	1.2986	−6.3926	3.0782	51.8508
−1.0620	−17.3390	−1.0620	−17.3390	0.3397	−1.8518	−15.0643	289.5843
−7.7023	7.1017	−7.7023	7.1017	0.0597	−7.7023	7.1017	59.6867
−9.4174	16.5765	−9.4174	16.5765	0.1399	−9.4174	16.5765	139.9408
−17.2619	37.6492	−17.2619	37.6492	0.7136	−17.2619	37.6492	713.6158
7.2354	10.8843	7.2354	10.8843	0.3807	7.2685	10.8333	380.5035
15.7304	−0.4265	15.7304	−0.4265	0.4817	12.2490	3.1412	386.8949
2.9303	−3.8191	2.9303	−3.8191	0.0094	2.9303	−3.8191	9.3767
30.8975	−31.2906	30.8975	−31.2906	0.9548	30.8975	−31.2906	954.8079

Generation: 3

Initial population		Initial population cost
−0.4343	−0.4773	1.0195
−5.3763	−0.4773	1.0195
2.9303	−3.8191	9.3767
−6.3926	3.0782	51.8508
−7.7023	7.1017	59.6867
−9.4174	16.5765	139.9408
−1.8518	−15.0643	289.5843
7.2685	10.8333	380.5035
12.2490	3.1412	386.8949
−17.2619	37.6492	713.6158

Population after teacher phase		Population after checking feasibility		Population cost after checking feasibility	Population after greedy selection		Population cost after greedy selection
−0.4066	−6.3459	−0.4066	−6.3459	45.7621	−0.4343	−0.4773	1.0195
−5.3486	−6.3459	−5.3486	−6.3459	165.3693	−5.3763	−0.4773	1.0195
2.9580	−9.6877	2.9580	−9.6877	54.0386	2.9303	−3.8191	9.3767
−6.3650	−2.7904	−6.3650	−2.7904	124.334	−6.3926	3.0782	51.8508
−7.6747	1.2331	−7.6747	1.2331	100.3955	−7.7023	7.1017	59.6867
−9.3897	10.7079	−9.3897	10.7079	89.9049	−9.3897	10.7079	89.9049
−1.8242	−20.9329	−1.8242	−20.9329	521.2133	−1.8518	−15.0643	289.5843
7.2961	4.9646	7.2961	4.9646	203.5592	7.2961	4.9646	203.5592
12.2766	−2.7274	12.2766	−2.7274	241.9022	12.2766	−2.7274	241.9022
−17.2343	31.7806	−17.2343	31.7806	508.6154	−17.2343	31.7806	508.6154

Population after learner phase		Population after checking feasibility		Population cost after checking feasibility	Population after greedy selection		Population cost after greedy selection
0.7780	11.9969	0.7780	11.9969	163.8018	−0.4343	−0.4773	1.0195
−3.6096	−5.4008	−3.6096	−5.4008	94.2164	−5.3763	−0.4773	1.0195
7.5402	−9.2547	7.5402	−9.2547	59.7945	2.9303	−3.8191	9.3767
−6.2181	2.9491	−6.2181	2.9491	49.3509	−6.2181	2.9491	49.3509
−19.6893	8.8097	−19.6893	8.8097	506.0354	−7.7023	7.1017	59.6867
−1.9004	−9.4107	−1.9004	−9.4107	131.5524	−9.3897	10.7079	89.9049
−1.8057	−14.5892	−1.8057	−14.5892	272.0528	−1.8057	−14.5892	272.0528
23.9429	−13.2332	23.9429	−13.2332	687.961	7.2961	4.9646	203.5592
−5.0283	8.0034	−5.0283	8.0034	34.1352	−5.0283	8.0034	34.1352
−17.0308	31.4214	−17.0308	31.4214	497.1375	−17.0308	31.4214	497.1375

A1.4 Example 4: Schwefel 2.21 Function

$$f(x) = \max_i\{|x_i|,\ 1 \leq i \leq n\}$$
$$-100 \leq x_i \leq 100, \quad \min(f) = f(0, 0\ldots0) = 0$$

Generation: 1

Initial population		Initial population cost
−27.8563	25.7134	27.8563
31.0886	3.0272	31.0886
7.6171	46.8397	46.8397
10.5190	60.8093	60.8093
77.8636	−15.7001	77.8636
41.0139	−79.8978	79.8978
−83.0201	−28.1705	83.0201
71.2215	93.6970	93.697
1.4503	−97.5835	97.5835
98.2282	−72.6985	98.2282

Population after teacher phase		Population after checking feasibility		Population cost after checking feasibility	Population after greedy selection		Population cost after greedy selection
−77.3135	45.9463	−77.3135	45.9463	77.3135	−27.8563	25.7134	27.8563
−18.3686	23.2601	−18.3686	23.2601	23.2601	−18.3686	23.2601	23.2601
−41.8401	67.0726	−41.8401	67.0726	67.0726	7.6171	46.8397	46.8397
−38.9382	81.0423	−38.9382	81.0423	81.0423	10.5190	60.8093	60.8093
28.4064	4.5328	28.4064	4.5328	28.4064	28.4064	4.5328	28.4064
−8.4434	−59.6648	−8.4434	−59.6648	59.6648	−8.4434	−59.6648	59.6648
−132.4773	−7.9376	−100.0000	−7.9376	100	−83.0201	−28.1705	83.0201
21.7643	113.9299	21.7643	100.0000	100	71.2215	93.6970	93.697
−48.0069	−77.3505	−48.0069	−77.3505	77.3505	−48.0069	−77.3505	77.3505
48.7709	−52.4656	48.7709	−52.4656	52.4656	48.7709	−52.4656	52.4656

Population after learner phase		Population after checking feasibility		Population cost after checking feasibility	Population after greedy selection		Population cost after greedy selection
−3.3771	49.6246	−3.3771	49.6246	49.6246	−27.8563	25.7134	27.8563
−45.4060	−11.8840	−45.4060	−11.8840	45.406	−18.3686	23.2601	23.2601
26.0179	62.0679	26.0179	62.0679	62.0679	7.6171	46.8397	46.8397
−16.4441	25.7617	−16.4441	25.7617	25.7617	−16.4441	25.7617	25.7617
41.5531	−22.2211	41.5531	−22.2211	41.5531	28.4064	4.5328	28.4064
−14.4586	−33.2096	−14.4586	−33.2096	33.2096	−14.4586	−33.2096	33.2096
−64.2011	−9.7883	−64.2011	−9.7883	64.2011	−64.2011	−9.7883	64.2011
59.1589	84.8105	59.1589	84.8105	84.8105	59.1589	84.8105	84.8105
−16.9840	−120.9257	−16.9840	−100.0000	100	−48.0069	−77.3505	77.3505
27.9060	−2.1179	27.9060	−2.1179	27.906	27.9060	−2.1179	27.906

Generation: 2

Initial population		Initial population cost
−18.3686	23.2601	23.2601
−16.4441	25.7617	25.7617
−27.8563	25.7134	27.8563
95.3769	25.7134	27.8563
27.9060	−2.1179	27.906
28.4064	4.5328	28.4064
−14.4586	−33.2096	33.2096
7.6171	46.8397	46.8397
−64.2011	−9.7883	64.2011
−48.0069	−77.3505	77.3505

Population after teacher phase		Population after checking feasibility		Population cost after checking feasibility	Population after greedy selection		Population cost after greedy selection
−23.6413	30.3752	−23.6413	30.3752	30.3752	−18.3686	23.2601	23.2601
−21.7167	32.8768	−21.7167	32.8768	32.8768	−16.4441	25.7617	25.7617
−33.1289	32.8285	−33.1289	32.8285	33.1289	−27.8563	25.7134	27.8563
90.1043	32.8285	90.1043	32.8285	90.1043	95.3769	25.7134	27.8563
22.6334	4.9971	22.6334	4.9971	22.6334	22.6334	4.9971	22.6334
23.1338	11.6479	23.1338	11.6479	23.1338	23.1338	11.6479	23.1338
−19.7312	−26.0945	−19.7312	−26.0945	26.0945	−19.7312	−26.0945	26.0945
2.3445	53.9548	2.3445	53.9548	53.9548	7.6171	46.8397	46.8397
−69.4738	−2.6732	−69.4738	−2.6732	69.4738	−64.2011	−9.7883	64.2011
−53.2795	−70.2355	−53.2795	−70.2355	70.2355	−53.2795	−70.2355	70.2355

Population after learner phase		Population after checking feasibility		Population cost after checking feasibility	Population after greedy selection		Population cost after greedy selection
−18.3610	23.5371	−18.3610	23.5371	23.5371	−18.3686	23.2601	23.2601
−19.5752	23.0188	−19.5752	23.0188	23.0188	−19.5752	23.0188	23.0188
−21.8007	48.5674	−21.8007	48.5674	48.5674	−27.8563	25.7134	27.8563
34.1165	8.2674	34.1165	8.2674	34.1165	95.3769	25.7134	27.8563
31.4716	6.5020	31.4716	6.5020	31.4716	22.6334	4.9971	22.6334
53.3077	43.9818	53.3077	43.9818	53.3077	23.1338	11.6479	23.1338
−15.2776	−54.4926	−15.2776	−54.4926	54.4926	−19.7312	−26.0945	26.0945
66.6510	93.3872	66.6510	93.3872	93.3872	7.6171	46.8397	46.8397
−2.5209	3.9339	−2.5209	3.9339	3.9339	−2.5209	3.9339	3.9339
−28.7171	−23.0137	−28.7171	−23.0137	28.7171	−28.7171	−23.0137	28.7171

Generation: 3

Initial population		Initial population cost
−2.5209	3.9339	3.9339
22.6334	4.9971	22.6334
−19.5752	23.0188	23.0188
23.1338	11.6479	23.1338
−18.3686	23.2601	23.2601
−80.5678	23.2601	23.2601
−19.7312	−26.0945	26.0945
−27.8563	25.7134	27.8563
−28.7171	−23.0137	28.7171
7.6171	46.8397	46.8397

Population after teacher phase		Population after checking feasibility		Population cost after checking feasibility	Population after greedy selection		Population cost after greedy selection
−1.1494	−1.4837	−1.1494	−1.4837	1.4837	−1.1494	−1.4837	1.4837
24.0049	−0.4204	24.0049	−0.4204	24.0049	22.6334	4.9971	22.6334
−18.2037	17.6012	−18.2037	17.6012	18.2037	−18.2037	17.6012	18.2037
24.5053	6.2303	24.5053	6.2303	24.5053	23.1338	11.6479	23.1338
−16.9971	17.8425	−16.9971	17.8425	17.8425	−16.9971	17.8425	17.8425
−79.1963	17.8425	−79.1963	17.8425	79.1963	−80.5678	23.2601	23.2601
−18.3597	−31.5121	−18.3597	−31.5121	31.5121	−19.7312	−26.0945	26.0945
−26.4847	20.2958	−26.4847	20.2958	26.4847	−26.4847	20.2958	26.4847
−27.3456	−28.4313	−27.3456	−28.4313	28.4313	−27.3456	−28.4313	28.4313
8.9886	41.4221	8.9886	41.4221	41.4221	8.9886	41.4221	41.4221

Population after learner phase		Population after checking feasibility		Population cost after checking feasibility	Population after greedy selection		Population cost after greedy selection
4.0686	−5.9694	4.0686	−5.9694	5.9694	−1.1494	−1.4837	1.4837
47.6169	21.7074	47.6169	21.7074	47.6169	22.6334	4.9971	22.6334
−49.4296	27.2388	−49.4296	27.2388	49.4296	−18.2037	17.6012	18.2037
48.3652	33.8640	48.3652	33.8640	48.3652	23.1338	11.6479	23.1338
−56.9539	24.0103	−56.9539	24.0103	56.9539	−16.9971	17.8425	17.8425
−41.8319	16.4052	−41.8319	16.4052	41.8319	−80.5678	23.2601	23.2601
−18.1198	−0.1991	−18.1198	−0.1991	18.1198	−18.1198	−0.1991	18.1198
−23.2746	19.4657	−23.2746	19.4657	23.2746	−23.2746	19.4657	23.2746
−26.6040	−27.8425	−26.6040	−27.8425	27.8425	−26.6040	−27.8425	27.8425
5.2191	32.5603	5.2191	32.5603	32.5603	5.2191	32.5603	32.5603

A1.5 Example 5: Rosenbrock Function

$$f(x) = \sum_{i=1}^{n-1} \left[100(x_{i+1} - x_i^2)^2 + (x_i - 1)^2\right]$$

$$-30 \leq x_i \leq 30, \quad \min(f) = f(1, 1\ldots1) = 0$$

Generation: 1

Initial population		Initial population cost 1.0e + 005
−1.5086	−3.9379	0.0387
3.5159	−7.2050	0.3829
5.8352	8.2983	0.6633
5.8591	7.4743	0.7214
7.3052	−0.0448	2.8531
−7.9559	2.5339	3.6929
−8.3702	7.8102	3.8759
−8.4321	8.5955	3.9078
8.8144	−3.4011	6.5771
9.2051	−3.2960	7.75

Population after teacher phase		Population after checking feasibility		Population cost after checking feasibility 1.0e + 005	Population after greedy selection		Population cost after greedy selection 1.0e + 005
−1.7359	−6.9808	−1.7359	−6.9808	0.1	−1.5086	−3.9379	0.0387
3.2886	−10.2479	3.2886	−10.0000	0.4333	3.5159	−7.2050	0.3829
5.6079	5.2555	5.6079	5.2555	0.6863	5.8352	8.2983	0.6633
5.6318	4.4315	5.6318	4.4315	0.7447	5.8591	7.4743	0.7214
7.0779	−3.0877	7.0779	−3.0877	2.829	7.0779	−3.0877	2.829
−8.1832	−0.5090	−8.1832	−0.5090	4.5536	−7.9559	2.5339	3.6929
−8.5975	4.7673	−8.5975	4.7673	4.7826	−8.3702	7.8102	3.8759
−8.6594	5.5526	−8.6594	5.5526	4.8219	−8.4321	8.5955	3.9078
8.5871	−6.4439	8.5871	−6.4439	6.4299	8.5871	−6.4439	6.4299
8.9778	−6.3389	8.9778	−6.3389	7.5592	8.9778	−6.3389	7.5592

Population after learner phase		Population after checking feasibility		Population cost after checking feasibility 1.0e + 005	Population after greedy selection		Population cost after greedy selection 1.0e + 005
−8.8153	−4.6615	−8.8153	−4.6615	6.7859	−1.5086	−3.9379	0.0387
7.4325	−12.1527	7.4325	−10.0000	4.2569	3.5159	−7.2050	0.3829
8.5209	8.2424	8.5209	8.2424	4.1432	5.8352	8.2983	0.6633
5.8551	7.6133	5.8551	7.6133	0.7114	5.8551	7.6133	0.7114
6.0748	5.6054	6.0748	5.6054	0.9798	6.0748	5.6054	0.9798
−7.9269	2.1642	−7.9269	2.1642	3.6818	−7.9269	2.1642	3.6818
−8.3510	7.5664	−8.3510	7.5664	3.8662	−8.3510	7.5664	3.8662
−7.6155	7.5155	−7.6155	7.5155	2.549	−7.6155	7.5155	2.549
6.1846	6.4265	6.1846	6.4265	1.013	6.1846	6.4265	1.013
7.5826	0.1596	7.5826	0.1596	3.2878	7.5826	0.1596	3.2878

Generation: 2

Initial population		Initial population cost 1.0e + 005
−1.5086	−3.9379	0.0387
−4.7189	−3.9379	0.0387
3.5159	−7.2050	0.3829
5.8352	8.2983	0.6633
5.8551	7.6133	0.7114
6.0748	5.6054	0.9798
6.1846	6.4265	1.013
−7.6155	7.5155	2.549
7.5826	0.1596	3.2878
−7.9269	2.1642	3.6818

Population after teacher phase		Population after checking feasibility		Population cost after checking feasibility 1.0e + 005	Population after greedy selection		Population cost after greedy selection 1.0e + 005
−3.0480	−7.9918	−3.0480	−7.9918	0.2988	−1.5086	−3.9379	0.0387
−6.2582	−7.9918	−6.2582	−7.9918	2.2243	−4.7189	−3.9379	0.0387
1.9766	−11.2589	1.9766	−10.0000	0.1934	1.9766	−10.0000	0.1934
4.2958	4.2445	4.2958	4.2445	0.202	4.2958	4.2445	0.202
4.3157	3.5595	4.3157	3.5595	0.2271	4.3157	3.5595	0.2271
4.5354	1.5516	4.5354	1.5516	0.3618	4.5354	1.5516	0.3618
4.6452	2.3726	4.6452	2.3726	0.369	4.6452	2.3726	0.369
−9.1548	3.4617	−9.1548	3.4617	6.4571	−7.6155	7.5155	2.549
6.0432	−3.8942	6.0432	−3.8942	1.6336	6.0432	−3.8942	1.6336
−9.4663	−1.8897	−9.4663	−1.8897	8.3733	−7.9269	2.1642	3.6818

Population after learner phase		Population after checking feasibility		Population cost after checking feasibility 1.0e + 006	Population after greedy selection		Population cost after greedy selection 1.0e + 005
−5.8724	−3.9632	−5.8724	−3.9632	0.1479	−1.5086	−3.9379	0.0387
−10.7218	−7.9834	−10.0000	−7.9834	1.1662	−4.7189	−3.9379	0.0387
1.4977	−12.9410	1.4977	−10.0000	0.015	1.4977	−10.0000	0.1499
10.4861	5.2982	10.0000	5.2982	0.8969	4.2958	4.2445	0.202
9.8074	1.7386	9.8074	1.7386	0.8921	4.3157	3.5595	0.2271
16.4707	0.9649	10.0000	0.9649	0.9809	4.5354	1.5516	0.3618
3.9016	−1.0751	3.9016	−1.0751	0.0266	3.9016	−1.0751	0.2657
3.8791	−2.0864	3.8791	−2.0864	0.0294	3.8791	−2.0864	0.2936
3.6033	−7.5576	3.6033	−7.5576	0.0422	3.6033	−7.5576	0.422
−7.6210	7.4203	−7.6210	7.4203	0.2567	−7.6210	7.4203	2.5672

Generation: 3

Initial population		Initial population cost 1.0e + 004
−1.5086	−3.9379	0.3867
4.3759	−3.9379	0.3867
1.4977	−10.0000	1.499
4.2958	4.2445	2.0201
4.3157	3.5595	2.2709
3.9016	−1.0751	2.6569
3.8791	−2.0864	2.9364
4.5354	1.5516	3.6182
3.6033	−7.5576	4.22
−4.7189	−3.9379	6.8707

Population after teacher phase		Population after checking feasibility		Population cost after checking feasibility 1.0e + 005	Population after greedy selection		Population cost after greedy selection 1.0e + 004
−3.7475	−3.7049	−3.7475	−3.7049	0.3152	−1.5086	−3.9379	0.3867
2.1370	−3.7049	2.1370	−3.7049	0.0684	4.3759	−3.9379	0.3867
−0.7412	−9.7669	−0.7412	−9.7669	0.1065	−0.7412	−9.7669	1.0646
2.0569	4.4776	2.0569	4.4776	0.0001	2.0569	4.4776	0.0007
2.0768	3.7926	2.0768	3.7926	0.0003	2.0768	3.7926	0.0028
1.6627	−0.8420	1.6627	−0.8420	0.013	1.6627	−0.8420	0.1301
1.6401	−1.8533	1.6401	−1.8533	0.0206	1.6401	−1.8533	0.2065
2.2965	1.7847	2.2965	1.7847	0.0122	2.2965	1.7847	0.1219
1.3643	−7.3246	1.3643	−7.3246	0.0844	1.3643	−7.3246	0.8438
−6.9578	−3.7049	−6.9578	−3.7049	2.7167	−4.7189	−3.9379	6.8707

Population after learner phase		Population after checking feasibility		Population cost after checking feasibility 1.0e + 004	Population after greedy selection		Population cost after greedy selection 1.0e + 004
−1.1072	−3.9379	−1.1072	−3.9379	0.2671	−1.1072	−3.9379	0.2671
2.2244	−3.9379	2.2244	−3.9379	0.7898	4.3759	−3.9379	0.3867
0.0622	−8.8349	0.0622	−8.8349	0.7813	0.0622	−8.8349	0.7813
2.9022	8.7810	2.9022	8.7810	0.0016	2.0569	4.4776	0.0007
1.9784	4.6919	1.9784	4.6919	0.0061	2.0768	3.7926	0.0028
1.9911	2.8343	1.9911	2.8343	0.0129	1.9911	2.8343	0.0129
1.7769	0.8599	1.7769	0.8599	0.0528	1.7769	0.8599	0.0528
2.5341	3.1018	2.5341	3.1018	0.1105	2.5341	3.1018	0.1105
1.6604	−0.8920	1.6604	−0.8920	0.1332	1.6604	−0.8920	0.1332
−2.5688	−3.9379	−2.5688	−3.9379	1.1115	−2.5688	−3.9379	1.1115

A1.6 Example 6: Step Function

$$f(x) = \sum_{i=1}^{n} [x_i + 0.5]^2$$

$$-100 \le x_i \le 100, \quad \min(f) = f(0, 0\ldots0) = 0$$

Generation: 1

Initial population		Initial population cost
13.8974	43.9576	2132
−42.5778	−17.9871	2173
−2.8289	−47.6670	2313
43.9244	−34.4907	3092
74.8079	−18.2191	5949
88.2787	42.6736	9593
−65.7761	−78.7583	10,597
69.8714	−79.3297	11,141
94.8875	−56.9660	12,274
−96.6432	−64.7305	13,634

Population after teacher phase		Population after checking feasibility		Population cost after checking feasibility	Population after greedy selection		Population cost after greedy selection
12.3506	102.2130	12.3506	100.0000	10,144	13.8974	43.9576	2132
−44.1246	40.2683	−44.1246	40.2683	3536	−42.5778	−17.9871	2173
−4.3757	10.5883	−4.3757	10.5883	137	−4.3757	10.5883	137
42.3776	23.7647	42.3776	23.7647	2340	42.3776	23.7647	2340
73.2610	40.0363	73.2610	40.0363	6929	74.8079	−18.2191	5949
86.7319	100.9289	86.7319	100.0000	17,569	88.2787	42.6736	9593
−67.3229	−20.5029	−67.3229	−20.5029	4930	−67.3229	−20.5029	4930
68.3246	−21.0743	68.3246	−21.0743	5065	68.3246	−21.0743	5065
93.3407	1.2893	93.3407	1.2893	8650	93.3407	1.2893	8650
−98.1900	−6.4752	−98.1900	−6.4752	9640	−98.1900	−6.4752	9640

Population after learner phase		Population after checking feasibility		Population cost after checking feasibility	Population after greedy selection		Population cost after greedy selection
2.7204	23.5468	2.7204	23.5468	585	2.7204	23.5468	585
−35.3481	−17.2520	−35.3481	−17.2520	1514	−35.3481	−17.2520	1514
−66.5297	33.2003	−66.5297	33.2003	5578	−4.3757	10.5883	137
16.6746	68.1822	16.6746	68.1822	4913	42.3776	23.7647	2340
74.7944	−18.2797	74.7944	−18.2797	5949	74.8079	−18.2191	5949
157.4880	60.9155	100.0000	60.9155	13,721	88.2787	42.6736	9593
−46.8811	−4.2792	−46.8811	−4.2792	2225	−46.8811	−4.2792	2225
30.2676	24.3978	30.2676	24.3978	1476	30.2676	24.3978	1476
83.9911	−0.0367	83.9911	−0.0367	7056	83.9911	−0.0367	7056
24.2766	−1.5105	24.2766	−1.5105	580	24.2766	−1.5105	580

Generation: 2

Initial population		Initial population cost
−4.3757	10.5883	137
24.2766	−1.5105	580
2.7204	23.5468	585
30.2676	24.3978	1476
−35.3481	−17.2520	1514
13.8974	43.9576	2132
−46.8811	−4.2792	2225
42.3776	23.7647	2340
74.8079	−18.2191	5949
83.9911	−0.0367	7056

Population after teacher phase		Population after checking feasibility		Population cost after checking feasibility	Population after greedy selection		Population cost after greedy selection
−7.5953	7.7191	−7.5953	7.7191	128	−7.5953	7.7191	128
21.0570	−4.3797	21.0570	−4.3797	457	21.0570	−4.3797	457
−0.4992	20.6776	−0.4992	20.6776	441	−0.4992	20.6776	441
27.0480	21.5285	27.0480	21.5285	1213	27.0480	21.5285	1213
−38.5677	−20.1213	−38.5677	−20.1213	1921	−35.3481	−17.2520	1514
10.6778	41.0884	10.6778	41.0884	1802	10.6778	41.0884	1802
−50.1007	−7.1484	−50.1007	−7.1484	2549	−46.8811	−4.2792	2225
39.1580	20.8955	39.1580	20.8955	1962	39.1580	20.8955	1962
71.5883	−21.0883	71.5883	−21.0883	5625	71.5883	−21.0883	5625
80.7715	−2.9059	80.7715	−2.9059	6570	80.7715	−2.9059	6570

Population after learner phase		Population after checking feasibility		Population cost after checking feasibility	Population after greedy selection		Population cost after greedy selection
−20.8500	13.3161	−20.8500	13.3161	610	−7.5953	7.7191	128
17.4906	−19.8030	17.4906	−19.8030	689	21.0570	−4.3797	457
31.1394	55.1132	31.1394	55.1132	3986	−0.4992	20.6776	441
13.7922	21.1191	13.7922	21.1191	637	13.7922	21.1191	637
−9.1219	−11.2669	−9.1219	−11.2669	202	−9.1219	−11.2669	202
−34.7068	87.4164	−34.7068	87.4164	8794	10.6778	41.0884	1802
−14.7468	13.0113	−14.7468	13.0113	394	−14.7468	13.0113	394
18.4831	35.5543	18.4831	35.5543	1620	18.4831	35.5543	1620
14.7320	−0.4037	14.7320	−0.4037	225	14.7320	−0.4037	225
39.0889	9.1898	39.0889	9.1898	1602	39.0889	9.1898	1602

Generation: 3

Initial population		Initial population cost
−7.5953	7.7191	128
−4.3757	10.5883	137
−9.1219	−11.2669	202
14.7320	−0.4037	225
−14.7468	13.0113	394
−0.4992	20.6776	441
21.0570	−4.3797	457
13.7922	21.1191	637
39.0889	9.1898	1602
18.4831	35.5543	1620

Population after teacher phase		Population after checking feasibility		Population cost after checking feasibility	Population after greedy selection		Population cost after greedy selection
−10.3836	6.4476	−10.3836	6.4476	136	−7.5953	7.7191	128
−7.1640	9.3168	−7.1640	9.3168	130	−7.1640	9.3168	130
−11.9103	−12.5384	−11.9103	−12.5384	313	−9.1219	−11.2669	202
11.9437	−1.6752	11.9437	−1.6752	148	11.9437	−1.6752	148
−17.5352	11.7399	−17.5352	11.7399	468	−14.7468	13.0113	394
−3.2875	19.4061	−3.2875	19.4061	370	−3.2875	19.4061	370
18.2687	−5.6512	18.2687	−5.6512	360	18.2687	−5.6512	360
11.0039	19.8476	11.0039	19.8476	521	11.0039	19.8476	521
36.3005	7.9183	36.3005	7.9183	1360	36.3005	7.9183	1360
15.6948	34.2828	15.6948	34.2828	1412	15.6948	34.2828	1412

Population after learner phase		Population after checking feasibility		Population cost after checking feasibility	Population after greedy selection		Population cost after greedy selection
−26.6913	7.6325	−26.6913	7.6325	793	−7.5953	7.7191	128
−1.2993	6.4594	−1.2993	6.4594	37	−1.2993	6.4594	37
−9.0348	−10.1833	−9.0348	−10.1833	181	−9.0348	−10.1833	181
−2.7169	5.3736	−2.7169	5.3736	34	−2.7169	5.3736	34
1.4561	4.0956	1.4561	4.0956	17	1.4561	4.0956	17
−6.5490	10.5577	−6.5490	10.5577	170	−6.5490	10.5577	170
14.9469	−3.9340	14.9469	−3.9340	241	14.9469	−3.9340	241
16.0116	2.2711	16.0116	2.2711	260	16.0116	2.2711	260
15.3616	17.7926	15.3616	17.7926	549	15.3616	17.7926	549
3.8513	12.5447	3.8513	12.5447	185	3.8513	12.5447	185

A1.7 Example 7: Quartic Function

$$f(x) = \sum_{i=1}^{n} \left[ix_i^4\right]$$
$$-1.28 \le x_i \le 1.28, \quad \min(f) = f(0, 0\ldots0) = 0$$

Generation: 1

Initial population		Initial population cost
−0.0609	0.2479	0.0076
0.3663	−0.2011	0.0213
0.1746	0.5606	0.1984
0.9645	0.2074	0.8692
1.0304	−0.3204	1.1485
−0.7380	0.8122	1.1668
0.2091	−0.9694	1.7678
−1.1930	0.1806	2.028
1.2417	0.6089	2.6523
0.0332	1.1201	3.1482

Population after teacher phase		Population after checking feasibility		Population cost after checking feasibility	Population after greedy selection		Population cost after greedy selection
−0.2137	0.2500	−0.2137	0.2500	0.0099	−0.0609	0.2479	0.0076
0.2136	−0.1990	0.2136	−0.1990	0.0052	0.2136	−0.1990	0.0052
0.0218	0.5627	0.0218	0.5627	0.2005	0.1746	0.5606	0.1984
0.8117	0.2095	0.8117	0.2095	0.4381	0.8117	0.2095	0.4381
0.8777	−0.3183	0.8777	−0.3183	0.6139	0.8777	−0.3183	0.6139
−0.8908	0.8143	−0.8908	0.8143	1.509	−0.7380	0.8122	1.1668
0.0563	−0.9672	0.0563	−0.9672	1.7504	0.0563	−0.9672	1.7504
−1.3458	0.1827	−1.2800	0.1827	2.6866	−1.1930	0.1806	2.028
1.0889	0.6111	1.0889	0.6111	1.6849	1.0889	0.6111	1.6849
−0.1195	1.1222	−0.1195	1.1222	3.1724	0.0332	1.1201	3.1482

Population after learner phase		Population after checking feasibility		Population cost after checking feasibility	Population after greedy selection		Population cost after greedy selection
−0.1742	1.4224	−0.1742	1.2800	5.3696	−0.0609	0.2479	0.0076
0.3405	−0.4057	0.3405	−0.4057	0.0676	0.2136	−0.1990	0.0052
0.2479	1.5074	0.2479	1.2800	5.3725	0.1746	0.5606	0.1984
0.7815	0.4520	0.7815	0.4520	0.4564	0.8117	0.2095	0.4381
0.2791	0.4299	0.2791	0.4299	0.0744	0.2791	0.4299	0.0744
−0.8614	1.0886	−0.8614	1.0886	3.3587	−0.7380	0.8122	1.1668
0.1420	0.1393	0.1420	0.1393	0.0012	0.1420	0.1393	0.0012
−0.7340	0.2672	−0.7340	0.2672	0.3005	−0.7340	0.2672	0.3005
1.7824	1.6710	1.2800	1.2800	8.0531	1.0889	0.6111	1.6849
−0.2692	0.9993	−0.2692	0.9993	1.9999	−0.2692	0.9993	1.9999

Generation: 2

Initial population		Initial population cost
0.1420	0.1393	0.0012
0.2136	−0.1990	0.0052
−0.0609	0.2479	0.0076
−0.6771	0.2479	0.0076
0.2791	0.4299	0.0744
0.1746	0.5606	0.1984
−0.7340	0.2672	0.3005
0.8117	0.2095	0.4381
−0.7380	0.8122	1.1668
1.0889	0.6111	1.6849

Population after teacher phase		Population after checking feasibility		Population cost after checking feasibility	Population after greedy selection		Population cost after greedy selection
0.1493	0.0635	0.1493	0.0635	0.0005	0.1493	0.0635	0.0005
0.2209	−0.2749	0.2209	−0.2749	0.0138	0.2136	−0.1990	0.0052
−0.0536	0.1720	−0.0536	0.1720	0.0018	−0.0536	0.1720	0.0018
−0.6697	0.1720	−0.6697	0.1720	0.2029	−0.6771	0.2479	0.0076
0.2864	0.3541	0.2864	0.3541	0.0382	0.2864	0.3541	0.0382
0.1819	0.4847	0.1819	0.4847	0.1115	0.1819	0.4847	0.1115
−0.7267	0.1913	−0.7267	0.1913	0.2815	−0.7267	0.1913	0.2815
0.8191	0.1337	0.8191	0.1337	0.4508	0.8117	0.2095	0.4381
−0.7306	0.7363	−0.7306	0.7363	0.8727	−0.7306	0.7363	0.8727
1.0963	0.5352	1.0963	0.5352	1.6086	1.0963	0.5352	1.6086

Population after learner phase		Population after checking feasibility		Population cost after checking feasibility	Population after greedy selection		Population cost after greedy selection
−0.7053	−0.3622	−0.7053	−0.3622	0.2818	0.1493	0.0635	0.0005
0.0208	0.0688	0.0208	0.0688	0	0.0208	0.0688	0
−0.0472	0.1667	−0.0472	0.1667	0.0015	−0.0472	0.1667	0.0015
−0.3295	0.2056	−0.3295	0.2056	0.0154	−0.6771	0.2479	0.0076
0.2222	−0.1337	0.2222	−0.1337	0.0031	0.2222	−0.1337	0.0031
0.2004	0.0849	0.2004	0.0849	0.0017	0.2004	0.0849	0.0017
−0.2138	0.1165	−0.2138	0.1165	0.0025	−0.2138	0.1165	0.0025
0.6072	0.1644	0.6072	0.1644	0.1374	0.6072	0.1644	0.1374
−0.4936	0.5388	−0.4936	0.5388	0.2279	−0.4936	0.5388	0.2279
0.9253	0.4812	0.9253	0.4812	0.8404	0.9253	0.4812	0.8404

Generation: 3

Initial population		Initial population cost
0.0208	0.0688	0
0.1493	0.0635	0.0005
0.1420	0.1393	0.0012
−0.0472	0.1667	0.0015
0.2004	0.0849	0.0017
−0.2138	0.1165	0.0025
0.2222	−0.1337	0.0031
0.6072	0.1644	0.1374
−0.6771	0.2479	0.2177
−0.4936	0.5388	0.2279

Population after teacher phase		Population after checking feasibility		Population cost after checking feasibility	Population after greedy selection		Population cost after greedy selection
0.0212	0.0655	0.0212	0.0655	0	0.0212	0.0655	0
0.1498	0.0602	0.1498	0.0602	0.0005	0.1498	0.0602	0.0005
0.1424	0.1361	0.1424	0.1361	0.0011	0.1424	0.1361	0.0011
−0.0468	0.1634	−0.0468	0.1634	0.0014	−0.0468	0.1634	0.0014
0.2009	0.0817	0.2009	0.0817	0.0017	0.2009	0.0817	0.0017
−0.2134	0.1132	−0.2134	0.1132	0.0024	−0.2134	0.1132	0.0024
0.2226	−0.1370	0.2226	−0.1370	0.0032	0.2222	−0.1337	0.0031
0.6076	0.1612	0.6076	0.1612	0.1377	0.6072	0.1644	0.1374
−0.6766	0.2446	−0.6766	0.2446	0.2168	−0.6766	0.2446	0.2168
−0.4932	0.5355	−0.4932	0.5355	0.2236	−0.4932	0.5355	0.2236

Population after learner phase		Population after checking feasibility		Population cost after checking feasibility	Population after greedy selection		Population cost after greedy selection
−0.0378	0.0680	−0.0378	0.0680	0	0.0212	0.0655	0
0.0445	0.0646	0.0445	0.0646	0	0.0445	0.0646	0
0.1445	0.1143	0.1445	0.1143	0.0008	0.1445	0.1143	0.0008
0.0400	0.1509	0.0400	0.1509	0.001	0.0400	0.1509	0.001
0.1523	0.0613	0.1523	0.0613	0.0006	0.1523	0.0613	0.0006
−0.1543	0.1012	−0.1543	0.1012	0.0008	−0.1543	0.1012	0.0008
0.1408	−0.0876	0.1408	−0.0876	0.0005	0.1408	−0.0876	0.0005
0.4808	0.1387	0.4808	0.1387	0.0542	0.4808	0.1387	0.0542
−0.1710	0.1795	−0.1710	0.1795	0.0029	−0.1710	0.1795	0.0029
0.0987	0.1485	0.0987	0.1485	0.0011	0.0987	0.1485	0.0011

A1.8 Example 8: Schwefel 2.26 Function

$$f(x) = -\sum_{i=1}^{30}\left(x_i \sin\left(\sqrt{|x_i|}\right)\right)$$

$$-500 \leq x_i \leq 500,$$

$$\min(f) = f(420.9687, 420.9687\ldots420.9687) = -12,569.5$$

Generation: 1

Initial population		Initial population cost
414.6404	−266.7841	−570.1683
402.8729	169.3761	−452.0135
−477.2336	414.1168	−343.9093
458.1882	−53.3658	−207.7937
236.6814	−89.3017	−77.4884
6.7347	60.6747	−64.048
−347.8892	−231.8518	38.9746
55.6974	−214.8973	134.76
−230.8016	139.7935	208.4045
329.3055	−383.1430	466.7711

Population after teacher phase		Population after checking feasibility		Population cost after checking feasibility	Population after greedy selection		Population cost after greedy selection
633.0672	−268.3715	500.0000	−268.3715	13.0763	414.6404	−266.7841	−570.1683
621.2997	167.7886	500.0000	167.7886	117.2709	402.8729	169.3761	−452.0135
−258.8069	412.5294	−258.8069	412.5294	−505.9366	−258.8069	412.5294	−505.9366
676.6150	−54.9533	500.0000	−54.9533	230.2863	458.1882	−53.3658	−207.7937
455.1082	−90.8892	455.1082	−90.8892	−288.1424	455.1082	−90.8892	−288.1424
225.1614	59.0872	225.1614	59.0872	−203.7609	225.1614	59.0872	−203.7609
−129.4625	−233.4393	−129.4625	−233.4393	−22.9493	−129.4625	−233.4393	−22.9493
274.1242	−216.4847	274.1242	−216.4847	387.2395	55.6974	−214.8973	134.76
−12.3748	138.2060	−12.3748	138.2060	95.5799	−12.3748	138.2060	95.5799
547.7323	−384.7305	500.0000	−384.7305	447.0268	500.0000	−384.7305	447.0268

Population after learner phase		Population after checking feasibility		Population cost after checking feasibility	Population after greedy selection		Population cost after greedy selection
386.4948	−389.1203	386.4948	−389.1203	19.1494	414.6404	−266.7841	−570.1683
384.2759	244.2618	384.2759	244.2618	−282.1897	402.8729	169.3761	−452.0135
−187.8422	340.9466	−187.8422	340.9466	298.6035	−258.8069	412.5294	−505.9366
870.1382	−221.0755	500.0000	−221.0755	345.1163	458.1882	−53.3658	−207.7937
453.1223	−115.0839	453.1223	−115.0839	−404.4391	453.1223	−115.0839	−404.4391
229.9795	62.0773	229.9795	62.0773	−180.8679	225.1614	59.0872	−203.7609
388.0430	158.1544	388.0430	158.1544	−292.8517	388.0430	158.1544	−292.8517
168.3098	−169.7026	168.3098	−169.7026	8.7987	168.3098	−169.7026	8.7987
−52.3783	345.7117	−52.3783	345.7117	130.3465	−12.3748	138.2060	95.5799
−107.5797	253.6387	−107.5797	253.6387	−32.4618	−107.5797	253.6387	−32.4618

Generation: 2

Initial population		Initial population cost
414.6404	−266.7841	−570.1683
−157.8041	−266.7841	−570.1683
−258.8069	412.5294	−505.9366
402.8729	169.3761	−452.0135
453.1223	−115.0839	−404.4391
388.0430	158.1544	−292.8517
458.1882	−53.3658	−207.7937
225.1614	59.0872	−203.7609
−107.5797	253.6387	−32.4618
168.3098	−169.7026	8.7987

Population after teacher phase		Population after checking feasibility		Population cost after checking feasibility	Population after greedy selection		Population cost after greedy selection
480.5672	−464.2750	480.5672	−464.2750	166.1976	414.6404	−266.7841	−570.1683
−91.8773	−464.2750	−91.8773	−464.2750	184.795	−157.8041	−266.7841	−570.1683
−192.8801	215.0385	−192.8801	215.0385	1.0701	−258.8069	412.5294	−505.9366
468.7997	−28.1148	468.7997	−28.1148	−179.421	402.8729	169.3761	−452.0135
519.0490	−312.5748	500.0000	−312.5748	−107.1863	453.1223	−115.0839	−404.4391
453.9698	−39.3365	453.9698	−39.3365	−287.5118	388.0430	158.1544	−292.8517
524.1150	−250.8567	500.0000	−250.8567	147.9468	458.1882	−53.3658	−207.7937
291.0882	−138.4037	291.0882	−138.4037	184.7692	225.1614	59.0872	−203.7609
−41.6530	56.1478	−41.6530	56.1478	−45.456	−41.6530	56.1478	−45.456
234.2366	−367.1935	234.2366	−367.1935	21.0725	168.3098	−169.7026	8.7987

Population after learner phase		Population after checking feasibility		Population cost after checking feasibility	Population after greedy selection		Population cost after greedy selection
470.6109	−363.0436	470.6109	−363.0436	−64.3866	414.6404	−266.7841	−570.1683
−484.5232	−347.9122	−484.5232	−347.9122	−78.1957	−157.8041	−266.7841	−570.1683
−291.0984	436.4608	−291.0984	436.4608	−673.141	−291.0984	436.4608	−673.141
559.7970	396.2217	500.0000	396.2217	−164.233	402.8729	169.3761	−452.0135
698.5646	−68.0152	500.0000	−68.0152	243.4152	453.1223	−115.0839	−404.4391
547.7561	255.2944	500.0000	255.2944	248.6771	388.0430	158.1544	−292.8517
457.4066	−62.8890	457.4066	−62.8890	−197.1044	458.1882	−53.3658	−207.7937
283.8279	95.4960	283.8279	95.4960	290.3188	225.1614	59.0872	−203.7609
364.2781	−84.3367	364.2781	−84.3367	−65.2052	364.2781	−84.3367	−65.2052
291.4291	−120.2912	291.4291	−120.2912	164.9353	168.3098	−169.7026	8.7987

A1.9 Example 9: Rastrigin Function

$$f(x) = \sum_{i=1}^{n} \left[x_i^2 - 10\cos(2\pi x_i) + 10\right]$$
$$-5.12 \leq x_i \leq 5.12, \quad \min(f) = f(0, 0\ldots0) = 0$$

Generation: 1

Initial population		Initial population cost
3.9866	−0.8038	23.2555
2.0999	−4.0908	24.6331
1.5917	0.1550	25.3207
0.0743	−4.9963	26.0399
0.5386	3.1134	32.1257
−1.4262	1.3165	36.7718
0.3900	2.3982	41.6312
−4.2506	−1.4423	49.5382
5.0293	−3.7222	51.0571
3.6465	4.7973	59.4359

Population after teacher phase		Population after checking feasibility		Population cost after checking feasibility	Population after greedy selection		Population cost after greedy selection
6.7378	−1.1040	5.1200	−1.1040	32.2039	3.9866	−0.8038	23.2555
4.8511	−4.3909	4.8511	−4.3909	64.6209	2.0999	−4.0908	24.6331
4.3429	−0.1452	4.3429	−0.1452	38.2752	1.5917	0.1550	25.3207
2.8255	−5.2964	2.8255	−5.1200	42.3418	0.0743	−4.9963	26.0399
3.2898	2.8133	3.2898	2.8133	37.339	0.5386	3.1134	32.1257
1.3250	1.0164	1.3250	1.0164	17.3798	1.3250	1.0164	17.3798
3.1412	2.0980	3.1412	2.0980	19.7912	3.1412	2.0980	19.7912
−1.4994	−1.7425	−1.4994	−1.7425	35.7561	−1.4994	−1.7425	35.7561
7.7805	−4.0223	5.1200	−4.0223	45.202	5.1200	−4.0223	45.202
6.3978	4.4971	5.1200	4.4971	69.1473	3.6465	4.7973	59.4359

Population after learner phase		Population after checking feasibility		Population cost after checking feasibility	Population after greedy selection		Population cost after greedy selection
3.6115	0.4839	3.6115	0.4839	50.8717	3.9866	−0.8038	23.2555
3.9958	−3.2433	3.9958	−3.2433	36.0653	2.0999	−4.0908	24.6331
1.9063	0.5495	1.9063	0.5495	25.1404	1.9063	0.5495	25.1404
1.9650	−4.1511	1.9650	−4.1511	25.5118	1.9650	−4.1511	25.5118
1.2046	1.2426	1.2046	1.2426	19.7133	1.2046	1.2426	19.7133
0.5002	1.5804	0.5002	1.5804	41.4995	1.3250	1.0164	17.3798
2.8528	3.0880	2.8528	3.0880	23.1449	3.1412	2.0980	19.7912
−0.9132	−1.3826	−0.9132	−1.3826	21.5989	−0.9132	−1.3826	21.5989
3.3667	1.4005	3.3667	1.4005	48.0998	5.1200	−4.0223	45.202
2.6048	2.4436	2.6048	2.4436	50.0468	2.6048	2.4436	50.0468

Generation: 2

Initial population		Initial population cost
1.3250	1.0164	17.3798
1.2046	1.2426	19.7133
3.1412	2.0980	19.7912
−0.9132	−1.3826	21.5989
3.9866	−0.8038	23.2555
4.8833	−0.8038	23.2555
2.0999	−4.0908	24.6331
1.9063	0.5495	25.1404
1.9650	−4.1511	25.5118
5.1200	−4.0223	45.202

Population after teacher phase		Population after checking feasibility		Population cost after checking feasibility	Population after greedy selection		Population cost after greedy selection
−0.2184	2.2790	−0.2184	2.2790	25.084	1.3250	1.0164	17.3798
−0.3388	2.5053	−0.3388	2.5053	41.6801	1.2046	1.2426	19.7133
1.5978	3.3607	1.5978	3.3607	48.4254	3.1412	2.0980	19.7912
−2.4565	−0.1200	−2.4565	−0.1200	28.3868	−0.9132	−1.3826	21.5989
2.4432	0.4588	2.4432	0.4588	45.2183	3.9866	−0.8038	23.2555
3.3399	0.4588	3.3399	0.4588	46.3876	4.8833	−0.8038	23.2555
0.5565	−2.8281	0.5565	−2.8281	32.9716	2.0999	−4.0908	24.6331
0.3629	1.8121	0.3629	1.8121	26.1248	1.9063	0.5495	25.1404
0.4216	−2.8884	0.4216	−2.8884	29.6903	1.9650	−4.1511	25.5118
3.5766	−2.7596	3.5766	−2.7596	48.6656	5.1200	−4.0223	45.202

Population after learner phase		Population after checking feasibility		Population cost after checking feasibility	Population after greedy selection		Population cost after greedy selection
1.3206	1.0450	1.3206	1.0450	17.5269	1.3250	1.0164	17.3798
1.1133	1.3328	1.1133	1.3328	20.4122	1.2046	1.2426	19.7133
2.6699	3.5558	2.6699	3.5558	53.9872	3.1412	2.0980	19.7912
−5.0395	−1.8701	−5.0395	−1.8701	32.3523	−0.9132	−1.3826	21.5989
4.1924	−0.4632	4.1924	−0.4632	43.9825	3.9866	−0.8038	23.2555
4.7898	0.4671	4.7898	0.4671	50.4706	4.8833	−0.8038	23.2555
2.6707	−0.6984	2.6707	−0.6984	35.584	2.0999	−4.0908	24.6331
1.8581	4.4133	1.8581	4.4133	45.1996	1.9063	0.5495	25.1404
0.8525	−3.0810	0.8525	−3.0810	15.4835	0.8525	−3.0810	15.4835
3.8238	−2.1783	3.8238	−2.1783	30.542	3.8238	−2.1783	30.542

Generation: 3

Initial population		Initial population cost
0.8525	−3.0810	15.4835
1.3250	1.0164	17.3798
−4.1251	1.0164	17.3798
1.2046	1.2426	19.7133
3.1412	2.0980	19.7912
−0.9132	−1.3826	21.5989
3.9866	−0.8038	23.2555
2.0999	−4.0908	24.6331
1.9063	0.5495	25.1404
3.8238	−2.1783	30.542

Population after teacher phase		Population after checking feasibility		Population cost after checking feasibility	Population after greedy selection		Population cost after greedy selection
0.7973	−4.9201	0.7973	−4.9201	33.1468	0.8525	−3.0810	15.4835
1.2698	−0.8227	1.2698	−0.8227	19.1185	1.3250	1.0164	17.3798
−4.1802	−0.8227	−4.1802	−0.8227	29.4956	−4.1251	1.0164	17.3798
1.1494	−0.5965	1.1494	−0.5965	23.9859	1.2046	1.2426	19.7133
3.0860	0.2589	3.0860	0.2589	21.5785	3.1412	2.0980	19.7912
−0.9683	−3.2217	−0.9683	−3.2217	19.7466	−0.9683	−3.2217	19.7466
3.9314	−2.6429	3.9314	−2.6429	39.586	3.9866	−0.8038	23.2555
2.0447	−5.9299	2.0447	−5.1200	33.4982	2.0999	−4.0908	24.6331
1.8511	−1.2896	1.8511	−1.2896	21.6187	1.8511	−1.2896	21.6187
3.7686	−4.0174	3.7686	−4.0174	39.2356	3.8238	−2.1783	30.542

Population after learner phase		Population after checking feasibility		Population cost after checking feasibility	Population after greedy selection		Population cost after greedy selection
0.5956	−2.8731	0.5956	−2.8731	29.8749	0.8525	−3.0810	15.4835
1.0620	2.1691	1.0620	2.1691	11.7131	1.0620	2.1691	11.7131
0.0423	1.0164	0.0423	1.0164	1.4384	0.0423	1.0164	1.4384
−0.4330	2.4472	−0.4330	2.4472	44.7575	1.2046	1.2426	19.7133
1.2130	1.2463	1.2130	1.2463	20.4864	3.1412	2.0980	19.7912
−0.1076	−1.6310	−0.1076	−1.6310	21.6704	−0.9683	−3.2217	19.7466
3.4884	0.9065	3.4884	0.9065	34.6418	3.9866	−0.8038	23.2555
2.4522	−1.9968	2.4522	−1.9968	29.5556	2.0999	−4.0908	24.6331
1.8416	−1.2524	1.8416	−1.2524	19.6678	1.8416	−1.2524	19.6678
3.8451	−1.9979	3.8451	−1.9979	23.1493	3.8451	−1.9979	23.1493

A1.10 Example 10: Ackley Function

$$f(x) = \sum_{i=1}^{n} -20\exp\left(-0.2\sqrt{\frac{1}{30}\sum_{i=1}^{30} x_i^2}\right) - \exp\left(\frac{1}{30}\sum_{i=1}^{30}\cos 2\pi x_i\right)$$

$$-32 \le x_i \le 32, \quad \min(f) = f(0,0\ldots0) = 0$$

Generation: 1

Initial population		Initial population cost
−6.2542	9.9629	17.3238
11.2311	12.2108	19.6048
−2.5180	15.4019	20.1133
14.9231	26.8464	20.3887
−13.8749	−25.8242	20.6188
0.2642	26.1139	20.8256
−22.8499	15.3343	21.2683
25.0947	−12.4015	21.3296
−5.6740	23.3924	21.5154
−23.9329	−26.5706	21.5864

Population after teacher phase		Population after checking feasibility		Population cost after checking feasibility	Population after greedy selection		Population cost after greedy selection
−9.5662	10.6640	−9.5662	10.6640	19.5917	−6.2542	9.9629	17.3238
7.9190	12.9118	7.9190	12.9118	18.0024	7.9190	12.9118	18.0024
−5.8300	16.1030	−5.8300	16.1030	19.0472	−5.8300	16.1030	19.0472
11.6110	27.5475	11.6110	27.5475	22.0039	14.9231	26.8464	20.3887
−17.1870	−25.1232	−17.1870	−25.1232	20.714	−13.8749	−25.8242	20.6188
−3.0478	26.8149	−3.0478	26.8149	20.3122	−3.0478	26.8149	20.3122
−26.1619	16.0354	−26.1619	16.0354	20.3395	−26.1619	16.0354	20.3395
21.7827	−11.7004	21.7827	−11.7004	21.1624	21.7827	−11.7004	21.1624
−8.9861	24.0934	−8.9861	24.0934	19.6963	−8.9861	24.0934	19.6963
−27.2449	−25.8695	−27.2449	−25.8695	21.1904	−27.2449	−25.8695	21.1904

Population after learner phase		Population after checking feasibility		Population cost after checking feasibility	Population after greedy selection		Population cost after greedy selection
−21.7872	−2.4208	−21.7872	−2.4208	21.0938	−6.2542	9.9629	17.3238
16.6620	12.1105	16.6620	12.1105	20.5028	7.9190	12.9118	18.0024
−5.6426	17.0799	−5.6426	17.0799	20.0129	−5.8300	16.1030	19.0472
22.6249	40.9328	22.6249	30.0000	21.4625	14.9231	26.8464	20.3887
3.6604	5.3427	3.6604	5.3427	14.1334	3.6604	5.3427	14.1334
−8.1724	24.4663	−8.1724	24.4663	21.4222	−3.0478	26.8149	20.3122
−14.3131	21.5942	−14.3131	21.5942	21.6612	−26.1619	16.0354	20.3395
18.2850	−6.2751	18.2850	−6.2751	20.5901	18.2850	−6.2751	20.5901
1.4832	17.1687	1.4832	17.1687	20.1933	−8.9861	24.0934	19.6963
−20.8435	−14.9419	−20.8435	−14.9419	20.0818	−20.8435	−14.9419	20.0818

Generation: 2

Initial population		Initial population cost
3.6604	5.3427	14.1334
−6.2542	9.9629	17.3238
22.0843	9.9629	17.3238
7.9190	12.9118	18.0024
−5.8300	16.1030	19.0472
−8.9861	24.0934	19.6963
−20.8435	−14.9419	20.0818
−3.0478	26.8149	20.3122
−26.1619	16.0354	20.3395
14.9231	26.8464	20.3887

Population after teacher phase		Population after checking feasibility		Population cost after checking feasibility	Population cost after greedy selection		Population after greedy selection
9.5000	0.9839	9.5000	0.9839	16.5396	3.6604	5.3427	14.1334
−0.4147	5.6041	−0.4147	5.6041	13.2465	−0.4147	5.6041	13.2465
27.9239	5.6041	27.9239	5.6041	21.3138	22.0843	9.9629	17.3238
13.7586	8.5530	13.7586	8.5530	20.0547	7.9190	12.9118	18.0024
0.0095	11.7442	0.0095	11.7442	17.3014	0.0095	11.7442	17.3014
−3.1465	19.7346	−3.1465	19.7346	20.2438	−8.9861	24.0934	19.6963
−15.0039	−19.3007	−15.0039	−19.3007	20.6785	−20.8435	−14.9419	20.0818
2.7917	22.4561	2.7917	22.4561	21.1996	−3.0478	26.8149	20.3122
−20.3224	11.6766	−20.3224	11.6766	21.3487	−26.1619	16.0354	20.3395
20.7626	22.4876	20.7626	22.4876	21.8226	14.9231	26.8464	20.3887

Population after learner phase		Population after checking feasibility		Population cost after checking feasibility	Population after greedy selection		Population cost after greedy selection
8.5813	−1.9534	8.5813	−1.9534	15.9138	3.6604	5.3427	14.1334
4.1663	−4.2774	4.1663	−4.2774	12.9424	4.1663	−4.2774	12.9424
24.0841	9.7112	24.0841	9.7112	20.845	22.0843	9.9629	17.3238
13.5270	11.7444	13.5270	11.7444	20.5304	7.9190	12.9118	18.0024
0.5493	10.7978	0.5493	10.7978	17.6632	0.0095	11.7442	17.3014
−14.0628	21.7668	−14.0628	21.7668	20.5337	−8.9861	24.0934	19.6963
−31.6847	−40.3804	−30.0000	−30.0000	19.9504	−30.0000	−30.0000	19.9504
4.3116	17.4852	4.3116	17.4852	20.6483	−3.0478	26.8149	20.3122
−7.3898	24.7900	−7.3898	24.7900	21.4321	−26.1619	16.0354	20.3395
12.6976	26.2608	12.6976	26.2608	21.5728	14.9231	26.8464	20.3887

Generation: 3

Initial population		Initial population cost
4.1663	−4.2774	12.9424
3.6604	5.3427	14.1334
−6.9711	5.3427	14.1334
0.0095	11.7442	17.3014
7.9190	12.9118	18.0024
22.0843	9.9629	19.5641
−8.9861	24.0934	19.6963
−30.0000	−30.0000	19.9504
−3.0478	26.8149	20.3122
−26.1619	16.0354	20.3395

Population after teacher phase		Population after checking feasibility		Population cost after checking feasibility	Population after greedy selection		Population cost after greedy selection
11.7377	−9.5264	11.7377	−9.5264	19.7727	4.1663	−4.2774	12.9424
11.2318	0.0937	11.2318	0.0937	17.0292	3.6604	5.3427	14.1334
0.6003	0.0937	0.6003	0.0937	3.353	0.6003	0.0937	3.353
7.5809	6.4952	7.5809	6.4952	17.4522	0.0095	11.7442	17.3014
15.4905	7.6628	15.4905	7.6628	20.5142	7.9190	12.9118	18.0024
29.6557	4.7139	29.6557	4.7139	21.7561	22.0843	9.9629	19.5641
−1.4147	18.8444	−1.4147	18.8444	20.4762	−8.9861	24.0934	19.6963
−22.4286	−35.2490	−22.4286	−30.0000	21.5674	−30.0000	−30.0000	19.9504
4.5236	21.5659	4.5236	21.5659	21.4459	−3.0478	26.8149	20.3122
−18.5905	10.7864	−18.5905	10.7864	21.0262	−26.1619	16.0354	20.3395

Population after learner phase		Population after checking feasibility		Population cost after checking feasibility	Population after greedy selection		Population cost after greedy selection
0.7753	−19.8092	0.7753	−19.8092	20.2082	4.1663	−4.2774	12.9424
6.4353	1.2284	6.4353	1.2284	14.1238	6.4353	1.2284	14.1238
19.6206	−11.2363	19.6206	−11.2363	21.1751	0.6003	0.0937	3.353
−6.4373	10.7925	−6.4373	10.7925	18.6169	0.0095	11.7442	17.3014
15.4376	7.9388	15.4376	7.9388	19.9996	7.9190	12.9118	18.0024
37.6668	2.8761	30.0000	2.8761	20.0825	22.0843	9.9629	19.5641
−3.6246	21.6551	−3.6246	21.6551	21.292	−8.9861	24.0934	19.6963
−33.3490	−70.1699	−30.0000	−30.0000	19.9504	−30.0000	−30.0000	19.9504
4.1081	−4.0267	4.1081	−4.0267	11.4367	4.1081	−4.0267	11.4367
−16.6886	20.4534	−16.6886	20.4534	21.7267	−26.1619	16.0354	20.3395

A1.11 Example 11: Griewank Function

$$f(x) = \frac{1}{4000}\sum_{i=1}^{n} x_i^2 - \prod_{i=1}^{n} \cos\left(\frac{x_i}{\sqrt{i}}\right) + 1$$

$$-600 \le x_i \le 600, \quad \min(f) = f(0, 0\ldots0) = 0$$

Generation: 1

Initial population		Initial population cost
−83.8021	−112.9558	5.8218
−76.0445	−202.4999	12.5024
91.0900	216.9729	13.9735
−47.3955	−255.7436	18.1008
177.8341	300.7528	31.7063
−74.5236	412.5378	45.51
−422.4655	94.9685	47.904
−159.2422	−444.7944	57.3225
13.4666	−596.6565	89.6718
593.9634	−379.8330	125.2432

Population after teacher phase		Population after checking feasibility		Population cost after checking feasibility	Population after greedy selection		Population cost after greedy selection
−92.5892	−109.1180	−92.5892	−109.1180	6.1034	−83.8021	−112.9558	5.8218
−84.8316	−198.6621	−84.8316	−198.6621	12.0413	−84.8316	−198.6621	12.0413
82.3029	220.8107	82.3029	220.8107	14.4052	91.0900	216.9729	13.9735
−56.1826	−251.9058	−56.1826	−251.9058	18.1991	−47.3955	−255.7436	18.1008
169.0470	304.5906	169.0470	304.5906	31.4851	169.0470	304.5906	31.4851
−83.3108	416.3756	−83.3108	416.3756	46.1143	−74.5236	412.5378	45.51
−431.2526	98.8063	−431.2526	98.8063	50.4151	−422.4655	94.9685	47.904
−168.0293	−440.9566	−168.0293	−440.9566	56.6367	−168.0293	−440.9566	56.6367
4.6794	−592.8187	4.6794	−592.8187	88.8569	4.6794	−592.8187	88.8569
585.1762	−375.9952	585.1762	−375.9952	122.2134	585.1762	−375.9952	122.2134

Population after learner phase		Population after checking feasibility		Population cost after checking feasibility	Population after greedy selection		Population cost after greedy selection
−106.6735	−23.2532	−106.6735	−23.2532	4.7148	−106.6735	−23.2532	4.7148
−104.6342	−168.4678	−104.6342	−168.4678	11.386	−104.6342	−168.4678	11.386
242.3155	252.8992	242.3155	252.8992	30.7799	91.0900	216.9729	13.9735
−21.2245	−900.4456	−21.2245	−600.0000	90.4004	−47.3955	−255.7436	18.1008
252.8723	762.2572	252.8723	600.0000	107.0118	169.0470	304.5906	31.4851
−77.2177	252.7957	−77.2177	252.7957	18.2332	−77.2177	252.7957	18.2332
−218.0158	281.5714	−218.0158	281.5714	32.5819	−218.0158	281.5714	32.5819
69.0966	161.1294	69.0966	161.1294	8.0157	69.0966	161.1294	8.0157
−20.8935	−268.2113	−20.8935	−268.2113	19.2761	−20.8935	−268.2113	19.2761
422.2508	−332.8732	422.2508	−332.8732	73.5558	422.2508	−332.8732	73.5558

Generation: 2

Initial population		Initial population cost
−106.6735	−23.2532	4.7148
−83.8021	−112.9558	5.8218
69.0966	161.1294	8.0157
−104.6342	−168.4678	11.386
91.0900	216.9729	13.9735
−47.3955	−255.7436	18.1008
−77.2177	252.7957	18.2332
−20.8935	−268.2113	19.2761
169.0470	304.5906	31.4851
−218.0158	281.5714	32.5819

Population after teacher phase		Population after checking feasibility		Population cost after checking feasibility	Population after greedy selection		Population cost after greedy selection
−142.6412	−85.6309	−142.6412	−85.6309	7.7262	−106.6735	−23.2532	4.7148
−119.7697	−175.3335	−119.7697	−175.3335	12.3764	−83.8021	−112.9558	5.8218
33.1289	98.7517	33.1289	98.7517	3.8195	33.1289	98.7517	3.8195
−140.6018	−230.8455	−140.6018	−230.8455	19.9766	−104.6342	−168.4678	11.386
55.1224	154.5952	55.1224	154.5952	7.85	55.1224	154.5952	7.85
−83.3631	−318.1213	−83.3631	−318.1213	28.0727	−47.3955	−255.7436	18.1008
−113.1854	190.4180	−113.1854	190.4180	14.1676	−113.1854	190.4180	14.1676
−56.8612	−330.5890	−56.8612	−330.5890	28.8612	−20.8935	−268.2113	19.2761
133.0793	242.2129	133.0793	242.2129	20.117	133.0793	242.2129	20.117
−253.9834	219.1937	−253.9834	219.1937	28.7024	−253.9834	219.1937	28.7024

Population after learner phase		Population after checking feasibility		Population cost after checking feasibility	Population after greedy selection		Population cost after greedy selection
−108.1442	81.4733	−108.1442	81.4733	5.4671	−106.6735	−23.2532	4.7148
−210.3619	−356.6942	−210.3619	−356.6942	44.4925	−83.8021	−112.9558	5.8218
79.0257	181.8495	79.0257	181.8495	9.9653	33.1289	98.7517	3.8195
−98.5351	−424.4390	−98.5351	−424.4390	48.5062	−104.6342	−168.4678	11.386
114.6219	485.5366	114.6219	485.5366	63.25	55.1224	154.5952	7.85
−72.8956	−216.8618	−72.8956	−216.8618	13.4203	−72.8956	−216.8618	13.4203
−72.4972	181.7579	−72.4972	181.7579	9.6404	−72.4972	181.7579	9.6404
−24.8979	−256.7760	−24.8979	−256.7760	16.8609	−24.8979	−256.7760	16.8609
−79.5062	−105.9209	−79.5062	−105.9209	5.8836	−79.5062	−105.9209	5.8836
−109.1569	−113.7565	−109.1569	−113.7565	7.4383	−109.1569	−113.7565	7.4383

Generation: 3

Initial population		Initial population cost
33.1289	98.7517	3.8195
−106.6735	−23.2532	4.7148
−591.7034	−23.2532	4.7148
−83.8021	−112.9558	5.8218
−79.5062	−105.9209	5.8836
−109.1569	−113.7565	7.4383
55.1224	154.5952	7.85
−72.4972	181.7579	9.6404
−104.6342	−168.4678	11.386
−72.8956	−216.8618	13.4203

Population after teacher phase		Population after checking feasibility		Population cost after checking feasibility	Population after greedy selection		Population cost after greedy selection
241.3195	235.8807	241.3195	235.8807	28.6684	33.1289	98.7517	3.8195
101.5171	113.8758	101.5171	113.8758	6.5975	−106.6735	−23.2532	4.7148
−383.5128	113.8758	−383.5128	113.8758	40.6236	−591.7034	−23.2532	4.7148
124.3886	24.1732	124.3886	24.1732	5.068	124.3886	24.1732	5.068
128.6844	31.2081	128.6844	31.2081	4.3936	128.6844	31.2081	4.3936
99.0337	23.3725	99.0337	23.3725	3.6386	99.0337	23.3725	3.6386
263.3130	291.7242	263.3130	291.7242	39.2041	55.1224	154.5952	7.85
135.6935	318.8869	135.6935	318.8869	31.6504	−72.4972	181.7579	9.6404
103.5565	−31.3388	103.5565	−31.3388	2.9473	103.5565	−31.3388	2.9473
135.2950	−79.7328	135.2950	−79.7328	8.1303	135.2950	−79.7328	8.1303

Population after learner phase		Population after checking feasibility		Population cost after checking feasibility	Population after greedy selection		Population cost after greedy selection
169.0586	217.3769	169.0586	217.3769	20.769	33.1289	98.7517	3.8195
−111.9678	−55.0120	−111.9678	−55.0120	4.7361	−106.6735	−23.2532	4.7148
−88.2729	14.8062	−88.2729	14.8062	3.4813	−88.2729	14.8062	3.4813
317.0003	−129.9907	317.0003	−129.9907	29.6889	124.3886	24.1732	5.068
111.4613	26.6567	111.4613	26.6567	4.3487	111.4613	26.6567	4.3487
207.2233	−76.5259	207.2233	−76.5259	13.9554	99.0337	23.3725	3.6386
113.9628	55.9009	113.9628	55.9009	5.1935	113.9628	55.9009	5.1935
−52.2304	156.2537	−52.2304	156.2537	7.4549	−52.2304	156.2537	7.4549
149.1970	−206.5484	149.1970	−206.5484	17.2314	103.5565	−31.3388	2.9473
−104.2316	−23.8232	−104.2316	−23.8232	3.502	−104.2316	−23.8232	3.502

A1.12 Example 12: Penalty1 Function

$$f(x) = \frac{\pi}{30}\left[10\sin^2(\pi y_1) + \sum_{i=1}^{n-1}(y_i - 1)^2\{1 + 10\sin^2(\pi y_{i+1})\} + (y_n - 1)^2\right]$$

$$+ \sum_{i=1}^{n} u(x_i, 10, 100, 4)$$

$$-50 \le x_i \le 50, \quad \min(f) = f(1, 1\ldots1) = 0$$

Generation: 1

Initial population		Initial population cost 1.0e + 008
4.8414	−0.8334	0
18.7035	0.8076	0.0057
23.4069	6.5174	0.0323
23.2508	17.6024	0.0342
25.1938	14.0609	0.0536
32.5263	25.9016	0.3214
−8.4667	−41.3839	0.9701
31.8744	−48.5974	2.4483
−45.8290	−40.4443	2.507
−42.3172	49.4002	3.5006

Population after teacher phase		Population after checking feasibility 1.0e + 008		Population cost after checking feasibility 1.0e + 008	Population after greedy selection		Population cost after greedy selection
3.9880	−0.4485	3.9880	−0.4485	0	3.9880	−0.4485	0
17.8501	1.1925	17.8501	1.1925	0.0038	17.8501	1.1925	0.0038
22.5536	6.9023	22.5536	6.9023	0.0248	22.5536	6.9023	0.0248
22.3975	17.9874	22.3975	17.9874	0.0277	22.3975	17.9874	0.0277
24.3404	14.4459	24.3404	14.4459	0.0427	24.3404	14.4459	0.0427
31.6729	26.2865	31.6729	26.2865	0.291	31.6729	26.2865	0.291
−9.3200	−40.9990	−9.3200	−40.9990	0.9234	−9.3200	−40.9990	0.9234
31.0210	−48.2125	31.0210	−48.2125	2.3274	31.0210	−48.2125	2.3274
−46.6823	−40.0594	−46.6823	−40.0594	2.6271	−45.8290	−40.4443	2.507
−43.1705	49.7851	−43.1705	49.7851	3.7161	−42.3172	49.4002	3.5006

Population after learner phase		Population after checking feasibility		Population cost after checking feasibility 1.0e + 008	Population after greedy selection		Population cost after greedy selection 1.0e + 008
−5.7953	−10.2458	−5.7953	−10.2458	0	3.9880	−0.4485	0
44.3796	42.3892	44.3796	42.3892	2.4976	17.8501	1.1925	0.0038
21.3308	14.8613	21.3308	14.8613	0.017	21.3308	14.8613	0.017
22.0845	18.5578	22.0845	18.5578	0.0267	22.0845	18.5578	0.0267
24.0015	13.0150	24.0015	13.0150	0.0385	24.0015	13.0150	0.0385
29.4033	24.2558	29.4033	24.2558	0.1831	29.4033	24.2558	0.1831
13.7206	−5.2201	13.7206	−5.2201	0.0002	13.7206	−5.2201	0.0002
43.5544	−49.4794	43.5544	−49.4794	3.6969	31.0210	−48.2125	2.3274
−5.3588	−13.9828	−5.3588	−13.9828	0.0003	−5.3588	−13.9828	0.0003
−8.4297	22.2486	−8.4297	22.2486	0.0225	−8.4297	22.2486	0.0225

Generation: 2

Initial population		Initial population cost 1.0e + 007
3.9880	−0.4485	0
4.8414	−0.8334	0
13.7206	−5.2201	0.0019
−5.3588	−13.9828	0.0025
17.8501	1.1925	0.038
21.3308	14.8613	0.1704
−8.4297	22.2486	0.2251
22.0845	18.5578	0.2669
24.0015	13.0150	0.3852
29.4033	24.2558	1.8305

Population after teacher phase		Population after checking feasibility		Population cost after checking feasibility 1.0e + 007	Population after greedy selection		Population cost after greedy selection 1.0e + 006
−15.7988	−13.9518	−15.7988	−13.9518	0.0138	3.9880	−0.4485	0
−14.9455	−14.3368	−14.9455	−14.3368	0.0095	4.8414	−0.8334	0
−6.0663	−18.7235	−6.0663	−18.7235	0.0579	13.7206	−5.2201	0.0192
−25.1457	−27.4861	−25.1457	−27.4861	1.4612	−5.3588	−13.9828	0.0252
−1.9367	−12.3108	−1.9367	−12.3108	0.0003	−1.9367	−12.3108	0.0029
1.5440	1.3580	1.5440	1.3580	0	1.5440	1.3580	0
−28.2165	8.7453	−28.2165	8.7453	1.1013	−8.4297	22.2486	2.2509
2.2977	5.0545	2.2977	5.0545	0	2.2977	5.0545	0
4.2147	−0.4883	4.2147	−0.4883	0	4.2147	−0.4883	0
9.6165	10.7524	9.6165	10.7524	0	9.6165	10.7524	0.0001

Population after learner phase		Population after checking feasibility		Population cost after checking feasibility 1.0e + 006	Population after greedy selection		Population cost after greedy selection 1.0e + 006
14.3822	−19.4469	14.3822	−19.4469	0.8336	3.9880	−0.4485	0
11.8705	−13.0591	11.8705	−13.0591	0.01	4.8414	−0.8334	0
6.0388	−1.4539	6.0388	−1.4539	0	6.0388	−1.4539	0
0.1561	−6.8732	0.1561	−6.8732	0	0.1561	−6.8732	0
−0.9110	−8.1042	−0.9110	−8.1042	0	−0.9110	−8.1042	0
2.5853	0.5883	2.5853	0.5883	0	2.5853	0.5883	0
−6.9070	21.2786	−6.9070	21.2786	1.6182	−6.9070	21.2786	1.6182
2.8041	7.5381	2.8041	7.5381	0	2.8041	7.5381	0
3.9283	−0.3307	3.9283	−0.3307	0	3.9283	−0.3307	0
7.7652	7.0683	7.7652	7.0683	0	7.7652	7.0683	0

Generation: 3

Initial population		Initial population cost 1.0e + 006
−0.9110	−8.1042	5.0356
2.8041	7.5381	11.3368
2.5853	0.5883	14.4636
3.9880	−0.4485	14.4824
41.5507	−0.4485	14.4824
0.1561	−6.8732	14.5782
3.9283	−0.3307	15.4044
6.0388	−1.4539	18.1844
7.7652	7.0683	19.1733
4.8414	−0.8334	19.3882

Population after teacher phase		Population after checking feasibility		Population cost after checking feasibility 1.0e + 007	Population after greedy selection		Population cost after greedy selection 1.0e + 006
−12.2771	−11.6788	−12.2771	−11.6788	0.0004	−0.9110	−8.1042	5.0356
−8.5620	3.9635	−8.5620	3.9635	0.0000	2.8041	7.5381	11.3368
−8.7808	−2.9863	−8.7808	−2.9863	0.0000	2.5853	0.5883	14.4636
−7.3781	−4.0231	−7.3781	−4.0231	0.0000	3.9880	−0.4485	14.4824
30.1846	−4.0231	30.1846	−4.0231	1.6599	41.5507	−0.4485	14.4824
−11.2100	−10.4479	−11.2100	−10.4479	0.0000	0.1561	−6.8732	14.5782
−7.4378	−3.9053	−7.4378	−3.9053	0.0000	3.9283	−0.3307	15.4044
−5.3273	−5.0286	−5.3273	−5.0286	0.0000	−5.3273	−5.0286	4.4565
−3.6009	3.4937	−3.6009	3.4937	0.0000	−3.6009	3.4937	16.0575
−6.5247	−4.4081	−6.5247	−4.4081	0.0000	4.8414	−0.8334	19.3882

Population after learner phase		Population after checking feasibility		Population cost after checking feasibility 1.0e + 003	Population after greedy selection		Population cost after greedy selection 1.0e + 006
−3.9851	−5.9633	−3.9851	−5.9633	0.0155	−0.9110	−8.1042	5.0356
5.9199	9.5056	5.9199	9.5056	0.0644	2.8041	7.5381	11.3368
2.3448	0.7529	2.3448	0.7529	0.0158	2.5853	0.5883	14.4636
−0.0955	−6.8298	−0.0955	−6.8298	0.0109	−0.0955	−6.8298	10.8868
12.5503	−5.6772	12.5503	−5.6772	4.3104	41.5507	−0.4485	14.4824
−0.9086	−8.1014	−0.9086	−8.1014	0.0050	−0.9086	−8.1014	5.0361
9.2820	−3.0501	9.2820	−3.0501	0.1294	3.9283	−0.3307	15.4044
−5.8927	−4.6348	−5.8927	−4.6348	0.0121	−5.3273	−5.0286	4.4565
−3.3078	2.2301	−3.3078	2.2301	0.0180	−3.6009	3.4937	16.0575
4.5613	0.3172	4.5613	0.3172	0.0396	4.8414	−0.8334	19.3882

A1.13 Example 13: Penalty2 Function

$$f(x) = 0.1\begin{bmatrix} \sin^2(\pi 3x_1) \\ +\sum_{i=1}^{n-1}(x_i-1)^2\{1+\sin^2(3\pi x_{i+1})\}+(x_n-1)^2(1+\sin^2(2\pi x_{30}) \end{bmatrix}$$

$$+\sum_{i=1}^{n} u(x_i, 5, 100, 4)$$

$$-50 \le x_i \le 50, \quad \min(f) = f(1, 1\ldots1) = 0$$

$$u(x_i, a, k, m) = \begin{cases} k(x_i-a)^m & x_i > a, \\ 0, & -a \le x_i \le a, \\ k(-x_i-a)^m, x_i < -a \end{cases}$$

$$y_i = 1 + 1/4(x_i+1)$$

Generation: 1

Initial population		Initial population cost 1.0e + 008
0.8752	20.4562	0.0571
−22.1707	−1.4415	0.0869
26.3507	−3.0696	0.2078
13.1101	27.0698	0.2416
20.7721	−26.6315	0.2808
8.8088	32.3612	0.5607
33.8588	31.7896	1.2087
1.4301	41.5484	1.7843
−43.9512	−0.5404	2.3019
37.7645	−47.4794	4.4086

Population after teacher phase		Population after checking feasibility		Population cost after checking feasibility 1.0e + 008	Population after greedy selection		Population cost after greedy selection 1.0e + 008
−2.7834	28.0601	−2.7834	28.0601	0.2828	0.8752	20.4562	0.0571
−25.8293	6.1624	−25.8293	6.1624	0.1882	−22.1707	−1.4415	0.0869
22.6921	4.5343	22.6921	4.5343	0.098	22.6921	4.5343	0.098
9.4515	34.6737	9.4515	34.6737	0.7757	13.1101	27.0698	0.2416
17.1135	−19.0276	17.1135	−19.0276	0.0603	17.1135	−19.0276	0.0603
5.1502	39.9651	5.1502	39.9651	1.4947	8.8088	32.3612	0.5607
30.2002	39.3936	30.2002	39.3936	1.8026	33.8588	31.7896	1.2087
−2.2284	49.1523	−2.2284	49.1523	3.8003	1.4301	41.5484	1.7843
−47.6097	7.0635	−47.6097	7.0635	3.2964	−43.9512	−0.5404	2.3019
34.1060	−39.8754	34.1060	−39.8754	2.1971	34.1060	−39.8754	2.1971

Population after learner phase		Population after checking feasibility		Population cost after checking feasibility	Population after greedy selection		Population cost after greedy selection
				1.0e + 008			1.0e + 008
−7.3903	15.9882	−7.3903	15.9882	0.0146	−7.3903	15.9882	0.0146
−8.0161	−2.0271	−8.0161	−2.0271	0.0001	−8.0161	−2.0271	0.0001
58.4801	7.2595	50.0000	7.2595	4.1007	22.6921	4.5343	0.098
24.7843	12.5982	24.7843	12.5982	0.1565	24.7843	12.5982	0.1565
18.2706	−26.1874	18.2706	−26.1874	0.2325	17.1135	−19.0276	0.0603
17.2958	15.3503	17.2958	15.3503	0.0343	17.2958	15.3503	0.0343
32.1251	31.1939	32.1251	31.1939	1.0121	32.1251	31.1939	1.0121
0.9638	23.8227	0.9638	23.8227	0.1255	0.9638	23.8227	0.1255
−4.2656	15.9489	−4.2656	15.9489	0.0144	−4.2656	15.9489	0.0144
23.4409	−13.2996	23.4409	−13.2996	0.1204	23.4409	−13.2996	0.1204

Generation: 2

Initial population		Initial population cost
		1.0e + 007 *
−8.0161	−2.0271	0.0008
−4.2656	15.9489	0.1437
−7.3903	15.9882	0.1461
17.2958	15.3503	0.3433
0.8752	20.4562	0.5707
17.1135	−19.0276	0.6025
22.6921	4.5343	0.9798
23.4409	−13.2996	1.2039
0.9638	23.8227	1.2553
24.7843	12.5982	1.5654

Population after teacher phase		Population after checking feasibility		Population cost after checking feasibility 1.0e + 006	Population after greedy selection		Population cost after greedy selection 1.0e + 006
−10.6007	−3.1703	−10.6007	−3.1703	0.0984	−8.0161	−2.0271	0.0083
−6.8502	14.8057	−6.8502	14.8057	0.9258	−6.8502	14.8057	0.9258
−9.9749	14.8451	−9.9749	14.8451	1.0008	−9.9749	14.8451	1.0008
14.7113	14.2071	14.7113	14.2071	1.6081	14.7113	14.2071	1.6081
−1.7094	19.3130	−1.7094	19.3130	4.1969	−1.7094	19.3130	4.1969
14.5290	−20.1708	14.5290	−20.1708	6.1217	17.1135	−19.0276	6.0253
20.1076	3.3911	20.1076	3.3911	5.2094	20.1076	3.3911	5.2094
20.8564	−14.4428	20.8564	−14.4428	7.1166	20.8564	−14.4428	7.1166
−1.6208	22.6795	−1.6208	22.6795	9.7699	−1.6208	22.6795	9.7699
22.1998	11.4550	22.1998	11.4550	8.9254	22.1998	11.4550	8.9254

Population after learner phase		Population after checking feasibility		Population cost after checking feasibility 1.0e + 008	Population after greedy selection		Population cost after greedy selection 1.0e + 006
−10.2235	−1.0779	−10.2235	−1.0779	0.0007	−8.0161	−2.0271	0.0083
−6.9988	14.5819	−6.9988	14.5819	0.0084	−6.9988	14.5819	0.8446
−10.4290	14.5996	−10.4290	14.5996	0.0094	−10.4290	14.5996	0.9361
13.8006	14.5418	13.8006	14.5418	0.0143	13.8006	14.5418	1.4289
−4.1552	20.1169	−4.1552	20.1169	0.0522	−1.7094	19.3130	4.1969
22.2417	−30.4442	22.2417	−30.4442	0.5075	17.1135	−19.0276	6.0253
19.7500	11.9068	19.7500	11.9068	0.0496	19.7500	11.9068	4.961
34.2232	−36.5188	34.2232	−36.5188	1.7162	20.8564	−14.4428	7.1166
18.4052	4.9023	18.4052	4.9023	0.0323	18.4052	4.9023	3.2292
20.8353	6.1957	20.8353	6.1957	0.0629	20.8353	6.1957	6.2881

Generation: 3

Initial population		Initial population cost 1.0e + 006
−8.0161	−2.0271	0.0083
−33.1205	−2.0271	0.0083
−6.9988	14.5819	0.8446
−10.4290	14.5996	0.9361
13.8006	14.5418	1.4289
18.4052	4.9023	3.2292
−1.7094	19.3130	4.1969
19.7500	11.9068	4.961
17.1135	−19.0276	6.0253
20.8353	6.1957	6.2881

Population after teacher phase		Population after checking feasibility		Population cost after checking feasibility 1.0e + 008	Population after greedy selection		Population cost after greedy selection 1.0e + 006
−21.7822	−15.9326	−21.7822	−15.9326	0.0936	−8.0161	−2.0271	0.0083
−46.8865	−15.9326	−46.8865	−15.9326	3.0925	−33.1205	−2.0271	0.0083
−20.7649	0.6765	−20.7649	0.6765	0.0618	−6.9988	14.5819	0.8446
−24.1951	0.6941	−24.1951	0.6941	0.1358	−10.4290	14.5996	0.9361
0.0346	0.6364	0.0346	0.6364	0	0.0346	0.6364	0
4.6391	−9.0031	4.6391	−9.0031	0.0003	4.6391	−9.0031	0.0257
−15.4754	5.4075	−15.4754	5.4075	0.012	−15.4754	5.4075	1.2042
5.9840	−1.9987	5.9840	−1.9987	0	5.9840	−1.9987	0.0001
3.3475	−32.9331	3.3475	−32.9331	0.6088	17.1135	−19.0276	6.0253
7.0692	−7.7097	7.0692	−7.7097	0.0001	7.0692	−7.7097	0.0072

Population after learner phase		Population after checking feasibility		Population cost after checking feasibility 1.0e + 007	Population after greedy selection		Population cost after greedy selection 1.0e + 006
−18.7106	3.8681	−18.7106	3.8681	0.3534	−8.0161	−2.0271	0.0083
−29.3857	−2.0271	−29.3857	−2.0271	3.5362	−33.1205	−2.0271	0.0083
−1.5266	3.7317	−1.5266	3.7317	0	−1.5266	3.7317	0
0.3295	3.7196	0.3295	3.7196	0	0.3295	3.7196	0
10.6663	−2.6342	10.6663	−2.6342	0.0103	0.0346	0.6364	0
5.2837	−5.6458	5.2837	−5.6458	0	5.2837	−5.6458	0
−16.7934	6.3957	−16.7934	6.3957	0.1935	−15.4754	5.4075	1.2042
15.4114	−14.0387	15.4114	−14.0387	0.1842	5.9840	−1.9987	0.0001
7.5548	−5.7040	7.5548	−5.7040	0.0004	7.5548	−5.7040	0.0043
11.5215	−14.7647	11.5215	−14.7647	0.109	7.0692	−7.7097	0.0072

A1.14 Example 14: Unconstrained Himmelblau Function

Minimize

$$f(X) = (x_1^2 + x_2 - 11)^2 + (x_1 + x_2^2 - 7)^2$$

Min $(f) = f\,(3,2) = 0$

Generation: 1

Initial population		Initial population cost (1e3*)
3.7569	1.4016	0.022
1.9407	2.1356	0.0262
2.5188	0.2674	0.0387
1.7994	1.4796	0.0485
0.3837	2.4598	0.0708
2.2672	3.6877	0.0833
2.9300	3.9305	0.1318
0.1654	0.7948	0.1421
1.2639	4.3798	0.206
4.4856	5.7037	1.1209

Population after teacher phase		Population after checking feasibility		Population cost after checking feasibility	Population after greedy selection		Population cost after greedy selection
3.4887	1.2798	3.4887	1.2798	9.5166	3.4887	1.2798	9.5166
1.6725	2.0137	1.6725	2.0137	39.924	1.9407	2.1356	26.2395
2.2505	0.1456	2.2505	0.1456	55.8745	2.5188	0.2674	38.7036
1.5311	1.3577	1.5311	1.3577	66.4021	1.7994	1.4796	48.5411
0.1154	2.3379	0.1154	2.3379	76.8138	0.3837	2.4598	70.7635
1.9990	3.5658	1.9990	3.5658	71.3303	1.9990	3.5658	71.3303
2.6618	3.8087	2.6618	3.8087	103.3982	2.6618	3.8087	103.3982
−0.1029	0.6729	0	0.6729	149.5145	0.1654	0.7948	142.0667
0.9957	4.2579	0.9957	4.2579	180.1031	0.9957	4.2579	180.1031
4.2174	5.5819	4.2174	5.5819	958.1119	4.2174	5.5819	958.1119

Population after learner phase		Population after checking feasibility		Population cost after checking feasibility	Population after greedy selection		Population cost after greedy selection
3.5405	1.2736	3.5405	1.2736	11.2667	3.4887	1.2798	9.5166
1.4675	1.0377	1.4675	1.0377	80.8305	1.9407	2.1356	26.2395
3.0446	−1.1102	3.0446	0	18.6401	3.0446	0	18.6401
2.9958	1.9810	2.9958	1.9810	0.0083	2.9958	1.9810	0.0083
−0.1390	0.9240	0	0.9240	139.3015	0.3837	2.4598	70.7635
2.0424	3.5359	2.0424	3.5359	67.7702	2.0424	3.5359	67.7702
2.3536	2.9763	2.3536	2.9763	23.9146	2.3536	2.9763	23.9146
0.3226	1.9939	0.3226	1.9939	86.5442	0.3226	1.9939	86.5442
3.2022	1.6221	3.2022	1.6221	2.1286	3.2022	1.6221	2.1286
4.1818	5.5413	4.1818	5.5413	922.4125	4.1818	5.5413	922.4125

Generation: 2

Initial population		Initial population cost
2.9958	1.9810	0.0083
3.2022	1.6221	2.1286
3.4887	1.2798	9.5166
3.0446	0	18.6401
3.7569	1.4016	22.0302
2.3536	2.9763	23.9146
1.9407	2.1356	26.2395
2.0424	3.5359	67.7702
0.3837	2.4598	70.7635
0.3226	1.9939	86.5442

Population after teacher phase		Population after checking feasibility		Population cost after checking feasibility	Population after greedy selection		Population cost after greedy selection
3.6045	2.0117	3.6045	2.0117	16.4594	2.9958	1.9810	0.0083
3.8109	1.6527	3.8109	1.6527	26.9961	3.2022	1.6221	2.1286
4.0974	1.3104	4.0974	1.3104	51.8049	3.4887	1.2798	9.5166
3.6533	0.0306	3.6533	0.0306	16.8452	3.6533	0.0306	16.8452
4.3656	1.4322	4.3656	1.4322	90.4207	3.7569	1.4016	22.0302
2.9623	3.0070	2.9623	3.0070	25.6535	2.3536	2.9763	23.9146
2.5494	2.1662	2.5494	2.1662	5.5075	2.5494	2.1662	5.5075
2.6511	3.5665	2.6511	3.5665	70.2417	2.0424	3.5359	67.7702
0.9924	2.4904	0.9924	2.4904	56.6607	0.9924	2.4904	56.6607
0.9313	2.0246	0.9313	2.0246	69.6218	0.9313	2.0246	69.6218

Population after learner phase		Population after checking feasibility		Population cost after checking feasibility	Population after zgreedy selection		Population cost after greedy selection
3.0384	1.9151	3.0384	1.9151	0.1079	2.9958	1.9810	0.0083
3.0139	1.9496	3.0139	1.9496	0.0354	3.0139	1.9496	0.0354
4.1386	0.2660	4.1386	0.2660	48.6737	3.4887	1.2798	9.5166
4.7411	−2.3364	4.7411	0	136.8488	3.6533	0.0306	16.8452
4.4876	0.5817	4.4876	0.5817	99.2097	3.7569	1.4016	22.0302
2.7838	2.3097	2.7838	2.3097	2.1362	2.7838	2.3097	2.1362
1.8935	2.5815	1.8935	2.5815	25.7848	2.5494	2.1662	5.5075
2.0425	3.5356	2.0425	3.5356	67.7364	2.0425	3.5356	67.7364
1.7954	2.3232	1.7954	2.3232	29.7753	1.7954	2.3232	29.7753
3.4538	1.2899	3.4538	1.2899	8.4656	3.4538	1.2899	8.4656

Generation: 3

Initial population		Initial population cost
2.9958	1.9810	0.0083
2.6911	1.9810	0.0083
3.0139	1.9496	0.0354
2.7838	2.3097	2.1362
2.5494	2.1662	5.5075
3.4538	1.2899	8.4656
3.4887	1.2798	9.5166
3.6533	0.0306	16.8452
3.7569	1.4016	22.0302
1.7954	2.3232	29.7753

Population after teacher phase		Population after checking feasibility		Population cost after checking feasibility	Population after greedy selection		Population cost after greedy selection
2.9824	2.0207	2.9824	2.0207	0.0115	2.9958	1.9810	0.0083
2.6776	2.0207	2.6776	2.0207	3.3319	2.6911	1.9810	0.0083
3.0005	1.9892	3.0005	1.9892	0.0019	3.0005	1.9892	0.0019
2.7703	2.3493	2.7703	2.3493	2.6155	2.7838	2.3097	2.1362
2.5360	2.2058	2.5360	2.2058	5.7457	2.5494	2.1662	5.5075
3.4404	1.3296	3.4404	1.3296	7.9007	3.4404	1.3296	7.9007
3.4753	1.3194	3.4753	1.3194	8.927	3.4753	1.3194	8.927
3.6398	0.0702	3.6398	0.0702	16.6337	3.6398	0.0702	16.6337
3.7435	1.4412	3.7435	1.4412	21.2363	3.7435	1.4412	21.2363
1.7820	2.3628	1.7820	2.3628	29.9643	1.7954	2.3232	29.7753

Population after learner phase		Population after checking feasibility		Population cost after checking feasibility	Population after greedy selection		Population cost after greedy selection
2.9976	1.9842	2.9976	1.9842	0.0052	2.9976	1.9842	0.0052
2.7338	1.9253	2.7338	1.9253	2.8771	2.6911	1.9810	0.0083
2.8501	2.1001	2.8501	2.1001	0.6713	3.0005	1.9892	0.0019
2.8136	2.2655	2.8136	2.2655	1.5646	2.8136	2.2655	1.5646
1.8716	2.5777	1.8716	2.5777	26.4996	2.5494	2.1662	5.5075
3.3809	1.4187	3.3809	1.4187	5.9991	3.3809	1.4187	5.9991
3.2994	1.4678	3.2994	1.4678	4.2234	3.2994	1.4678	4.2234
3.5007	1.1263	3.5007	1.1263	10.646	3.5007	1.1263	10.646
2.5994	2.1358	2.5994	2.1358	4.4661	2.5994	2.1358	4.4661
2.8554	1.6829	2.8554	1.6829	3.0768	2.8554	1.6829	3.0768

A1.15 Example 15: Constrained Himmelblau Function

Minimize

$$f(X) = (x_1^2 + x_2 - 11)^2 + (x_1 + x_2^2 - 7)^2$$

Subjected to:

$$g_1(X) = 4.84 - (x_1 - 0.05)^2 - (x_2 - 2.5)^2$$

$$g_2(X) = x_1^2 + (x_2 - 2.5)^2 - 4.84$$

Generation: 1

Initial population		Initial population cost (1e5)	Contraints	
			g1	g2
2.1196	2.3256	0.0112	0.5263	−0.3169
0.7140	4.2950	0.0241	1.1771	−1.1082
0.5930	3.6448	0.1120	3.2346	−3.1778
1.0476	2.1611	0.1422	3.7299	−3.6277
1.4024	5.2129	0.2039	−4.3488	4.4866
3.4453	2.0708	0.4824	−6.8723	7.2143
3.9228	1.2542	1.3817	−11.7106	12.1004
4.4953	1.7961	2.3877	−15.4162	15.8632
4.4893	4.9126	4.2963	−20.6880	21.1345
5.3575	5.0954	9.0602	−30.0657	30.5989

Population After Teacher Phase		Population After Checking Feasibility		Population Cost After Checking Feasibility (1e5)	Population After greedy selection		Population Cost After greedy selection (1e5)	Contraints	
								g1	g2
1.1208	−1.6613	1.1208	0	0.0767	2.1196	2.3256	0.0112	0.5263	−0.3169
−0.2848	0.3081	0	0.3081	0.0116	0	0.3081	0.0116	0.0331	−0.0356
−0.4058	−0.3422	0	0	0.0317	0	0	0.0317	1.4125	1.4100
0.0489	−1.8259	0.0489	0	0.0316	0.0489	0	0.0316	−1.4100	1.4124
0.4037	1.2259	0.4037	1.2259	0.1044	0.4037	1.2259	0.1044	3.0916	−3.0537
2.4466	−1.9162	2.4466	0	0.5222	3.4453	2.0708	0.4824	−6.8723	7.2143
2.9240	−2.7328	2.9240	0	0.9453	2.9240	0	0.9453	−9.6699	9.9598
3.4965	−2.1909	3.4965	0	1.7760	3.4965	0	1.7760	−13.2884	13.6355
3.4905	0.9256	3.4905	0.9256	0.9080	3.4905	0.9256	0.9080	−9.4758	9.8223
4.3587	1.1084	4.3587	1.1084	2.4636	4.3587	1.1084	2.4636	15.6614	16.0948

Population after learner phase		Population after checking feasibility		Population cost after checking feasibility	Population after greedy selection		Population cost after greedy selection	Contraints	
								g1	g2
3.0395	3.3588	3.0395	3.3588	0.2443	2.1196	2.3256	0.0112	0.5263	−0.3169
−2.2672	0.5469	0	0.5469	0.0221	0	0.3081	0.0116	0.0331	−0.0356
−0.1656	−0.5031	0	0	0.0317	0	0	0.0317	−1.4125	1.4100
0.0853	0	0.0853	0	0.0316	0.0489	0	0.0316	−1.4100	1.4124
1.6862	2.0478	1.6862	2.0478	0.0425	1.6862	2.0478	0.0425	1.9584	−1.7922
3.1658	1.9931	3.1658	1.9931	0.2727	3.1658	1.9931	0.2727	−5.1252	5.4392
2.7571	0	2.7571	0	0.7739	2.7571	0	0.7739	−8.7384	9.0116
3.4811	0.6229	3.4811	0.6229	1.1035	3.4811	0.6229	1.1035	−10.4560	10.8016
3.2066	1.2155	3.2066	1.2155	0.4689	3.2066	1.2155	0.4689	−6.7741	7.0922
3.8129	1.6835	3.8129	1.6835	1.0075	3.8129	1.6835	1.0075	−9.9861	10.3649

Generation: 2

Initial population		Initial population cost (1e5)
2.1196	2.3256	0.0112
3.4353	2.3256	0.0112
0	0.3081	0.0116
0.0489	0	0.0316
0	0	0.0317
1.6862	2.0478	0.0425
3.1658	1.9931	0.2727
3.2066	1.2155	0.4689
2.7571	0	0.7739
3.8129	1.6835	1.0075

Population after teacher phase		Population after checking feasibility		Population cost after checking feasibility (1e4)	Population after greedy selection		Population cost after greedy selection (1e4)
1.5631	2.3201	1.5631	2.3201	0.6629	2.1196	2.3256	0.1118
2.8788	2.3201	2.8788	2.3201	1.1208	3.4353	2.3256	0.1118
−0.5565	0.3026	0	0.3026	0.1162	0	0.3026	0.1162
−0.5076	−0.0055	0	0	0.3165	0.0489	0	0.3157
−0.5565	−0.0055	0	0	0.3165	0	0	0.3165
1.1297	2.0424	1.1297	2.0424	1.2314	1.6862	2.0478	0.4251
2.6093	1.9876	2.6093	1.9876	0.4896	2.6093	1.9876	0.4896
2.6501	1.2100	2.6501	1.2100	1.3865	2.6501	1.2100	1.3865
2.2006	−0.0055	2.2006	0	3.7484	2.2006	0	3.7484
3.2564	1.6780	3.2564	1.6780	3.8419	3.2564	1.6780	3.8419

Population after learner phase		Population after checking feasibility		Population cost after checking feasibility (1e4)	Population after greedy selection		Population cost after greedy selection (1e4)
2.1104	2.5902	2.1104	2.5902	0.1162	2.1196	2.3256	0.1118
2.1348	2.3256	2.1348	2.3256	0.1081	2.1348	2.3256	0.1081
−2.0526	0.5848	0	0.5848	0.2526	0	0.3026	0.1162
−0.0737	0	0	0	0.3165	0.0489	0	0.3157
−0.0231	−0.0281	0	0	0.3165	0	0	0.3165
0.7766	0.9432	0.7766	0.9432	0.4405	1.6862	2.0478	0.4251
3.0526	2.1690	3.0526	2.1690	1.9364	2.6093	1.9876	0.4896
2.5473	1.2994	2.5473	1.2994	0.9071	2.5473	1.2994	0.9071
1.8217	−0.6023	1.8217	0	2.1778	1.8217	0	2.1778
2.2895	2.2288	2.2895	2.2288	0.1075	2.2895	2.2288	0.1075

Generation: 3

Initial population		Initial population cost 1.0e + 003
2.2895	2.2288	1.0746
2.1348	2.3256	1.0809
2.1196	2.3256	1.1183
3.4049	2.3256	1.1183
0	0.3026	1.1623
0.0489	0	3.1574
0	0	3.1652
1.6862	2.0478	4.2514
2.6093	1.9876	4.8957
2.5473	1.2994	9.0705

Population after teacher phase		Population after checking feasibility		Population cost after checking feasibility 1.0e + 005	Population after greedy selection		Population cost after greedy selection 1.0e + 003
2.8845	2.8934	2.8845	2.8934	0.1224	2.2895	2.2288	1.0746
2.7298	2.9902	2.7298	2.9902	0.0769	2.1348	2.3256	1.0809
2.7146	2.9902	2.7146	2.9902	0.0727	2.1196	2.3256	1.1183
3.9999	2.9902	3.9999	2.9902	1.2214	3.4049	2.3256	1.1183
0.5950	0.9672	0.5950	0.9672	0.0569	0	0.3026	1.1623
0.6439	0.6646	0.6439	0.6646	0.0225	0.6439	0.6646	2.25
0.5950	0.6646	0.5950	0.6646	0.0238	0.5950	0.6646	2.3834
2.2811	2.7124	2.2811	2.7124	0.0105	2.2811	2.7124	1.05
3.2043	2.6522	3.2043	2.6522	0.2736	2.6093	1.9876	4.8957
3.1423	1.9640	3.1423	1.9640	0.2609	2.5473	1.2994	9.0705

Population after learner phase		Population after checking feasibility		Population cost after checking feasibility 1.0e + 004	Population after greedy selection		Population cost after greedy selection 1.0e + 003	Contraints	
								g1	g2
2.2842	2.5355	2.2842	2.5355	0.1037	2.2842	2.5355	1.0369	−0.1529	0.3788
3.3939	3.5188	3.3939	3.5188	5.5549	2.1348	2.3256	1.0809	0.4632	−0.2522
2.2221	2.5710	2.2221	2.5710	0.0016	2.2221	2.5710	0.0155	0.1169	0.1028
3.6784	2.6528	3.6784	2.6528	7.0736	3.4049	2.3256	1.1183	−6.4458	6.7838
−1.2465	−0.5024	0	0	0.3165	0	0.3026	1.1623	0.0089	−0.0114
0.9779	1.0367	0.9779	1.0367	0.4142	0.6439	0.6646	2.25	1.1186	−1.0567
1.6999	1.3177	1.6999	1.3177	0.1364	1.6999	1.3177	1.3642	0.7200	−0.5525
2.2774	2.9260	2.2774	2.9260	0.1115	2.2811	2.7124	1.05	−0.1829	0.4085
1.3415	1.1549	1.3415	1.1549	0.2599	1.3415	1.1549	2.599	1.3627	−1.2311
2.0939	1.1220	2.0939	1.1220	0.2572	2.0939	1.1220	2.5722	−1.2364	1.4433

BEST solutions obtained by using the TLBO algorithm for different number of population sizes and generations

Example	A	B	C	D	E	F
1	0.000007	0	0	0	0	0
2	0.010012	0.000015	0.000001	0	0	0
3	0.000217	0	0	0	0	0
4	0.011658	0.001001	0.00003	0	0	0
5	0.031595	0	0.000005	0	0	0
6	0	0	0	0	0	0
7	0	0	0	0	0	0
8	−769.158	−837.964	−837.966	−837.965	−837.966	−837.966
9	0.00367	0.000016	0.000005	0	0	0
10	0.090041	0.000772	0	0	0	0
11	0.020971	0.014999	0.007453	0.005157	0.000001	0
12	0.012046	0.000002	0	0	0	0
13	0.000454	0.000034	0	0	0	0
14	0.001171	0.000002	0	0	0	0
15	14.16304	13.60063	13.59097	13.59087	13.59087	13.59084

A = Population 10; Generations 10

B = Population 10; Generations 20

C = Population 10; Generations 30

D = Population 10; Generations 40

E = Population 10; Generations 50

F = Population 10; Generations 100

Appendix 2
Sample Codes

The codes for TLBO, ABC, BBO and PSO algorithms are given below. Readers are to note that a blank Microsoft access file "Matrices.mat" is to be created for output storage while running these codes. Even though elitism is mentioned in the TLBO code (i.e. the word 'keep' in thecode), it is not actually implemented in the results presented in this book.However, interested readers may make use of elitism concept in TLBO code.

R. V. Rao and V. J. Savsani, *Mechanical Design Optimization Using Advanced Optimization Techniques*, Springer Series in Advanced Manufacturing,
DOI: 10.1007/978-1-4471-2748-2,

A2.1 TLBO Code

```
function ateacher(ProblemFunction, DisplayFlag, ProbFlag)
global ll
if ~exist('DisplayFlag', 'var')
  DisplayFlag = true;
end
if ~exist('ProbFlag', 'var')
  ProbFlag = true;
end
[OPTIONS, MinCost, AvgCost, InitFunction, CostFunction, FeasibleFunction, ...
  MaxParValue, MinParValue, Population,FeasibleFunction_1,CostFunction_1] = Init(DisplayFlag,
ProblemFunction);
Keep=2;
TF=1;
MR=1;
for GenIndex = 1 : OPTIONS.Maxgen
  for i = 1 : Keep
    chromKeep(i,:) = Population(i).chrom;
    costKeep(i) = Population(i).cost;
  end
  for i=1:length(Population)
  pp(i,:)=Population(i).chrom;
  pp_cost(i)=Population(i).cost;
  end
  pp_new=pp;
  pp
  Mean_pop=mean(pp);
  Stdev_pop=std(pp);
   %TF=round(1+rand*(1));
 for i=1:size(pp,2)
   New_Mean(i)=((pp(1,i)));
   Diff_Mean(i)=rand*(New_Mean(i)-TF*Mean_pop(i));
   for j=1:size(pp,1)
     F=(size(pp,1)-j)/size(pp,1);
      pp_new(j,i)=pp(j,i)+1*(1)*(Diff_Mean(i)); %%%% nor cor %%%%
   end
   end
  pp_new
    pp_new = FeasibleFunction_1(OPTIONS, pp_new);
     pp_new_cost = CostFunction_1(OPTIONS, pp_new);
    %%%%%%%%%% Greedy selection
   %for i = 1 : length(Population)-Keep
     for i = 1 : length(Population)
      if pp_new_cost(i)<Population(i).cost
        Population(i).chrom =pp_new(i,:);
        Population(i).cost=pp_new_cost(i);%%%%%%%
      end
     end
 for i = 1 : 1 : length(Population)
     ii=ceil(rand*(length(Population)));
     while ii==i
          ii=ceil(rand*(length(Population)));
     end
    if Population(i).cost<Population(ii).cost
      Island_1(i,:) = (Population(i).chrom + 1*rand*(Population(i).chrom-Population(ii).chrom));
      else
      Island_1(i,:) = (Population(i).chrom+1*rand*(Population(ii).chrom-Population(i).chrom));
     end
 end
     Island_1 = FeasibleFunction_1(OPTIONS, Island_1);
     Island_1_cost = CostFunction_1(OPTIONS, Island_1);
     for i = 1 : length(Population)
      if Island_1_cost(i)<Population(i).cost
        Population(i).chrom =Island_1(i,:);
```

```
            Population(i).cost=Island_1_cost(i);%%%%%%%
          end
        end
    Population = FeasibleFunction(OPTIONS, Population);
    Population = CostFunction(OPTIONS, Population);
    Population = PopSort(Population);
    n = length(Population);
    for i = 1 : Keep
        Population(n-i+1).chrom = chromKeep(i,:);
        Population(n-i+1).cost = costKeep(i);
    end
     Population = ClearDups(Population, MaxParValue, MinParValue);
    Population = PopSort(Population);
    [AverageCost, nLegal] = ComputeAveCost(Population);
     MinCost = [MinCost Population(1).cost];
    AvgCost = [AvgCost AverageCost];
    if DisplayFlag
            %disp([num2str(MinCost(end))]);
    end
end
fprintf('\n %f',MinCost(end));
Conclude(DisplayFlag, OPTIONS, Population, nLegal, MinCost);
```

```
function [Population] = ClearDups(Population, MaxParValue, MinParValue)
global ll
% Make sure there are no duplicate individuals in the population.
% This logic does not make 100% sure that no duplicates exist, but any duplicates that are found are
% randomly mutated, so there should be a good chance that there are no duplicates after this procedure.
for i = 1 : length(Population)
    Chrom1 = sort(Population(i).chrom);
    for j = i+1 : length(Population)
        Chrom2 = sort(Population(j).chrom);
        if isequal(Chrom1, Chrom2)
```

```
function [AveCost, nLegal] = ComputeAveCost(Population)
% Compute the average cost of all legal individuals in the population.
% OUTPUTS: AveCost = average cost
%          nLegal = number of legal individuals in population
% Save valid population member fitnesses in temporary array
Cost = [];
nLegal = 0;
for i = 1 : length(Population)
    if Population(i).cost < inf
        Cost = [Cost Population(i).cost];
        nLegal = nLegal + 1;
    end
end
% Compute average cost.
AveCost = mean(Cost);
return;
```

```
function Conclude(DisplayFlag, OPTIONS, Population, nLegal, MinCost)
% Output results of population-based optimization algorithm.
if DisplayFlag
    % Count the number of duplicates
    NumDups = 0;
    for i = 1 : OPTIONS.popsize
        Chrom1 = sort(Population(i).chrom);
        for j = i+1 : OPTIONS.popsize
            Chrom2 = sort(Population(j).chrom);
            if isequal(Chrom1, Chrom2)
```

```
            NumDups = NumDups + 1;
         end
      end
   end
   %disp([num2str(NumDups), ' duplicates in final population.']);
   %disp([num2str(nLegal), ' legal individuals in final population.']);
   % Display the best solution
   Chrom = sort(Population(1).chrom);

   %disp(['Best chromosome = ', num2str(Chrom)]);

   %{
   % Plot some results
   close all;
   plot([0:OPTIONS.Maxgen], MinCost, 'r');
   xlabel('Generation');
   ylabel('Minimum Cost');
%}
end
return;
```

```
function [OPTIONS, MinCost, AvgCost, InitFunction, CostFunction, FeasibleFunction, ...
   MaxParValue, MinParValue, Population,FeasibleFunction_1,CostFunction_1] = Init(DisplayFlag,
ProblemFunction, RandSeed)
OPTIONS.popsize = 50000;
OPTIONS.Maxgen = 500000;
OPTIONS.numVar = 5;
OPTIONS.pmutate = 0;
if ~exist('RandSeed', 'var')
   RandSeed = round(sum(100*clock));
end
rand('state', RandSeed); % initialize random number generator
if DisplayFlag
   %disp(['random # seed = ', num2str(RandSeed)]);
end
[InitFunction, CostFunction, FeasibleFunction,FeasibleFunction_1,CostFunction_1] = ProblemFunction();
[MaxParValue, MinParValue, Population, OPTIONS] = InitFunction(OPTIONS);
 Population = ClearDups(Population, MaxParValue, MinParValue);
 Population = CostFunction(OPTIONS, Population);
Population = PopSort(Population);
AverageCost = ComputeAveCost(Population);
MinCost = [Population(1).cost];
AvgCost = [AverageCost];
return;
```

```
function [Population, indices] = PopSort(Population)
% Sort the population members from best to worst
popsize = length(Population);
Cost = zeros(1, popsize);
indices = zeros(1, popsize);
for i = 1 : popsize
   Cost(i) = Population(i).cost;
end
[Cost, indices] = sort(Cost, 2, 'ascend');
Chroms = zeros(popsize, length(Population(1).chrom));
for i = 1 : popsize
   Chroms(i, :) = Population(indices(i)).chrom;
end
for i = 1 : popsize
   Population(i).chrom = Chroms(i, :);
   Population(i).cost = Cost(i);
End
```

```
function [InitFunction, CostFunction, FeasibleFunction,FeasibleFunction_1,CostFunction_1] = SphereCont
InitFunction = @SphereInit;
CostFunction = @SphereCost;
CostFunction_1 = @SphereCost_1;
FeasibleFunction = @SphereFeasible;
FeasibleFunction_1 = @SphereFeasible_1;
return
%%%%%%%%%%%%%%%%%%%%%%%%%%%%%%%%%%%%%%%%%%%%%%%%%%%%%%%%%%%%%%%%%%%%%%%%
%%%%%%%%%%%%%%%
function [MaxParValue, MinParValue, Population, OPTIONS] = SphereInit(OPTIONS)
global MinParValue MaxParValue
MinParValue = -5.12;
MaxParValue = 5.12;
% Initialize population
for popindex = 1 : OPTIONS.popsize
   chrom = MinParValue + (MaxParValue - MinParValue) * rand(1,OPTIONS.numVar);
   Population(popindex).chrom = chrom;
end
OPTIONS.OrderDependent = false;
return
%%%%%%%%%%%%%%%%%%%%%%%%%%%%%%%%%%%%%%%%%%%%%%%%%%%%%%%%%%%%%%%%%%%%%%%%
%%%%%%%%%%%%%%%%%%
function [Population] = SphereCost(OPTIONS,Population)
% Compute the cost of each member in Population
global MinParValue MaxParValue
%p = length(Population(1).chrom);
for popindex = 1 : OPTIONS.popsize
   Population(popindex).cost = 0;
   for i = 1 : OPTIONS.numVar
      x = Population(popindex).chrom(i);
      Population(popindex).cost = Population(popindex).cost + x^2;
   end
end
return
%%%%%%%%%%%%%%%%%%%%%%%%%%%%%%%%%%%%%%%%%%%%%%%%%%%%%%%%%%%%%%%%%%%%%%%%
%%%%%%%%%%%%%%%%%%
function [Population_1] = SphereCost_1(OPTIONS,Population)
% Compute the cost of each member in Population
global MinParValue MaxParValue
%p = length(Population(1).chrom);
for popindex = 1 : OPTIONS.popsize
   Population_1(popindex) = 0;
  x=[];
   for i = 1 : OPTIONS.numVar
      x = Population(popindex,i);
      Population_1(popindex) = Population_1(popindex) + x^2;
   end
end
return
%%%%%%%%%%%%%%%%%%%%%%%%%%%%%%%%%%%%%%%%%%%%%%%%%%%%%%%%%%%%%%%%%%%%%%%%
%%%%%%%%%%
function [Population] = SphereFeasible(OPTIONS,Population)
global MinParValue MaxParValue
for i = 1 : OPTIONS.popsize
   for k = 1 : OPTIONS.numVar
      Population(i).chrom(k) = max(Population(i).chrom(k), MinParValue);
      Population(i).chrom(k) = min(Population(i).chrom(k), MaxParValue);
   end
end
return
%%%%%%%%%%%%%%%%%%%%%%%%%%%%%%%%%%%%%%%%%%%%%%%%%%%%%%%%%%%%%%%%%%%%%%%%
%%%%%%%%%
```

```
function [Population] = SphereFeasible_1(OPTIONS,Population)
global MinParValue MaxParValue
for i = 1 : OPTIONS.popsize
   for k = 1 : OPTIONS.numVar
      Population(i,k) = max(Population(i,k), MinParValue);
      Population(i,k) = min(Population(i,k), MaxParValue);
   end
end
return
```

A2.2 ABC Code

```
function ABC_cont(ProblemFunction, DisplayFlag, ProbFlag)
global ll
if ~exist('DisplayFlag', 'var')
   DisplayFlag = true;
end
if ~exist('ProbFlag', 'var')
   ProbFlag = true;
end
[OPTIONS, MinCost, AvgCost, InitFunction, CostFunction, FeasibleFunction, ...
   MaxParValue, MinParValue, Population,FeasibleFunction_1,CostFunction_1] = Init(DisplayFlag,
ProblemFunction);
OPTIONS.pmodify = 1;
OPTIONS.pmutate = 0.05;
Keep = 5; % elitism parameter: how many of the best habitats to keep from one generation to the next
lambdaLower = 0.0; % lower bound for immigration probabilty per gene
lambdaUpper = 1; % upper bound for immigration probabilty per gene
dt = 0.1; % step size used for numerical integration of probabilities
I = 1; % max immigration rate for each island
E = 1; % max emigration rate, for each island
P = OPTIONS.popsize; % max species count, for each island
pmutate=0.05;
for i = 1 : length(Population)
   Prob(i) = 1 / length(Population);
end
for GenIndex = 1 : OPTIONS.Maxgen
   for i = 1 : Keep
      chromKeep(i,:) = Population(i).chrom;
      costKeep(i) = Population(i).cost;
   end
   % Map cost values to species counts.
   [Population] = GetSpeciesCounts(Population, P);
   [lambda, mu] = GetLambdaMu(Population, I, E, P);
   if ProbFlag
      % Compute the time derivative of Prob(i) for each habitat i.
      for i = 1 : length(Population)
         % Compute lambda for one less than the species count of habitat i.
         lambdaMinus = I * (1 - (Population(i).SpeciesCount - 1) / P);
         % Compute mu for one more than the species count of habitat i.
         muPlus = E * (Population(i).SpeciesCount + 1) / P;
         % Compute Prob for one less than and one more than the species count of habitat i.
         % Note that species counts are arranged in an order opposite to that presented in
         % MacArthur and Wilson's book - that is, the most fit
         % habitat has index 1, which has the highest species count.
         if i < length(Population)
            ProbMinus = Prob(i+1);
         else
            ProbMinus = 0;
         end
         if i > 1
            ProbPlus = Prob(i-1);
         else
```

```
            ProbPlus = 0;
          end
          ProbDot(i) = -(lambda(i) + mu(i)) * Prob(i) + lambdaMinus * ProbMinus + muPlus * ProbPlus;
        end
        % Compute the new probabilities for each species count.
        Prob = Prob + ProbDot * dt;
        Prob = max(Prob, 0);
        Prob = Prob / sum(Prob);
    end
    % Now use lambda and mu to decide how much information to share between habitats.
    lambdaMin = min(lambda);
    lambdaMax = max(lambda);
    %for i = 1 : 1 : length(Population)-Keep
        for i = 1 : 1 : length(Population)
        if rand > OPTIONS.pmodify
          continue;
        end
        % Normalize the immigration rate.
        lambdaScale = lambdaLower + (lambdaUpper - lambdaLower) * (lambda(i) - lambdaMin) / (lambdaMax -
lambdaMin);
        % Probabilistically input new information into habitat i
        for j = 1 : OPTIONS.numVar
          if rand < lambdaScale
            % Pick a habitat from which to obtain a feature
            RandomNum = rand * sum(mu);
            Select = mu(1);
            SelectIndex = 1;
            while (RandomNum > Select) & (SelectIndex < OPTIONS.popsize)
              SelectIndex = SelectIndex + 1;
              Select = Select + mu(SelectIndex);
            end
            Island(i,j) = Population(SelectIndex).chrom(j);
          else
            Island(i,j) = Population(i).chrom(j);
          end
        end
        end
        for i = 1 : 1 : length(Population)
          for j = 1 : OPTIONS.numVar
            Population_old(i,j) = Population(i).chrom(j);
          end
        end
  if ProbFlag
        for i = 1 : 1 : length(Population)
          ii=ceil(rand*(length(Population)));
          if ii==i
            ii=ceil(rand*(length(Population)));
          end
          Island_1(i,:) = (Population(i).chrom+(-1+rand*(2))*(Population(i).chrom-Population(ii).chrom));
        end
    end
     Island_1 = FeasibleFunction_1(OPTIONS, Island_1);
     Island_1_cost = CostFunction_1(OPTIONS, Island_1);
    for i = 1 : length(Population)
        if Island_1_cost(i)<Population(i).cost
          Population(i).chrom =Island_1(i,:);
          Population(i).cost=Island_1_cost(i);%%%%%%%
        end
    end
       for i=1:length(Population)
            xx(i)=1/(Population(i).cost+1);
        end
        for i=1:length(Population)
            fit(i)=xx(i)/sum(xx);
```

```
    end
  y=zeros(1,length(Population));
    for i=1:length(Population)
     while y(i)==0
         r=rand;
       for j=1:length(Population)
         if r<fit(j)
            Island(i,:)=Population(i).chrom;
            y(i)=1;
            break;
         end
       end
       end
    end
  for i = 1 : 1 : length(Population)
      ii=ceil(rand*(length(Population)));
     if ii==i
        ii=ceil(rand*(length(Population)));
     end
     for j = 1 : OPTIONS.numVar
      Island_1(i,j) = (Island(i,j)+(-1+rand*(2))*(Island(i,j)-Population(ii).chrom(j)));
           end
   end
 end
   Island_1 = FeasibleFunction_1(OPTIONS, Island_1);
   Island_1_cost = CostFunction_1(OPTIONS, Island_1);

   for i = 1 : length(Population)
    if Island_1_cost(i)<Population(i).cost
       Population(i).chrom =Island_1(i,:);
       Population(i).cost=Island_1_cost(i);%%%%%%
    end
  end
  Population = FeasibleFunction(OPTIONS, Population);
  Population = CostFunction(OPTIONS, Population);
  Population = PopSort(Population);
  n = length(Population);
  for i = 1 : Keep
     Population(n-i+1).chrom = chromKeep(i,:);
     Population(n-i+1).cost = costKeep(i);
  end
  Population = ClearDups(Population, MaxParValue, MinParValue);
  Population = PopSort(Population);
  [AverageCost, nLegal] = ComputeAveCost(Population);
  MinCost = [MinCost Population(1).cost];
  AvgCost = [AvgCost AverageCost];
  if DisplayFlag
     %disp([num2str(MinCost(end))]);
  end
end
disp([num2str(MinCost(end))]);
Conclude(DisplayFlag, OPTIONS, Population, nLegal, MinCost);
%%%%%%%%%%%%%%%%%%%%%%%%%%%%%%%%%%%%%%%%%%%%%%
%%%%%%%%%%%%
function [Population] = GetSpeciesCounts(Population, P)
for i = 1 : length(Population)
   if Population(i).cost < inf
      Population(i).SpeciesCount = P - i;
   else
      Population(i).SpeciesCount = 0;
   end
end
return;
```

```
%%%%%%%%%%%%%%%%%%%%%%%%%%%%%%%%%%%%%%%%%%%%%%%%%%%%%%%%%%%%%%%%%%%%%%%%%%%%%%%%%%%%%%%
%%%%%%%%%%%%%%%
function [lambda, mu] = GetLambdaMu(Population, I, E, P)
for i = 1 : length(Population)
   lambda(i) = I * (1 - Population(i).SpeciesCount / P);
   mu(i) = E * Population(i).SpeciesCount / P;
end
return;
----------------------------------------------------------------------------------------------------------------------------------------
----
function [Population] = ClearDups(Population, MaxParValue, MinParValue)
global ll
% Make sure there are no duplicate individuals in the population.
% This logic does not make 100% sure that no duplicates exist, but any duplicates that are found are
% randomly mutated, so there should be a good chance that there are no duplicates after this procedure.
for i = 1 : length(Population)
   Chrom1 = sort(Population(i).chrom);
   for j = i+1 : length(Population)
      Chrom2 = sort(Population(j).chrom);
      if isequal(Chrom1, Chrom2)
         %parnum = ceil(length(Population(j).chrom) * rand);
         parnum = floor(1+(length(Population(j).chrom)-1) * rand);
         %Population(j).chrom(parnum) = floor(MinParValue + (MaxParValue - MinParValue ) * rand);
          Population(j).chrom(parnum) = (ll(parnum) + (MaxParValue(parnum) - ll(parnum)) * rand);
          %Population(j).chrom(parnum) = (MinParValue + (MaxParValue - MinParValue ) * rand);
          %Population(j).chrom(parnum)   =   (Population(j).chrom(parnum)   +   (MaxParValue(parnum)   -
Population(j).chrom(parnum)) * rand/2);
         end
   end
end
return;
----------------------------------------------------------------------------------------------------------------------------------------
function [AveCost, nLegal] = ComputeAveCost(Population)
% Compute the average cost of all legal individuals in the population.
% OUTPUTS: AveCost = average cost
% nLegal = number of legal individuals in population
% Save valid population member fitnesses in temporary array
Cost = [];
nLegal = 0;
for i = 1 : length(Population)
   if Population(i).cost < inf
      Cost = [Cost Population(i).cost];
      nLegal = nLegal + 1;
   end
end
% Compute average cost.
AveCost = mean(Cost);
return;
----------------------------------------------------------------------------------------------------------------------------------------
function Conclude(DisplayFlag, OPTIONS, Population, nLegal, MinCost)
% Output results of population-based optimization algorithm.
if DisplayFlag
   % Count the number of duplicates
   NumDups = 0;
   for i = 1 : OPTIONS.popsize
      Chrom1 = sort(Population(i).chrom);
      for j = i+1 : OPTIONS.popsize
         Chrom2 = sort(Population(j).chrom);
         if isequal(Chrom1, Chrom2)
            NumDups = NumDups + 1;
         end
      end
```

```
  end
  %disp([num2str(NumDups), ' duplicates in final population.']);
  %disp([num2str(nLegal), ' legal individuals in final population.']);
  % Display the best solution
  Chrom = sort(Population(1).chrom);
  %disp(['Best chromosome = ', num2str(Chrom)]);
  %{
  % Plot some results
  close all;
  plot([0:OPTIONS.Maxgen], MinCost, 'r');
  xlabel('Generation');
  ylabel('Minimum Cost');
%}
end
return;
-------------------------------------------------------------------------------------------------------------------------------

function [OPTIONS, MinCost, AvgCost, InitFunction, CostFunction, FeasibleFunction, ...
   MaxParValue,    MinParValue,    Population,FeasibleFunction_1,CostFunction_1]    =    Init(DisplayFlag,
ProblemFunction, RandSeed)
OPTIONS.popsize = 50000;
OPTIONS.Maxgen = 500000;
OPTIONS.numVar = 5;
OPTIONS.pmutate = 0;
if ~exist('RandSeed', 'var')
   RandSeed = round(sum(100*clock));
end
rand('state', RandSeed); % initialize random number generator
if DisplayFlag
   %disp(['random # seed = ', num2str(RandSeed)]);
end
[InitFunction, CostFunction, FeasibleFunction,FeasibleFunction_1,CostFunction_1] = ProblemFunction();
[MaxParValue, MinParValue, Population, OPTIONS] = InitFunction(OPTIONS);
 Population = ClearDups(Population, MaxParValue, MinParValue);
 Population = CostFunction(OPTIONS, Population);
Population = PopSort(Population);
AverageCost = ComputeAveCost(Population);
MinCost = [Population(1).cost];
AvgCost = [AverageCost];
return;
-------------------------------------------------------------------------------------------------------------------------------

function [Population, indices] = PopSort(Population)
% Sort the population members from best to worst
popsize = length(Population);
Cost = zeros(1, popsize);
indices = zeros(1, popsize);
for i = 1 : popsize
   Cost(i) = Population(i).cost;
end
[Cost, indices] = sort(Cost, 2, 'ascend');
Chroms = zeros(popsize, length(Population(1).chrom));
for i = 1 : popsize
   Chroms(i, :) = Population(indices(i)).chrom;
end
for i = 1 : popsize
   Population(i).chrom = Chroms(i, :);
   Population(i).cost = Cost(i);
End
-------------------------------------------------------------------------------------------------------------------------------

function [InitFunction, CostFunction, FeasibleFunction,FeasibleFunction_1,CostFunction_1] = SphereCont
InitFunction = @SphereInit;
CostFunction = @SphereCost;
```

```
CostFunction_1 = @SphereCost_1;
FeasibleFunction = @SphereFeasible;
FeasibleFunction_1 = @SphereFeasible_1;
return
%%%%%%%%%%%%%%%%%%%%%%%%%%%%%%%%%%%%%%%%%%%%%%%%%%%
%%%%%%%%%%%%
function [MaxParValue, MinParValue, Population, OPTIONS] = SphereInit(OPTIONS)
global MinParValue MaxParValue
MinParValue = -5.12;
MaxParValue = 5.12;
% Initialize population
for popindex = 1 : OPTIONS.popsize
  chrom = MinParValue + (MaxParValue - MinParValue) * rand(1,OPTIONS.numVar);
  Population(popindex).chrom = chrom;
end
OPTIONS.OrderDependent = false;
return
%%%%%%%%%%%%%%%%%%%%%%%%%%%%%%%%%%%%%%%%%%%%%%%%%%%
%%%%%%%%%%%%%%%
function [Population] = SphereCost(OPTIONS,Population)
% Compute the cost of each member in Population
global MinParValue MaxParValue
%p = length(Population(1).chrom);
for popindex = 1 : OPTIONS.popsize
  Population(popindex).cost = 0;
  for i = 1 : OPTIONS.numVar
    x = Population(popindex).chrom(i);
    Population(popindex).cost = Population(popindex).cost + x^2;
  end
end
return
%%%%%%%%%%%%%%%%%%%%%%%%%%%%%%%%%%%%%%%%%%%%%%%%%%%
%%%%%%%%%%%%%%%
function [Population_1] = SphereCost_1(OPTIONS,Population)
% Compute the cost of each member in Population
global MinParValue MaxParValue
%p = length(Population(1).chrom);
for popindex = 1 : OPTIONS.popsize
  Population_1(popindex) = 0;
 x=[];
  for i = 1 : OPTIONS.numVar
    x = Population(popindex,i);
    Population_1(popindex) = Population_1(popindex) + x^2;
  end
end
return
%%%%%%%%%%%%%%%%%%%%%%%%%%%%%%%%%%%%%%%%%%%%%%%%%%%
%%%%%%%
function [Population] = SphereFeasible(OPTIONS,Population)
global MinParValue MaxParValue
for i = 1 : OPTIONS.popsize
  for k = 1 : OPTIONS.numVar
    Population(i).chrom(k) = max(Population(i).chrom(k), MinParValue);
    Population(i).chrom(k) = min(Population(i).chrom(k), MaxParValue);
  end
end
return
%%%%%%%%%%%%%%%%%%%%%%%%%%%%%%%%%%%%%%%%%%%%%%%%%%%
%%%%%%%
function [Population] = SphereFeasible_1(OPTIONS,Population)
global MinParValue MaxParValue
for i = 1 : OPTIONS.popsize
  for k = 1 : OPTIONS.numVar
    Population(i,k) = max(Population(i,k), MinParValue);
```

```
    Population(i,k) = min(Population(i,k), MaxParValue);
  end
end
return
```

A2.3 BBO Code

```
%function [MinCost] =BBO_cont(ProblemFunction, DisplayFlag, ProbFlag, RandSeed)
function BBO_cont(ProblemFunction, DisplayFlag, ProbFlag, RandSeed)
if ~exist('DisplayFlag', 'var')
  DisplayFlag = true;
end
if ~exist('ProbFlag', 'var')
  ProbFlag = false;
end
if ~exist('RandSeed', 'var')
  RandSeed = round(sum(100*clock));
end
[OPTIONS, MinCost, AvgCost, InitFunction, CostFunction, FeasibleFunction, ...
  MaxParValue, MinParValue, Population] = Init(DisplayFlag, ProblemFunction, RandSeed);
Population = CostFunction(OPTIONS, Population);
OPTIONS.pmodify = 1; % habitat modification probability
OPTIONS.pmutate = 0.005; % initial mutation probability
Keep = 2; % elitism parameter: how many of the best habitats to keep from one generation to the next
lambdaLower = 0.0; % lower bound for immigration probabilty per gene
lambdaUpper = 1; % upper bound for immigration probabilty per gene
dt = 1; % step size used for numerical integration of probabilities
I = 1; % max immigration rate for each island
E = 1; % max emigration rate, for each island
P = OPTIONS.popsize; % max species count, for each island
% Initialize the species count probability of each habitat
% Later we might want to initialize probabilities based on cost
for j = 1 : length(Population)
  Prob(j) = 1 / length(Population);
end
% Begin the optimization loop
for GenIndex = 1 : OPTIONS.Maxgen
  % Save the best habitats in a temporary array.
  for j = 1 : Keep
    chromKeep(j,:) = Population(j).chrom;
    costKeep(j) = Population(j).cost;
  end
  % Map cost values to species counts.
  [Population] = GetSpeciesCounts(Population, P);
  % Compute immigration rate and emigration rate for each species count.
  % lambda(i) is the immigration rate for habitat i.
  % mu(i) is the emigration rate for habitat i.
  [lambda, mu] = GetLambdaMu(Population, I, E, P);
  if ProbFlag
    % Compute the time derivative of Prob(i) for each habitat i.
    for j = 1 : length(Population)
      % Compute lambda for one less than the species count of habitat i.
      lambdaMinus = I * (1 - (Population(j).SpeciesCount - 1) / P);
      % Compute mu for one more than the species count of habitat i.
      muPlus = E * (Population(j).SpeciesCount + 1) / P;
      % Compute Prob for one less than and one more than the species count of habitat i.
      % Note that species counts are arranged in an order opposite to that presented in
      % MacArthur and Wilson's book - that is, the most fit
      % habitat has index 1, which has the highest species count.
      if j < length(Population)
        ProbMinus = Prob(j+1);
      else
        ProbMinus = 0;
```

```
        end
        if j > 1
            ProbPlus = Prob(j-1);
        else
            ProbPlus = 0;
        end
        ProbDot(j) = -(lambda(j) + mu(j)) * Prob(j) + lambdaMinus * ProbMinus + muPlus * ProbPlus;
    end
    % Compute the new probabilities for each species count.
    Prob = Prob + ProbDot * dt;
    Prob = max(Prob, 0);
    Prob = Prob / sum(Prob);
  end
  % Now use lambda and mu to decide how much information to share between habitats.
  lambdaMin = min(lambda);
  lambdaMax = max(lambda);
  for k = 1 : length(Population)
    if rand > OPTIONS.pmodify
        continue;
    end
    % Normalize the immigration rate.
    lambdaScale = lambdaLower + (lambdaUpper - lambdaLower) * (lambda(k) - lambdaMin) / (lambdaMax -
lambdaMin);
    % Probabilistically input new information into habitat i
    for j = 1 : OPTIONS.numVar
        if rand < lambdaScale
            % Pick a habitat from which to obtain a feature
            RandomNum = rand * sum(mu);
            Select = mu(1);
            SelectIndex = 1;
            while (RandomNum > Select) & (SelectIndex < OPTIONS.popsize)
                SelectIndex = SelectIndex + 1;
                Select = Select + mu(SelectIndex);
            end
            Island(k,j) = Population(SelectIndex).chrom(j);
        else
            Island(k,j) = Population(k).chrom(j);
        end
    end
  end
  if ProbFlag
    % Mutation
    Pmax = max(Prob);
    MutationRate = OPTIONS.pmutate * (1 - Prob / Pmax);
    % Mutate only the worst half of the solutions
    Population = PopSort(Population);
    for k = round(length(Population)/2) : length(Population)
        for parnum = 1 : OPTIONS.numVar
            if MutationRate(k) > rand
                Island(k,parnum) = (MinParValue + (MaxParValue - MinParValue) * rand);
            end
        end
    end
  end
  % Replace the habitats with their new versions.
  for k = 1 : length(Population)
    Population(k).chrom = Island(k,:);
  end
  % Make sure each individual is legal.
  Population = FeasibleFunction(OPTIONS, Population);
  % Calculate cost
  Population = CostFunction(OPTIONS, Population);
  % Sort from best to worst
  Population = PopSort(Population);
```

```
  % Replace the worst with the previous generation's elites.
  n = length(Population);
  for k = 1 : Keep
     Population(n-k+1).chrom = chromKeep(k,:);
     Population(n-k+1).cost = costKeep(k);
  end
  % Make sure the population does not have duplicates.
  Population = ClearDups(Population, MaxParValue, MinParValue);
  % Sort from best to worst
  Population = PopSort(Population);
  % Compute the average cost
  [AverageCost, nLegal] = ComputeAveCost(Population);
  % Display info to screen
  MinCost = [MinCost Population(1).cost];
  AvgCost = [AvgCost AverageCost];
  if DisplayFlag
     %disp([num2str(MinCost(end))]);
  end
end
disp([num2str(MinCost(end))]);
Conclude(DisplayFlag, OPTIONS, Population, nLegal, MinCost);
function [Population] = GetSpeciesCounts(Population, P)
for i = 1 : length(Population)
   if Population(i).cost < inf
      Population(i).SpeciesCount = P - i;
   else
      Population(i).SpeciesCount = 0;
   end
end
return;
%%%%%%%%%%%%%%%%%%%%%%%%%%%%%%%%%%%%%%%%%%%%%%%%%%%%%%%%%%%%
%%%%%%%%%%%
function [lambda, mu] = GetLambdaMu(Population, I, E, P)
% Compute immigration rate and extinction rate for each species count.
% lambda(i) is the immigration rate for individual i.
% mu(i) is the extinction rate for individual i.
for i = 1 : length(Population)
   lambda(i) = I * (1 - Population(i).SpeciesCount / P);
   mu(i) = E * Population(i).SpeciesCount / P;
end
return;
------------------------------------------------------------------------------------------------------------------------

function [Population] = ClearDups(Population, MaxParValue, MinParValue)
global ll
% Make sure there are no duplicate individuals in the population.
% This logic does not make 100% sure that no duplicates exist, but any duplicates that are found are
% randomly mutated, so there should be a good chance that there are no duplicates after this procedure.
for i = 1 : length(Population)
   Chrom1 = sort(Population(i).chrom);
   for j = i+1 : length(Population)
      Chrom2 = sort(Population(j).chrom);
      if isequal(Chrom1, Chrom2)
         %parnum = ceil(length(Population(j).chrom) * rand);
         parnum = floor(1+(length(Population(j).chrom)-1) * rand);
         %Population(j).chrom(parnum) = floor(MinParValue + (MaxParValue - MinParValue ) * rand);
          Population(j).chrom(parnum) = (ll(parnum) + (MaxParValue(parnum) - ll(parnum)) * rand);
          %Population(j).chrom(parnum) = (MinParValue + (MaxParValue - MinParValue ) * rand);
                 %Population(j).chrom(parnum) = (Population(j).chrom(parnum) + (MaxParValue(parnum) -
Population(j).chrom(parnum)) * rand/2);
             end
   end
end
return;
------------------------------------------------------------------------------------------------------------------------
```

```
function [AveCost, nLegal] = ComputeAveCost(Population)
% Compute the average cost of all legal individuals in the population.
% OUTPUTS: AveCost = average cost
%         nLegal = number of legal individuals in population
% Save valid population member fitnesses in temporary array
Cost = [];
nLegal = 0;
for i = 1 : length(Population)
   if Population(i).cost < inf
      Cost = [Cost Population(i).cost];
      nLegal = nLegal + 1;
   end
end
% Compute average cost.
AveCost = mean(Cost);
return;
-----------------------------------------------------------------------------------------------------------------
function Conclude(DisplayFlag, OPTIONS, Population, nLegal, MinCost)
% Output results of population-based optimization algorithm.
if DisplayFlag
   % Count the number of duplicates
   NumDups = 0;
   for i = 1 : OPTIONS.popsize
      Chrom1 = sort(Population(i).chrom);
      for j = i+1 : OPTIONS.popsize
         Chrom2 = sort(Population(j).chrom);
         if isequal(Chrom1, Chrom2)
            NumDups = NumDups + 1;
         end
      end
   end
  %disp([num2str(NumDups), ' duplicates in final population.']);
   %disp([num2str(nLegal), ' legal individuals in final population.']);
   % Display the best solution
   Chrom = sort(Population(1).chrom);
  %disp(['Best chromosome = ', num2str(Chrom)]);
   %{
   % Plot some results
   close all;
   plot([0:OPTIONS.Maxgen], MinCost, 'r');
   xlabel('Generation');
   ylabel('Minimum Cost');
%}
end
return;
-----------------------------------------------------------------------------------------------------------------
function [OPTIONS, MinCost, AvgCost, InitFunction, CostFunction, FeasibleFunction, ...
   MaxParValue, MinParValue, Population,FeasibleFunction_1,CostFunction_1] = Init(DisplayFlag,
ProblemFunction, RandSeed)
OPTIONS.popsize = 50000;
OPTIONS.Maxgen = 500000;
OPTIONS.numVar = 5;
OPTIONS.pmutate = 0;
if ~exist('RandSeed', 'var')
   RandSeed = round(sum(100*clock));
end
rand('state', RandSeed); % initialize random number generator
if DisplayFlag
   %disp(['random # seed = ', num2str(RandSeed)]);
end
```

```
[InitFunction, CostFunction, FeasibleFunction,FeasibleFunction_1,CostFunction_1] = ProblemFunction();
[MaxParValue, MinParValue, Population, OPTIONS] = InitFunction(OPTIONS);
 Population = ClearDups(Population, MaxParValue, MinParValue);
 Population = CostFunction(OPTIONS, Population);
Population = PopSort(Population);
AverageCost = ComputeAveCost(Population);
MinCost = [Population(1).cost];
AvgCost = [AverageCost];
return;
--------------------------------------------------------------------------------------------------------------------

function [Population, indices] = PopSort(Population)
% Sort the population members from best to worst
popsize = length(Population);
Cost = zeros(1, popsize);
indices = zeros(1, popsize);
for i = 1 : popsize
   Cost(i) = Population(i).cost;
end
[Cost, indices] = sort(Cost, 2, 'ascend');
Chroms = zeros(popsize, length(Population(1).chrom));
for i = 1 : popsize
   Chroms(i, :) = Population(indices(i)).chrom;
end
for i = 1 : popsize
   Population(i).chrom = Chroms(i, :);
   Population(i).cost = Cost(i);
End
--------------------------------------------------------------------------------------------------------------------

function [InitFunction, CostFunction, FeasibleFunction,FeasibleFunction_1,CostFunction_1] = SphereCont
InitFunction = @SphereInit;
CostFunction = @SphereCost;
CostFunction_1 = @SphereCost_1;
FeasibleFunction = @SphereFeasible;
FeasibleFunction_1 = @SphereFeasible_1;
return
%%%%%%%%%%%%%%%%%%%%%%%%%%%%%%%%%%%%%%%%%%%%%%%%%%%%%%%%%%%%%%%%
%%%%%%%%%%%%%%
function [MaxParValue, MinParValue, Population, OPTIONS] = SphereInit(OPTIONS)
global MinParValue MaxParValue
MinParValue = -5.12;
MaxParValue = 5.12;
% Initialize population
for popindex = 1 : OPTIONS.popsize
   chrom = MinParValue + (MaxParValue - MinParValue) * rand(1,OPTIONS.numVar);
   Population(popindex).chrom = chrom;
end
OPTIONS.OrderDependent = false;
return
%%%%%%%%%%%%%%%%%%%%%%%%%%%%%%%%%%%%%%%%%%%%%%%%%%%%%%%%%%%%%%%%
%%%%%%%%%%%%%%%%
function [Population] = SphereCost(OPTIONS,Population)
% Compute the cost of each member in Population
global MinParValue MaxParValue
%p = length(Population(1).chrom);
for popindex = 1 : OPTIONS.popsize
   Population(popindex).cost = 0;
   for i = 1 : OPTIONS.numVar
      x = Population(popindex).chrom(i);
      Population(popindex).cost = Population(popindex).cost + x^2;
   end
end
return
```

```
%%%%%%%%%%%%%%%%%%%%%%%%%%%%%%%%%%%%%%%%%%%%%%%%%%%%%%%%%%%%
%%%%%%%%%%%%%%%%%%
function [Population_1] = SphereCost_1(OPTIONS,Population)
% Compute the cost of each member in Population
global MinParValue MaxParValue
%p = length(Population(1).chrom);
for popindex = 1 : OPTIONS.popsize
  Population_1(popindex) = 0;
  x=[];
  for i = 1 : OPTIONS.numVar
    x = Population(popindex,i);
    Population_1(popindex) = Population_1(popindex) + x^2;
  end
end
return
%%%%%%%%%%%%%%%%%%%%%%%%%%%%%%%%%%%%%%%%%%%%%%%%%%%%%%%%%%%%
%%%%%%%%%
function [Population] = SphereFeasible(OPTIONS,Population)
global MinParValue MaxParValue
for i = 1 : OPTIONS.popsize
  for k = 1 : OPTIONS.numVar
    Population(i).chrom(k) = max(Population(i).chrom(k), MinParValue);
    Population(i).chrom(k) = min(Population(i).chrom(k), MaxParValue);
  end
end
return
%%%%%%%%%%%%%%%%%%%%%%%%%%%%%%%%%%%%%%%%%%%%%%%%%%%%%%%%%%%%
%%%%%%%%%
function [Population] = SphereFeasible_1(OPTIONS,Population)
global MinParValue MaxParValue
for i = 1 : OPTIONS.popsize
  for k = 1 : OPTIONS.numVar
    Population(i,k) = max(Population(i,k), MinParValue);
    Population(i,k) = min(Population(i,k), MaxParValue);
  end
end
return
```

A2.1 PSO Code

```
function PSO(ProblemFunction, DisplayFlag)
if ~exist('DisplayFlag', 'var')
  DisplayFlag = true;
end
[OPTIONS, MinCost, AvgCost, InitFunction, CostFunction, FeasibleFunction, ...
  MaxParValue, MinParValue, Population] = Init(DisplayFlag, ProblemFunction);
OPTIONS.Keep = 2; % elitism parameter: how many of the best particles to keep from one iteration to the next
OPTIONS.neighbors = 0; % size of particle swarm neighborhood
OPTIONS.w = 0.4; % inertial constant
OPTIONS.c1 = 2; % cognitive constant
OPTIONS.c2 = 2; % social constant for swarm interaction
OPTIONS.c3 = 0; % social constant for neighborhood interaction
vel = zeros(OPTIONS.popsize, OPTIONS.numVar); % velocities
pbest = Population; % personal best of each particle
nbest = Population; % neighborhood best of each particle
gbest = Population(1); % global best

for GenIndex = 1 : OPTIONS.Maxgen
  if ~OPTIONS.OrderDependent
    for i = 1 : OPTIONS.popsize
      [chrom, indices] = sort(Population(i).chrom);
      Population(i).chrom = chrom;
      VelTemp = vel(i, :);
      for j = 1 : OPTIONS.numVar
```

```
            vel(i, j) = VelTemp(indices(j));
         end
      end
   end
   % Update the global best if needed
   if Population(1).cost < gbest.cost
      gbest = Population(1);
   end
   % Update personal best and neighborhood best for each particle
   for i = 1 : OPTIONS.popsize
      % Update each personal best if needed
      if Population(i).cost < pbest(i).cost
         pbest(i) = Population(i);
      end
      % Update each neighborhood best if needed
      Distance = zeros(OPTIONS.popsize, 1);
      for j = 1 : OPTIONS.popsize
         Distance(j) = norm(Population(i).chrom-Population(j).chrom);
      end
      [Distance, indices] = sort(Distance);
      nbest(i).cost = inf;
      for j = 2 : OPTIONS.neighbors+1
         nindex = indices(j);
         if Population(nindex).cost < nbest(i).cost
            nbest(i) = Population(nindex);
         end
      end
   end
   % Update the position and velocity of each particle (except the elites)
   for i = OPTIONS.Keep+1 : OPTIONS.popsize
      r = rand(3, OPTIONS.numVar);
      x = Population(i).chrom;
      deltaVpersonal = OPTIONS.c1 * r(1,:) .* (pbest(i).chrom - x);
      deltaVswarm = OPTIONS.c2 * r(2,:) .* (gbest.chrom - x);
      deltaVneighborhood = OPTIONS.c3 * r(3,:) .* (nbest(i).chrom - x);
      vel(i,:) = OPTIONS.w * vel(i,:) + deltaVpersonal + deltaVswarm + deltaVneighborhood;
      Population(i).chrom = (x + vel(i,:));%%%% chane for interger value
   end
   Population = ClearDups(Population, MaxParValue, MinParValue);
      Population = FeasibleFunction(OPTIONS, Population);
      Population = CostFunction(OPTIONS, Population);
     Population = PopSort(Population);
      [AverageCost, nLegal] = ComputeAveCost(Population);
     MinCost = [MinCost Population(1).cost];
   AvgCost = [AvgCost AverageCost];
end
disp([num2str(MinCost(end))]);
Conclude(DisplayFlag, OPTIONS, Population, nLegal, MinCost);
return;
```

```
function [Population] = ClearDups(Population, MaxParValue, MinParValue)
global ll
% Make sure there are no duplicate individuals in the population.
% This logic does not make 100% sure that no duplicates exist, but any duplicates that are found are
% randomly mutated, so there should be a good chance that there are no duplicates after this procedure.
for i = 1 : length(Population)
   Chrom1 = sort(Population(i).chrom);
   for j = i+1 : length(Population)
      Chrom2 = sort(Population(j).chrom);
      if isequal(Chrom1, Chrom2)
         %parnum = ceil(length(Population(j).chrom) * rand);
         parnum = floor(1+(length(Population(j).chrom)-1) * rand);
         %Population(j).chrom(parnum) = floor(MinParValue + (MaxParValue - MinParValue ) * rand);
```

```
        Population(j).chrom(parnum) = (ll(parnum) + (MaxParValue(parnum) - ll(parnum)) * rand);
        %Population(j).chrom(parnum) = (MinParValue + (MaxParValue - MinParValue ) * rand);
        %Population(j).chrom(parnum)  =  (Population(j).chrom(parnum)  +  (MaxParValue(parnum)  -
Population(j).chrom(parnum)) * rand/2);
            end
    end
end
return;
---------------------------------------------------------------------------------------------------------------------------------
function [AveCost, nLegal] = ComputeAveCost(Population)
% Compute the average cost of all legal individuals in the population.
% OUTPUTS: AveCost = average cost
%        nLegal = number of legal individuals in population
% Save valid population member fitnesses in temporary array
Cost = [];
nLegal = 0;
for i = 1 : length(Population)
    if Population(i).cost < inf
        Cost = [Cost Population(i).cost];
        nLegal = nLegal + 1;
    end
end
% Compute average cost.
AveCost = mean(Cost);
return;
---------------------------------------------------------------------------------------------------------------------------------
function Conclude(DisplayFlag, OPTIONS, Population, nLegal, MinCost)
% Output results of population-based optimization algorithm.
if DisplayFlag
    % Count the number of duplicates
    NumDups = 0;
    for i = 1 : OPTIONS.popsize
        Chrom1 = sort(Population(i).chrom);
        for j = i+1 : OPTIONS.popsize
            Chrom2 = sort(Population(j).chrom);
            if isequal(Chrom1, Chrom2)
                NumDups = NumDups + 1;
            end
        end
    end
        %disp([num2str(NumDups), ' duplicates in final population.']);
    %disp([num2str(nLegal), ' legal individuals in final population.']);
    % Display the best solution
    Chrom = sort(Population(1).chrom);
        %disp(['Best chromosome = ', num2str(Chrom)]);
        %{
    % Plot some results
    close all;
    plot([0:OPTIONS.Maxgen], MinCost, 'r');
    xlabel('Generation');
    ylabel('Minimum Cost');
%}
end
return;
---------------------------------------------------------------------------------------------------------------------------------
function [OPTIONS, MinCost, AvgCost, InitFunction, CostFunction, FeasibleFunction, ...
    MaxParValue,    MinParValue,    Population,FeasibleFunction_1,CostFunction_1]   =   Init(DisplayFlag,
ProblemFunction, RandSeed)
OPTIONS.popsize = 50000;
OPTIONS.Maxgen = 500000;
OPTIONS.numVar = 5;
```

```
OPTIONS.pmutate = 0;

if ~exist('RandSeed', 'var')
   RandSeed = round(sum(100*clock));
end
rand('state', RandSeed); % initialize random number generator
if DisplayFlag
   %disp(['random # seed = ', num2str(RandSeed)]);
end
[InitFunction, CostFunction, FeasibleFunction,FeasibleFunction_1,CostFunction_1] = ProblemFunction();
[MaxParValue, MinParValue, Population, OPTIONS] = InitFunction(OPTIONS);
 Population = ClearDups(Population, MaxParValue, MinParValue);
 Population = CostFunction(OPTIONS, Population);
Population = PopSort(Population);
AverageCost = ComputeAveCost(Population);
MinCost = [Population(1).cost];
AvgCost = [AverageCost];
return;
-------------------------------------------------------------------------------------------------------------------------------
function [Population, indices] = PopSort(Population)
% Sort the population members from best to worst
popsize = length(Population);
Cost = zeros(1, popsize);
indices = zeros(1, popsize);
for i = 1 : popsize
   Cost(i) = Population(i).cost;
end
[Cost, indices] = sort(Cost, 2, 'ascend');
Chroms = zeros(popsize, length(Population(1).chrom));
for i = 1 : popsize
   Chroms(i, :) = Population(indices(i)).chrom;
end
for i = 1 : popsize
   Population(i).chrom = Chroms(i, :);
   Population(i).cost = Cost(i);
End
-------------------------------------------------------------------------------------------------------------------------------
function [InitFunction, CostFunction, FeasibleFunction,FeasibleFunction_1,CostFunction_1] = SphereCont
InitFunction = @SphereInit;
CostFunction = @SphereCost;
CostFunction_1 = @SphereCost_1;
FeasibleFunction = @SphereFeasible;
FeasibleFunction_1 = @SphereFeasible_1;
return
%%%%%%%%%%%%%%%%%%%%%%%%%%%%%%%%%%%%%%%%%%%%%%%%%%%%%%%%%%%%%%
%%%%%%%%%%%%%
function [MaxParValue, MinParValue, Population, OPTIONS] = SphereInit(OPTIONS)
global MinParValue MaxParValue
MinParValue = -5.12;
MaxParValue = 5.12;
% Initialize population
for popindex = 1 : OPTIONS.popsize
   chrom = MinParValue + (MaxParValue - MinParValue) * rand(1,OPTIONS.numVar);
   Population(popindex).chrom = chrom;
end
OPTIONS.OrderDependent = false;
return
%%%%%%%%%%%%%%%%%%%%%%%%%%%%%%%%%%%%%%%%%%%%%%%%%%%%%%%%%%%%%%
%%%%%%%%%%%%%%%
function [Population] = SphereCost(OPTIONS,Population)
% Compute the cost of each member in Population
global MinParValue MaxParValue
```

```
%p = length(Population(1).chrom);
for popindex = 1 : OPTIONS.popsize
   Population(popindex).cost = 0;
   for i = 1 : OPTIONS.numVar
      x = Population(popindex).chrom(i);
      Population(popindex).cost = Population(popindex).cost + x^2;
   end
end
return
%%%%%%%%%%%%%%%%%%%%%%%%%%%%%%%%%%%%%%%%%%%%%%%%%%%%%%%%%%
%%%%%%%%%%%%%%%%
function [Population_1] = SphereCost_1(OPTIONS,Population)
% Compute the cost of each member in Population
global MinParValue MaxParValue
%p = length(Population(1).chrom);
for popindex = 1 : OPTIONS.popsize
   Population_1(popindex) = 0;
  x=[];
   for i = 1 : OPTIONS.numVar
      x = Population(popindex,i);
      Population_1(popindex) = Population_1(popindex) + x^2;
   end
end
return
%%%%%%%%%%%%%%%%%%%%%%%%%%%%%%%%%%%%%%%%%%%%%%%%%%%%%%%%%%
%%%%%%%
function [Population] = SphereFeasible(OPTIONS,Population)
global MinParValue MaxParValue
for i = 1 : OPTIONS.popsize
   for k = 1 : OPTIONS.numVar
      Population(i).chrom(k) = max(Population(i).chrom(k), MinParValue);
      Population(i).chrom(k) = min(Population(i).chrom(k), MaxParValue);
   end
end
return
%%%%%%%%%%%%%%%%%%%%%%%%%%%%%%%%%%%%%%%%%%%%%%%%%%%%%%%%%%
%%%%%%%%
function [Population] = SphereFeasible_1(OPTIONS,Population)
global MinParValue MaxParValue
for i = 1 : OPTIONS.popsize
   for k = 1 : OPTIONS.numVar
      Population(i,k) = max(Population(i,k), MinParValue);
      Population(i,k) = min(Population(i,k), MaxParValue);
   end
end
return
```

Authors' Biographies

Dr. R. VenkataRao is a Professor in the Department of Mechanical Engineering, S.V. National Institute of Technology, Surat, India. He has more than 20 years of teaching and research experience. He was deputed twice on deputation to the Asian Institute of Technology, Bangkok, Thailand as a visiting Professor. He gained his B.Tech in 1988, M.Tech in 1991, and Ph.D. in 2002. Dr. Rao's research interests include: CAD/CAM, CIMS, advanced optimization methods, and fuzzy multiple attribute decision making methods. He has published more than 200 research papers in national and international journals and conference proceedings and received national and international awards for best research work. He has been a reviewer to many national and international journals and on the editorial boards of few International journals. He had already authored two books entitled "Decision Making in the Manufacturing Environment Using Graph Theory and Fuzzy Multiple Attribute Decision Making Methods" and "Advanced Modeling and Optimization of Manufacturing Processes: International Research and Development" and these books were published by Springer Verlag, UK in 2007 and 2011 respectively.

Mr. Vimal J. Savsani is an Associate Professor in the Department of Mechanical Engineering of B. H. Gardi College of Engineering and Technology, Rajkot, India. He has about 10 years of teaching and research experience. Mr. Savsani's research interests include: CAD/CAM and advanced optimization methods. He has published about 25 research papers in national and international journals and conference proceedings.

R. V. Rao and V. J. Savsani, *Mechanical Design Optimization Using Advanced Optimization Techniques*, Springer Series in Advanced Manufacturing,
DOI: 10.1007/978-1-4471-2748-2, © Springer-Verlag London 2012

Authors' Addresses

Dr. R. VENKATARAO
Department of Mechanical Engineering
S.V. National Institute of Technology
Ichchhanath, Surat
Gujarat 395 007
India

and

VIMAL J. SAVSANI
Department of Mechanical Engineering
B. H. Gardi College of Engineering and Technology
Rajkot
Gujarat
India

Index

R. V. Rao and V. J. Savsani, *Mechanical Design Optimization Using Advanced Optimization Techniques*, Springer Series in Advanced Manufacturing,
DOI: 10.1007/978-1-4471-2748-2, © Springer-Verlag London 2012